LIMITLESS
THE AUTOBIOGRAPHY

Tim Peake is a European Space Agency astronaut. He finished his 186-day Principia mission working on the International Space Station for Expedition 46/47 when he landed back on Earth 18 June 2016. He is also a test pilot and served in the British Army Air Corps. Tim is a Fellow of a number of UK science, aviation and space-based organisations. He is also a STEM ambassador. He is married with two sons.

'Awesome'
Jason Fox

'Fasten your seatbelt for an exhilarating read.'
Express

'Full of courage, camaraderie and daring escapades, this reads like a Boys' Own adventure.'
Mirror

'Awe-inspiring stuff'
Evening Standard

'Exhilarating'
Good Housekeeping

Also by Tim Peake

Hello, Is This Planet Earth?
Ask and Astronaut
The Astronaut Selection Test Book

TIM PEAKE

LIMITLESS
THE AUTOBIOGRAPHY

arrow books

1 3 5 7 9 10 8 6 4 2

Arrow Books
20 Vauxhall Bridge Road
London SW1V 2SA

Arrow Books is part of the Penguin Random House group of companies
whose addresses can be found at global.penguinrandomhouse.com.

Penguin
Random House
UK

First published by Century in 2020
First published in paperback by Arrow Books in 2021

www.penguin.co.uk

A CIP catalogue record for this book is available from the British Library.

ISBN 9781787465961

Printed and bound in Great Britain by Clays Ltd, Elcograf S.p.A.

The authorised representative in the EEA is Penguin Random House
Ireland, Morrison Chambers, 32 Nassau Street, Dublin D02 YH68.

Twenty years from now you will be more disappointed by the things that you didn't do than by the ones you did. So throw off the bowlines. Sail away from the safe harbour. Catch the trade winds in your sails. Explore. Dream. Discover.

Sarah Frances Brown

To Rebecca, for sailing away from the safe harbour with me.

CONTENTS

MY JOURNEY TO SPACE: A TIMELINE

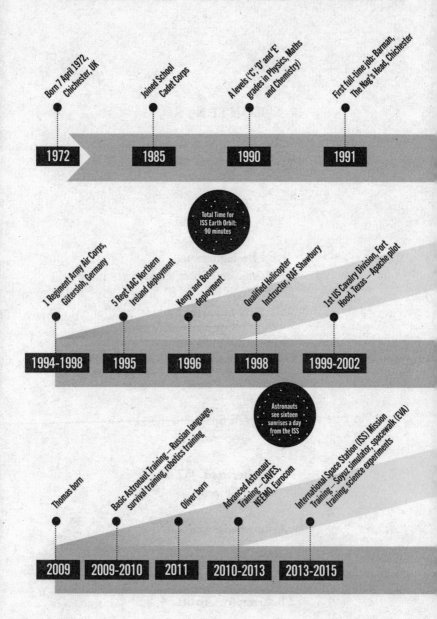

Born 7 April 1972, Chichester, UK
1972

Joined School Cadet Corps
1985

A levels ('C', 'D' and 'E' grades in Physics, Maths and Chemistry)
1990

First full-time job: Barman, The Nag's Head, Chichester
1991

Total Time for ISS Earth Orbit: 90 minutes

1 Regiment Army Air Corps, Gütersloh, Germany
1994–1998

5 Regt AAC Northern Ireland deployment
1995

Kenya and Bosnia deployment
1996

Qualified Helicopter Instructor, RAF Shawbury
1998

1st US Cavalry Division, Fort Hood, Texas — Apache pilot
1999–2002

Astronauts see sixteen sunrises a day from the ISS

Thomas born
2009

Basic Astronaut Training – Russian language, survival training, robotics training
2009–2010

Oliver born
2011

Advanced Astronaut Training – CAVES, NEEMO, Eurocom
2010–2013

International Space Station (ISS) Mission Training – Soyuz simulator, spacewalk (EVA) training, science experiments
2013–2015

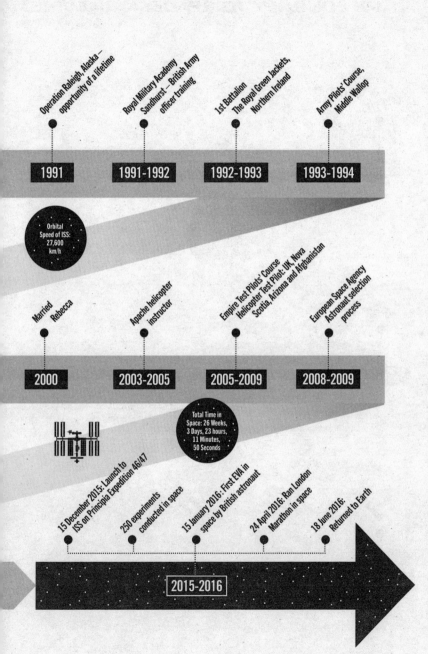

Operation Raleigh, Alaska – opportunity of a lifetime

1991

Royal Military Academy Sandhurst – British Army officer training

1991-1992

1st Battalion The Royal Green Jackets, Northern Ireland

1992-1993

Army Pilots' Course, Middle Wallop

1993-1994

Orbital Speed of ISS: 27,600 km/h

Married Rebecca

2000

Apache helicopter instructor

2003-2005

Empire Test Pilots' Course Helicopter Test Pilot: UK, Nova Scotia, Arizona and Afghanistan

2005-2009

European Space Agency Astronaut selection process

2008-2009

Total Time in Space: 26 Weeks, 3 Days, 23 hours, 11 Minutes, 50 Seconds

15 December 2015: Launch to ISS on Principia Expedition 46/47

250 experiments conducted in space

15 January 2016: First EVA in space by British astronaut

24 April 2016: Ran London Marathon in space

18 June 2016: Returned to Earth

2015-2016

PROLOGUE

Knocking on heaven's door

Yuri Malenchenko is one of the most accomplished Russian cosmonauts in history. By December 2015 he is already the veteran of five separate missions and has logged just over 641 days in orbit, which means he has spent more time off the planet than almost anybody. He is also the calmest man I have ever met. A quietly spoken, undemonstrative fifty-three-year-old, Yuri is one of those naturally composed people who can reassure you with the smallest gesture, the slightest look. All in all, I could not have hoped for a steadier commander to be strapped in beside on my first voyage into space.

Which is why I know it's a bad sign when I glance across at Yuri and notice that, with his adrenaline surging, there's a tremor in his hands.

And it had all been going so well. For six hours our tiny Soyuz capsule, blasted into orbit by rocket from Kazakhstan, had travelled through space at twenty-five times the speed of sound towards our destination, some 400 kilometres from Earth. With just 400 metres to go until contact, the spacecraft, in automated mode, began its standard fly-around of the International Space Station before locking on to the docking port for the final approach.

As we slowly closed in from below, I glanced up at the tiny window just above my head. There, gliding by, right outside, filling the view, were the huge golden foils of the Space Station's solar panels. The first time you see something big, close to you, in space is not a moment you forget. Especially when that something is the width of a football pitch, utterly dwarfing the little vehicle you are in.

Awed by such a sight, I suddenly felt compelled to share my reaction with Tim Kopra, my American colleague, on his second trip to space and seated on the other side of Yuri.

'Tim!' I said, eagerly. 'There are some bloody great solar panels out there!'

Tim shot me a look. I had forgotten the hot microphone. It had been switched off for the last few hours but now, for this critical end-phase, it was back on, meaning everything we said was being relayed with crystal clarity to Mission Control in Russia. Every stupid remark. I winced.

Suddenly, though, we all had something more important to occupy us.

Creeping softly forward, we were less than 20 metres away now, close enough for me to see the Space Station's cluttered docking area. There, practically alongside us, beneath its gold umbrella of solar panels, was the Cygnus resupply craft. It had arrived a few days earlier and was carrying, I knew, the suit in which I would, at some point in this mission if all went to plan, venture out and walk in space.

By this point, the camera view of our port had begun to loom large on the monitor in front of Yuri. And unbeknown to me, our orbit had just begun to carry us over the United Kingdom which would be a nice piece of synchronicity with which to kick off this whole adventure – me arriving at the entrance to my temporary home for the next six months in full view of my permanent one.

And then, at 17 metres from contact, the master alarm went off and red lights flashed on the console.

In truth, this isn't the first time on this journey that we have heard the alarm's loud, intermittent tone and seen the console light up. The alarm has been triggered three or four times, in fact, and always the same minor problem: abnormal moisture levels in the cabin's atmosphere. It was my simple task, in that case, to restore the balance by pumping away condensate which had collected in our module.

So when the alarm sounds this time, I assume it's the same deal and sure enough, the monitor confirms this.

But something is different. Yuri's demeanour immediately changes. Then I notice that our gentle forward crawl has stopped and that we are backing away from the Space Station.

There's a moment of confusion. The Soyuz is aborting. Why are we aborting? There has never been an automated docking which aborted from this close to the Space Station. Even for Yuri, this is new territory.

I'm rather feebly getting ready to pump condensate. Tim is digging down through the emergency signals and trying to figure out exactly what's going on. It's a thruster sensor failure, apparently. It looks like we got two alarms right on top of one another: one for the moisture level, one for the thruster sensor. The second of those has caused the Soyuz to abandon its plans and retreat.

But no worries. There's a procedure. In spaceflight, there's always a procedure. In this case, the procedure is that Yuri will switch to manual control and fly us in by hand.

Simple.

Well, in theory.

The Soyuz gently backs out to 90 metres from the Space Station and then Yuri gets ready to bring us forward again. Our orbit is taking us from day to night and the lighting conditions are less than ideal. At this distance, in the pitch darkness of space, the searchlight on our little craft suddenly seems to have the feeble range of a bedroom torch. Yuri hunches forward and peers into the periscope in front of him. I don't know what he can see out there but I imagine it's not much. There are visuals from the cameras, too, on screens in front of us, but one of those screens chooses this moment to develop a fault, limiting Yuri still further.

And it's all on him. There is very little that Tim and I can do at this point. In the absence of forward-facing windows, we can't see much that will help Yuri with the approach; we can't gauge the closure rate as the Soyuz draws back in. The one useful act I can perform is to grab the communication column, which sits, a bit like a pilot's joystick, between Yuri's legs, and hold it out of his way because I can tell that it's really bugging him at this point.

And that's when I notice the trembling hands and think, well, if Yuri is anxious, maybe this isn't such a great situation.

We'll see it later on the video images: coming in practically blind, the Soyuz has yawed by about 30 degrees and is off target, drifting towards the back of the Space Station. Had we drifted in the other direction, we would have been certain to clatter into the Cygnus resupply vehicle. But for now, as far as Tim and I are concerned, there are just voices from Mission Control, all sorts of people jumping on the radio down there to offer suggestions, none of which seems particularly conducive to Yuri's calm. His levels of anxiety are increasing, and Tim and I are catching it, too.

Maybe it would be useful if, at this point in the story, I paused to outline some of the consequences of crashing a Soyuz capsule into the International Space Station. The same space station that cost $100 billion to be built, took ten years of multi-national collaboration to be constructed in low Earth orbit, and is arguably the most advanced structure humanity has ever made. The worst-case scenario would be a rupture to the Space Station, or a rupture to the Soyuz, or both, leading to a rapid depressurisation – survivable, possibly, but without doubt a major catastrophe.

Alternatively, bumping into a solar panel or radiator could damage multiple systems in both the Space Station and Soyuz, creating a situation in which the loss of a thruster sensor would be the least of one's problems. Furthermore, glancing against a solid object could send the craft into a tumble which might prove impossible to arrest. The Space Station is not able to grapple a Soyuz capsule with the robotic arm – unlike with cargo vehicles, it has no means to reach out mechanically and grab you. You would be left rolling through space forever.

In conclusion, crashing a Soyuz capsule into the International Space Station is to be avoided if at all possible.

This moment is where Yuri's experience really tells. Maybe a more junior cosmonaut would have ploughed on in the hope of correcting the manoeuvre and getting the docking done, with possibly disastrous consequences. Not Yuri. He takes his own reading of the situation and, despite the pressure and the voices in his ears, he has the composure and clarity to knock that approach on the head – and gently back the Soyuz away out into space. Then he settles himself and prepares to try again.

While this drama continues to develop, there are lots of things that are running through my mind and many things that I am very determinedly not thinking about. And maybe it's just my habitual mindset as a helicopter pilot, which is to be constantly checking over your systems and contingency plans; but uppermost among my concerns at this point is our fuel level. There needs to be enough fuel in the spacecraft to get us home safely – and at this stage, yes, there's plenty. Yet all this unplanned flying around on manual control is eating into our precisely budgeted fuel supply. If we stay out here long enough, it's not impossible that the fuel level will drop to 'Bingo fuel' – the point where our only option will be to cut our losses, give up on the docking altogether and head back to Earth.

Which, I probably don't need to say, would be a somewhat anticlimactic outcome. To leave my former life behind and spend six years training for the expedition of a lifetime, to get blasted into space, to reach the front door, practically, of the International Space Station, and then to turn round and trudge all the way back again ...

This after the build-up, the hype and fanfare, the ceremonial farewells, the emotional partings down there at the launch site from Rebecca, my wife, our two boys, and my parents, the big departure in clouds of smoke with the nation looking on ... Suffice to say, I'm going to feel a bit sheepish if I return about twelve hours later.

'Hi, everyone, I'm back.'

Assuming we get back. To a far greater extent than docking procedures, emergency re-entry attempts are fraught with peril.

All in all, then, as Yuri once more lowers his face to the periscope and begins to line up the Soyuz for its third attempt at docking, I'm feeling quite keen that he should get it right.

After all, I've come all this way ...

PART 1: EARTH

CHAPTER ONE

First adventures in flight, what space travel and caravanning have in common, and the lesson of the exploding Portaloo

Even now, the mere memory can make my stomach turn and my palms glow hot. Most of all, it's the feeling of absolute powerlessness – the sickening realisation that, as the nose drops and the plane begins its terrible vertical plummet towards the ground, I have tried everything that I know and am entirely out of options. This aircraft is going down and the sole question now is not whether this ends badly, but how badly this ends.

Good job it was only a radio-controlled one, really. But even so, the mortification ... holy smokes. I still shudder. With half the school watching, too, and parents and teachers craning their necks, and with a brigadier in full regalia looking on as guest of honour – a big crowd for the annual Chichester High School for Boys Cadet parade, all primed for this feature item on the day's printed programme: '*Radio-Controlled Aircraft Display by Cadet Sergeant Timothy Peake*'.

Well, it started off OK. Using skills honed over a number of Saturdays at the Horndean remote-controlled flying club, I sent my hand-painted, 3-foot-long, mid-wing stunt model (complete with stick-on lightning flashes, obviously) skittering off across the playground, got it smoothly airborne and sent it out to perform a preliminary wide circuit over the school playing fields.

Things were still looking good as the aircraft, its mighty engines screaming (or, at any rate, its single engine whining thinly, like a trapped mosquito) gained altitude ahead of my big opening number – a loop. (Solid gold crowd-pleaser, the loop: guaranteed, surely, to get the spectators eating out of your hand.)

Unfortunately that's where things stopped looking good. Somewhere early in the manoeuvre, for reasons I will never understand, the plane went rogue, veered wildly away in the direction of the High School for Girls school next door, lost power entirely, and then, with me by now beetroot red and jabbing pointlessly at the controller, simply fell out of the sky, as though it were a pigeon and someone had just shot it.

A few seconds later came the noise of splintering balsa wood as Cadet Sergeant Timothy Peake's prized model aircraft trashed itself against the Girls school roof.

Display over. Plane over, too, for that matter. How that brigadier didn't wet himself laughing, I have no idea. Military self-control, I guess – my first real experience of it in action.

This is the autobiography of an astronaut, so I guess by rights the childhood scenes ought to be packed full of space stuff – early dreams of rockets and spaceflight, unignorable evidence of extreme Apollo fandom, tales of the little kid who parted the curtains each night before bed to stare longingly at the night sky and tell himself, one day, I'm going out there.

But I can't deceive you: that child wasn't me. Among my boyhood preoccupations, space barely figured at all. In December 1972, when Apollo 17 landed on the moon, the last of the Apollo missions to do so, I was just eight months old – born too late to witness that culture-changing adventure as it unfolded. Sometime in the early 1980s, I had a Mini Lego Space Shuttle Launcher, which I guess I played with a fair bit. I learned a few constellations: you did that on Cadet night-hikes. But that was about it.

By comparison with other astronauts I know, I can feel a bit of a fraud in this area. My European Space Agency colleague Luca Parmitano was so obsessed with spaceflight that his nickname in the Italian Air Force was 'NASA'. Andy Mogensen had a doctorate in aerospace engineering and had clearly devoted most of his waking life to thinking about rocketry and the physics of satellite navigation. Scott Kelly, who was my first commander on the International Space Station, puts it all down to reading Tom Wolfe's *The Right Stuff* as a young teenager. By the time I opened that book, I would be in my twenties and training to be an army pilot.

So, I can claim no youthful plans to go into space, I'm afraid. But I did very badly want a radio-controlled plane. Actually, what I *really* lusted after was a radio-controlled helicopter. I had longed for one of those since I was nine, when I saw a feature on *Blue Peter* about the opening sequence for the James Bond movie *For Your Eyes Only*. Maybe you remember it. It's the one where Roger Moore has to get a helicopter under control while it does its best to kill him by flinging itself around in a disused gasworks.

Spoiler alert: it doesn't succeed.

Spoiler alert again: a toy was used in the creation of this sequence.

And what a toy! *Blue Peter* shared some behind-the-scenes footage and, at the sight of that model helicopter buzzing into the air exactly like the real thing, something stirred deep inside my boyhood soul. This, surely, was the business, and clearly a massive step up from the green plastic helicopter in which I had been carrying my Action Man soldier into combat zones – mostly on the small strip of carpet beside the bed in my bedroom, but sometimes outside the door on the landing. The Action Man helicopter was a decent piece of kit in its own way, but as the label beside the picture on the box so disappointingly stated, 'NB Does not fly.'

The thing was, as I pretty quickly established, radio-controlled helicopters were prohibitively expensive, not to mention treacherously difficult to pilot. A simple radio-controlled aeroplane, though … that had to be within reach, didn't it?

It turned out that a few miles up the road from our West Sussex village, people were gathering in a field on Saturday mornings to fly model planes. Kevin Porter and Martin Fleet, two of my schoolfriends, knew about it and spread the word. We would cycle out to Horndean together and look on in awe. There were significantly more adults standing in that field than kids, one noticed – sticking their flags in the ground with their radio frequencies marked on them to avoid clashes. And these guys were operating some serious gear – vintage numbers, a Spitfire, even. A radio-controlled tail-dragger! Intimidatingly great. Still, if the skies weren't too full, nobody there seemed to mind young interlopers sending up their own, lesser aircraft.

Desperate to be part of this action, I acquired a cheap starter kit – a basic high-wing model, a little like a Piper Cub, with a tiny, single-cylinder, petrol engine; one of those planes where you had to flick the propeller over with a wooden stick to get it going and hope it didn't take your finger off when it fired. Witnessed reluctantly by my sister Fiona, this aircraft's big maiden flight took place near home, at Westbourne Common – a small patch of land, adjacent to a strip of houses and almost entirely unsuitable for the flying of model planes, but never mind. My aircraft flew high and steady that afternoon, a wonder to behold. Even Fiona was just about to become officially impressed when I did something wrong and watched my first proper flying machine go down heavily somewhere behind Westbourne youth club.

Still, as I found out once I made it to the crash site, the mechanical parts of those kit planes – the engines, the servos – are actually quite robust. It was only the wings and the fuselage that were entirely destroyed. Those, surely, I could easily and cheaply replace with bodywork built by myself. So I bought a stack of balsa wood and set to designing my own plane around the surviving bits. The result was pretty impressive, if I say so myself. It looked more or less like a plane, anyway. The question was, would it fly like one? I took it over to the club at Horndean to find out.

And the answer was no. No sooner had it raised itself to a respectable height than my hand-crafted special piled into the ground again, almost as if terrified of the air. Plenty of lessons there for me to absorb about centre of gravity and the importance of wing shape, as I picked the shards of balsa wood out of the grass and headed home.

I needed to up my game, clearly – both as a pilot and as an investor in remote-controlled hardware. In due course, a little bit of money for this project was available. Martin, Kevin and myself all had part-time jobs stacking shelves in Chichester Waitrose – Tuesdays and Thursdays after school and all day Saturday. My zone was the dairy aisle which, by the way, had something to recommend it to the slightly bored teenage shift-worker: access to eggs. An egg slipped gently into Kevin's or Martin's pocket while they weren't paying attention would most likely be a broken egg by the next time they put their hands in there. They would get me back by doctoring my trolley so that all the milk fell off it. And thus the shifts would go by.

Anyway, with funding courtesy of Waitrose, I was ultimately able to splash out on the mid-wing sporty number, mentioned earlier. I lavished hours on constructing it. Maybe I loved building those things almost as much as I loved flying them, in fact. Either way, I privately commended myself for a beautiful job on the paintwork: white for the fuselage and red for the wings. But that was another schoolboy error, of course: once that plane was up in the air, being a mid-wing you couldn't tell from its colour scheme whether it was upside down or not. This would turn out to matter – critically so above the playground of Chichester High School for Boys during the annual Cadets parade.

Whence this craving to fly things? My dad, Nigel, must shoulder a lot of the blame, for taking me to air shows. Dad was a journalist, a features editor at the *Portsmouth News*, and his interest wasn't really aircraft so much as the history they were a part of. Nevertheless he took me to the big display at Farnborough, which was only an hour from where we lived. And with even greater resolve, he carted me off in the car to North Weald Airfield, 100 miles away in Essex, for a show which they started staging regularly there in the mid-eighties, called the Fighter Meet. It became an annual outing for us. The field would be humming with aircraft – Second World War planes, a Vulcan Bomber from the 1950s, maybe a Harrier Jump Jet – and there would be fly-pasts, stunts, pyrotechnic displays. All of this left a deep imprint on my impressionable mind. Planes, explosions and a picnic: what's not to like?

Or maybe something had bitten even earlier than that – in the mid-1970s, when my parents twice treated our family to Spanish package holidays. The first of these, when I was still a baby and oblivious, involved a resort in Majorca which, by all accounts, was going to be very nice when it was finished. But the second was in 1975, this time in Ibiza, when I was three and old enough to absorb the excitements of Gatwick Airport and the mysteries of taxi and take-off and flight and landing.

One evening in the middle of that holiday, at the nightly Six O'Clock Club, the children's entertainer gathered the kids on stage and asked us all to tell the room what we wanted to be when we grew up. No doubt there were a few footballers among us, some doctors, some nurses, possibly the odd cowboy. When the microphone came to me, I solemnly replied, 'An airportist.'

The watching parents got a good chuckle out of that – much to my confusion. It was as good a word as any, wasn't it?

Either way, one thing was clear: I knew from the start that I wanted to fly. It took me a little longer to work out how far.

*   *   *

When asked about it, I frequently find myself describing my childhood as 'ordinary', or referring to my 'normal upbringing'. By which I mean that it was stable, secure, untroubled by extremes. Those phrases can sound dismissive, though. Only when I had children of my own did I realise how much effort goes into producing a 'normal upbringing', an 'ordinary childhood'. Normal and ordinary don't always get the credit they deserve.

For my mum and dad, the serious work of producing my 'ordinary childhood' got under way on 7 April 1972 at St Richard's Hospital in Chichester. I had been due to arrive on the 2nd, but eventually appeared five days late, thus missing the 5 April tax deadline and depriving my parents of the rebate they would have received if I had been thoughtful enough to show up on time. They were very nice about it, though. They still took me home.

Whereupon I rewarded them with thirteen months of broken nights. My sister Fiona had never given them any trouble over sleeping, so this was new territory. My poor mum, Angela, dazed with lack of sleep, would push me around the neighbourhood in the pram in the hope that I might consent to nod off for a few minutes. Apparently I was a large baby – my hands in particular.

So, on those walks, people would lean into the pram to coo over the new arrival and end up saying, 'Oooh, hasn't he got … big hands?' My mum had mixed feelings about this – as I probably would have, too, if I had been capable of understanding anything. (For the record, my hands in due course became standard sized relative to the rest of my body.)

But nothing my mum or dad tried could dissuade me from staging a one-infant protest against the concept of bedtime. Indeed, my campaign only intensified. As time went by, I was able to move on from simply lying there and bawling and could eventually pull myself up and rattle the bars of the cot as well. Then one day, magically, I could walk and it all stopped. That was apparently the breakthrough that I had been waiting for and, to the relief of my shattered parents, I slept soundly from that point on, restricting my demands for action to more sociable hours.

Still, even then those demands were fairly taxing. As a preschool toddler I would expect to be taken out, every day, without fail, whatever the weather. It seems I was already someone who felt reined in by walls and carpets, who needed to be outdoors.

In that respect, I was lucky to be growing up in Whitley Close – a cul-de-sac, tucked quite deep into a small, quiet, brand new housing estate, on the edge of the small, quiet village of Westbourne. There was no passing traffic, so from a fairly young age we could be safely turned out into the close to play. There were plenty of kids out there to play with, too. Because the estate was new, people had moved in at the same time, at more or less the same stage in their lives. They were having children at the same time, taking those children to the same playgroups, sending them to the same school. I was born into a ready-made multi-family community. There were the Smiths – Andrew, Neil and Claire. There were the Towles – Philip and Julian. Next door to us were the

Mayhews – Nic and Bev. And there were the Peakes – Fiona and me. We were the Whitley Close gang.

Ours was a typical semi-detached, three-bedroom, late-sixties new-build house, which my parents stretched themselves a little to buy. Skirting boards had been an optional extra, so we didn't have any. Garages cost extra, too, so we didn't have one of those, either. Our orange Datsun Sunny just had to lump it, out on the drive. The gap where the garage would have been, though, linking the drive with the back garden, made for another communal play area, open to the street, and provided a good strip for go-karting, so there were bonuses.

My mum was a midwife at the Havant Health Centre, over towards Portsmouth, and she brought to housekeeping the discipline that was essential to her work. The place was always scrupulously clean and magnificently uncluttered. Everything had its place and, in my small bedroom, toys were stored systematically under the bed. Minimalist, I suppose you could call it, but this was practical as much as aesthetic. There wasn't really room for it to be any other way. (Clutter still sets me on edge. We seem to have so much stuff now. How much of it do we really need?)

Those carefully stowed toys would come to include Lego and Scalextric – and Action Man with his helicopter, whom I already mentioned. And I will never forget the first air rifle I was allowed to own – a Stoeger ATAC TS2 with Gas-Ram spring piston, zoom scope and patented Dual-Stage noise reduction system. But that's because I only bought it in January 2020, at the age of forty-seven – a belated Christmas present for myself after a lifetime of yearning. My parents wouldn't let me have an air rifle in Whitley Close any more than they would have let me have an actual gun with bullets. In fairness to them, being not much bigger than a pocket handkerchief, our garden did lack many of the key features of a safe and workable shooting range. I pleaded, but those pleas fell on deaf ears. (I phoned my mum to tell her that I'd finally acquired one. She still didn't sound like she approved.)

Beyond our front door, life was lived very locally. Things for the most part were within a short walk from home. I went to the local village school, Westbourne Primary. I shopped for sweets on Saturday mornings in the local village shop. I went to the local village Scout hut, just behind the local village school, for Cubs on Tuesday nights. (First badge: First Aider. Second badge: Handyman, for which I think I had to bang a nail into a plank of wood.) In due course, in my mid-teens, the Westbourne youth club would become the venue where all the action was. We would gather for table tennis and discos, with music courtesy of Philip, who by then had turned DJ and was spending all his money on records and flashing lights. Philip spun Wham!, David Bowie and (my big musical heroes at the time) Madness, and the rest of us danced and tried to stay upright while the hut's battered wooden floor bounced up and down beneath our feet.

My world broadened significantly in the summer holidays. At some point my parents acquired a caravan, and a car strong enough to tow it. The Datsun Sunny was replaced on the drive by a brown Citroën DS, which opened up exotic summer holiday possibilities in southern France, the caravan hitched to the back; my mum in the passenger seat, navigating with a paper map and trying, on Dad's keen instructions, to keep us away from the steepest hills – and mostly succeeding, although there was at least one occasion when we had to turn around awkwardly in a French allotment to avoid defeat on the mountain that had 'unexpectedly' risen up ahead.

It was on one of those holidays that I learned an important – and indeed unforgettable – lesson about physics. We had pulled into a lay-by for a brew and Fiona took the opportunity to avail herself of the caravan's Portaloo. This would have been an uneventful experience, but for the fact that our journey had taken us halfway up a mountain (one that our Citroën definitely could manage), en route for Briançon in the Alps.

Now, this may not be news to you, but it turns out that there is a marked difference between the air pressure at sea level and the air pressure at 2,000 metres. Astride the Portaloo, Fiona unsuspectingly pulled the lever which would ordinarily have opened the trapdoor beneath her, permitting gravity to do the rest. But not on this occasion. From inside the caravan came the sound of a muffled explosion, followed by a scream, followed by the sight of my sister running out dressed in three days' worth of Portaloo contents. If that were not funny enough, the indelible blue dye from the loo's chemical freshener acted as a constant reminder of the incident for the duration of the holiday. I'm not sure but I think that may have been the moment my early interest in science became a fully fledged passion.

We would meet up in France with the Smiths and the Towles, transplanting Whitley Close to the Ardèche. Living in close proximity to people, in a confined and rudimentary domestic setting in which everything has to be locked down for travel … caravanning may have more in common with spaceflight than is often remarked. In any case, I absolutely loved it, from sleeping in the awning to the opportunities to run off for the day and go exploring and float around on rafts and canoes. In the morning, there would be Nutella (a substance entirely alien to the UK in those days) to spread on your bread, and in the evening you would carry your saucepan over to the camp kitchens and get it filled up with French fries (also alien) to take back to your pitch for supper. That was heaven to me.

But so, for that matter, were the holidays closer to home, in the Lake District and the Yorkshire Dales. Anywhere outside seemed to work as far as I was concerned. By the time I was seventeen, and we were away in the Dales, my parents were picking a pub and telling me to meet them there. And then, on my own, I would yomp off across the wilderness with a map, a compass and some waterproofs while they drove round in the car.

I have to give my mum and dad credit; they weren't potholers, kayakers, rock climbers or orienteers, but they were happy to encourage me in all of that reckless stuff when I showed an interest in it. It's quite something, really, to let your son go stomping off across the Dales on his own and have faith that he will join you in one piece for lunch. But maybe the times helped a little. Instilling and stimulating a taste for adventure in your kids, from the remote safety of your car, doesn't seem such an easy move these days.

In 1983, aged eleven, I left Westbourne Primary and went to Chichester High School for Boys, a non-selective state school. A place there was considered something of a golden ticket, especially given that Westbourne was right out on the fringes of the catchment area. When I think about the directions my life has taken, I realise that a lot pivoted on that one piece of administrative luck. A different school and my life could have headed off down another track altogether.

As it was, I put on the green tie and the green blazer with the West Sussex crest on its pocket, and set off on the nine-mile daily journey. Fiona was at Chichester High School for Girls, and sometimes we would get a lift in the morning with Philip and Julian, whose mum, Val, was a primary teacher in Chichester. When Val's chauffeuring wasn't available, I would take the train from Emsworth, and walk the mile and a half home from the station at the end of the day, entering the house with fingers that were practically ready to drop off after lugging up the road a holdall stuffed with schoolbooks.

Why a holdall and not a rucksack, which would have been more sensible? Because it was the eighties. You had to have your school things in a holdall in the eighties. It was the law.

Apart from being a village boy (rather than a Chichester city boy), which wasn't deemed to be especially cool, life had dealt me two potential disadvantages for coping in the jungle which is the playground at an all-boys school. First, I was small (the promise that I had shown, size-wise, as a baby didn't last beyond the romper-suit phase). And second, I was ginger. Arriving on the scene with those twin blessings, I had to develop a few survival skills, or certainly a thick skin for banter. But I got by. I made friends and I was quickly happy there.

When it came to schoolwork, however, nobody could claim I broke any records. My grades were consistently below average, across the full range of available subjects, with the solitary exception of what went by the impressive title of 'Rural & Environmental Science', and which mostly seemed to involve getting a pig out of a pen during a field trip. I absolutely nailed that. None of my more disappointing grades were for the want of trying, though. My end-of-year reports were nearly always a set of variations on the theme of: '*Tim struggles but he perseveres.*' Effort I could do, it seems; achievement less so.

Those first couple of years at secondary school, I went through a bit of an aimless phase, which might have slightly concerned my parents, who saw me bouncing around from one thing to the next and not sticking at anything. I took up swimming for a bit, which was Fiona's speciality and which I convinced myself very briefly was going to be mine. The truth became emphatically clear at a sponsored swim for the British Heart Foundation, where I was still grimly ploughing up and down the pool for the allotted distance long after practically everyone else, including my sister, had finished and got changed. At least I stayed the course, I guess. But I abandoned the swimming club soon after.

Then I tried horse riding – again enthusiastically following Fiona over to the local stables. But my inner Princess Anne proved hard to awaken, and the enthusiasm didn't last long there, either.

So what about acting, maybe? I was picked to play Oliver in the lower-school play, and did so to great acclaim – or at least to a relatively positive review from Miss Griffiths, the drama teacher, who encouraged me to join an out-of-school drama club. But I soon found I couldn't really see the point and dumped it.

I was even, for a while, a chorister at Westbourne church. My family only did church at Christmas and Easter, but my mate Richard North's mum played the organ every week, and the church was a big part of their family. Soon enough, I found myself going along, donning the red robe and the white cassock, and joining Richard in the choir stalls. We messed around so much that we had to be separated. The choirmaster put us on opposite sides of the aisle, but actually that was even better because then we could ping rolled-up bits of A4 hymn sheet at each other.

I wasn't there long.

Altogether I was floating from one interest to the next and failing to settle at anything. But that's OK, isn't it? I mean, it worked out all right eventually. It's why I have come to feel slightly uncomfortable about the idea of the 'lifetime dream'. I'm sure it works that way for some people – that they find their purpose, their burning interest, right at the beginning of their lives and pursue it single-mindedly from that point onwards. But it certainly wasn't like that for me. And when you think about it, there's no rule that says the dream of your lifetime should be the first dream you have, rather than, say, the second, or the ninth or even the fourteenth … Maybe the dream of your lifetime is a dream you haven't even had yet. No need to narrow your chances, then. Certainly no need to stop trying things.

For a couple of years I definitely had form as someone who took things up and then junked them. Indeed, after the swimming and the horse riding and the acting and the singing, I'm sure when I joined the school Cadet Corps, at the age of thirteen, my parents would have been wondering whether this was just the next thing that I was going to try on for size and soon discard.

But actually it didn't work out that way.

CHAPTER TWO

The Cadet that squeaked, a quick trip to the sky,
and the best car boot sale ever

There were probably more fashionable things to be wearing around Chichester in 1985 than the uniform of the Combined Cadet Force (Army division). At best the critics might have described the ensemble as 'retro' – or possibly 'museum piece'. The outfit we're talking about consisted of black Dr. Martens boots and square-cut combat trousers overlaid with puttees – strips of bandage, essentially, which you wrapped around your leg from your ankle to your knee. I don't think puttees had featured in official military uniform since the outbreak of the Second World War, but clearly they were still a big fashion item for the Cadets – a counterpart, maybe, for the Jane Fonda-style workout leggings which were also very much the rage at this particular moment in the eighties.

But the real monster in this standard-issue get-up from my point of view was the KF shirt. Again, this was an item of clothing which seemed to have come off the designer's sketchpad somewhere around 1939. But it wasn't the look of the thing that was the problem: it was the texture of it. KF stands for 'khaki flannel', which in this case described a material with all the seductive, body-hugging softness of a sheet of coarse-grained sandpaper.

Did I mention that I was ginger-haired? I think I did. And with that ginger hair came some fair and sensitive skin – the kind that finds direct contact with wool, for instance, to be an ordeal akin to getting tossed into an ants' nest. To button up the KF shirt was, for me, to enter a whole new world of irritation, like getting kidnapped in a hessian sack. At home, on the nights before CCF sessions, I found myself feverishly ironing the inside of this hostile garment over and over, in a mostly doomed attempt to take some of the edge off.

Ironically, it was the uniform that had been a big part of my initial attraction to the Cadets. In my first two years at Chichester High, I would see the older boys and girls going about in this clobber – which looked extremely grown-up, whatever else you wanted to say about it – and I would find myself feeling slightly star-struck at the sight. There were also the stories about what these people were getting up to in those uniforms: night hikes, apparently, and rock climbing and shooting ... So you got to dress up as a soldier AND you got to go in a kayak? That felt like an irrefusable offer to me. The minute I was eligible, at thirteen, I leapt at it.

And I wasn't disappointed by what I found when I joined, either. Apart from the khaki flannel, of course, but even that I could put up with in the circumstances. Over and above the dressing-up, CCF just seemed to be a whole bundle of things that appealed to my imagination.

For instance, very quickly, after just a few of the regular after-school sessions on Fridays, we were being taken to an indoor shooting range in Chichester for our first go at .22 rifle shooting. For a kid with thwarted dreams of air-rifle ownership, this was a huge break. And soon after that we were off to the Bordon and Longmoor Military Ranges, not far away in Hampshire, and shown how to handle a Lee-Enfield .303 rifle.

Now, that's a pretty large lump of weaponry for a thirteen-year-old to be carting around, especially if the thirteen-year-old in question happens to cut a relatively small and slight figure. When fired, it kicked like an elephant, threatening to throw my still diminutive person backwards every time I squeezed its trigger. And then there was the noise – literally deafening. Not everyone bothered to hand out ear-defenders in those days. Which seems quite shocking, on reflection. Yet of course, nobody thought kids needed seat belts in cars in that period either, and if you had suggested putting on a helmet for cycling, people would have thought you were eccentric to say the least. Different times and different attitudes. But bad luck for my ears, though. Those first shooting-range sessions were just the beginning of a battering for my eardrums which went on to include eighteen years of exposure to screaming helicopter engines, followed by six years of the constant pressurising and depressurising involved in spaceflight training. Little wonder, really, that I now suffer a bit with tinnitus. In many ways, I'm probably lucky I can hear at all.

I loved the shooting range, though. The accuracy it called for, the discipline of it, the sense of getting better at something. Here was the literal embodiment of 'targets to aim at'. I also loved the CCF weekends away: loading up the minibus, unpacking in some army barrack block, maybe doing some kind of night exercise, then getting up the next morning and doing skills and drills, or an orienteering course, or kayaking and learning to do Eskimo rolls, or maybe going sea-kayaking, but always being kept busy, and mostly outdoors. Even when the winter months forced us inside for sessions in the classroom on weapon cleaning, map reading and tactical studies, I found I enjoyed all of that, too. On all of these occasions, CCF seemed to me to be essentially saying, 'Here's a challenge – see if you can step up to it.'

I also liked the mix of ages. You don't do much at school out of your year group, but CCF lumped together people from Year 9 to Year 13, and that was great in both directions – whether you were young and getting to hang out with the older kids, or whether you were older and helping out the younger ones, making sure they were OK, that they had the right kit and so on. Very early on there was an opportunity to accept responsibility for other people, and I seemed to warm to that.

I soon earned my first stripe – Cadet Lance Corporal. But it was in the autumn half term of 1986, when I was fourteen, that I went off for the week to Langley Lines Cadet training centre at Longmoor to do the Cadet Sergeant's course and see if I could earn serious promotion. When I looked around, I realised I was the youngest student there. Also the smallest.

Here my tender ears took another beating from two unprotected days in the company of a Bren machine gun, firing balsa wood bullets through a shredder attachment so that, essentially, ribbons of wood plumed from the nozzle. It might not have been live ammo, but it still made one hell of a racket. It was like manning a wood-chipper for two days. There was also a Drill component; and an assault course exercise called Exercise Holdtight, which you did individually and then as a team; and there was a night navigation and Fieldcraft and Map & Compass work … I threw myself into all of it.

The typed-up course report from the Senior Cadet Instructor Cadre – on a single sheet of white A4 which I still have in my possession – alludes relatively gently to what we might call my lack of physical stature. The sergeant who marked me for Physical Training notes, '*Hampered by his size but very enthusiastic and competitive.*' Otherwise it goes unmentioned.

The far larger area of concern appears to have been my squeaky, as yet unbroken, voice – well-suited, possibly, at this stage in my life, for communicating with bats, but less so for bawling commands across a mocked-up battlefield. Accordingly, the instructor overseeing the assault course exercise felt obliged to write, '*Hampered by his voice but plenty of spirit, drive and enthusiasm.*' The sergeant in charge of Fieldcraft observed, '*Should project himself more (voice).*' There was a slightly more optimistic tone to the comment from the sergeant overseeing Drill: '*His voice will improve with age.*'

Overall, though, I was given a B-grade pass by the commanding officer ('A very good result. Well done!') and, on the strength of that, the school was able to make me an unusually young, and unusually pipe-voiced, cadet sergeant. After that, I became colour sergeant. The only position really available above that was warrant officer, and most school Cadet forces didn't have one of those, because your force had to be a certain size to qualify. But by the time I was in my final year at school, the number of people doing CCF had expanded, so the head of our Cadet force wrote and said, '*We'd like to have a Warrant Officer position.*' His application was accepted and I was promoted to the rank, which effectively made me senior cadet in charge of both of the school's Army and Air Force sections.

You'll be wondering, perhaps, how much this advancing military status improved my teenage social standing in and around school – how much kudos and respect it earned me from my fellow pupils beyond the circles of the CCF. And I would have to say, not very much at all, really. I don't think I was ever under any misconception that being in the Cadets could automatically lump you in with the cool kids. On the contrary. At Chichester, the 'cool' accolade seemed mostly to go to the sports crowd – the footballers, the rugby players. I was in the hockey team, but it didn't quite count in the same way. Those of us spending our weekends sleeping in barracks and doing night hikes definitely weren't considered cool. The cool kids tended to have a fair bit of scorn, all in all, for the CCF kids.

But it didn't bother me in the slightest. As far as I was concerned, I had stepped into a giant adventure playground – and, to some extent, found a second home among its many and various rides. Before all this, I had felt a little lost in this large and slightly intimidating school – a little unsure about where I fitted in. Now I felt more anchored, and a lot more sure of myself. I had a few people to thank for that, clearly. If you're lucky, your path crosses with some good teachers at some point in your life. My school's CCF was run by Anthony Forrest, a chemistry master, and RE and drama teacher Gina Griffiths, with some help from a senior master who had the excellent name, for a schoolteacher, of Mr Thrasher. I think back about the evenings and weekends which those staff members selflessly gave up, on top of their teaching commitments, so that we could have these extraordinary opportunities and I realise how fortunate I was that such people were around. I owe them a lot.

And I especially owe them for slightly bending the rules on my behalf one time. As passionate as I was about flying – and as determined as I remained to become some kind of 'airportist' at some point in the future – it didn't occur to me to join the Cadets' Royal Air Force section. The activities on the Army side were much more appealing to me. Nevertheless, when I heard that the Air Force Cadets were getting lined up for a trip to go gliding, I pleaded with the teachers to let me sneak across the forces divide and join them.

And so it was, one weekend, that I found myself at RAF Hullavington in Wiltshire, sitting in front of an instructor in a tandem-seat glider, getting launched up into the air by a winch, dragged through an aggressively steep climb, and then feeling the stomach-dropping lurch as the cable dropped away and left us to glide.

And I was instantly enchanted. Bear in mind that at this point in my life I hadn't even driven a car. So, unless you counted the lawnmower or the occasional remote-controlled plane, this was the first time I had come anywhere close to being in charge of a significant piece of equipment. And as I was shown the controls and eventually allowed to take them, the effect on me was immediate. To be up there in the sky, unconstrained by roads or paths, and to be able to feel the glider conforming to the air around you, felt magical and, at the same time, perfectly natural.

I was in my element, I suppose you could say. Certainly I knew straight away that, if I ever got to choose my element, the air would be the one.

* * *

Helicopters continued to be the grand passion, and my experience of them expanded greatly in these years. By the time I reached my mid-teens I had already logged simply hundreds of hours at the controls of an Apache, flying into hostile combat zones, identifying ground threats, locking on to targets and squeezing off rocket salvos – and very often doing all this while my parents watched television in the same room.

And, OK, the 1986 Tomahawk Helicopter Flight Simulation for my dad's Amstrad, with its simple flickering black and green graphics, would be regarded as a primitive cave painting by comparison with the entire worlds digitally conjured in today's computer sims. But actually it wasn't such a bad place to learn some of the basics. (About ten years later, in 1996, I acquired the Jane's Combat Simulations version of the AH-64D Apache Longbow, which I didn't play around with anywhere near as much, being older and busier. But as I would eventually

23

discover at the controls of the real thing, that sim offered a pretty decent approximation in terms of showing you where things were.)

I piloted that simulated Apache in the sitting room, by the way, because that was where my dad's recently acquired 'home computer' had been stationed. (Note to younger readers: at the dawn of domestic computing, many homes in the mid-1980s had only one computerised device, and the chances were that you had to share it with your parents and your sister. And if that sounds mind-blowingly restrictive, wait until you find out what used to go on with the phone.) Both the game and the Amstrad still reside in my parents' attic, and are no doubt well on their way to becoming priceless museum pieces.

One Sunday morning during this time my great schoolfriend and fellow cadet Paul Hunter and I were poking around at a car boot sale in a field somewhere outside Chichester when we found gold. Well, it was gold as far as we were concerned. Tucked between the usual china cat ornaments, incomplete board games and discarded Betamax videos was a tatty, cheaply printed booklet with a black cover – utterly unremarkable and yet somehow beckoning to us. Paul tugged it out for a closer look.

To our astonishment, he was now holding a 'how-to' guide to the manufacture of home-made explosives. Here was everything you needed to know about making *Improvised Black Powder*. (*'Black powder can be prepared in a simple, safe manner. It can be used as blasting or gun powder.'*) Here, too, was the easy-to-follow recipe for an *'Improvised White Smoke Munition'*. (*'Can be used for either signalling or screening.'*) And here, furthermore, was the complete low-down on putting together a *'Sawdust/Wax Incendiary'* complete with a *'time-delay igniter'*. (*'More effective than napalm for the destruction of heavy wood timbers and hard-to-ignite combustibles.'*)

Moreover, not only were all of the ingredients clearly specified, with details on where to find them, but each stage of the process was rendered painstakingly clear with the use of idiot-proof drawings. For example, where it said, '*Measure 9 teaspoons of hydrogen peroxide into a container,*' there was a drawing of a hand using a teaspoon to drop some bits of hydrogen peroxide into a container. And where, in the instructions for making an '*Improvised White Flare,*' it said, '*Measure 21 tablespoons of the powdered nitrate into a quart jar,*' there was a drawing of possibly the same hand using a tablespoon to drop powdered nitrate into a quart jar.

What a find this was! Ordinarily we were delighted if our car boot scavenging turned up any old bits of discarded military kit – webbing, water bottles, mess tins. A recipe book for explosives was another kind of treasure altogether. It's hard now to convey the rarity value of this kind of information in a time before the internet – just as it is hard, too, post 9/11, post 7/7, to stress how innocent and merely playful our interest in manufacturing munitions seemed to us. For Paul and me, this kind of stuff was simply a place where a certain kind of curiosity about chemistry met a teenage interest in things that went bang. Still, the plain fact is that, nowadays, if you wanted to know how to rustle up a slow-burning fuse using three shoelaces and some granulated sugar, there would be Google. Back then you had to ... well, go to car boot sales, clearly.

Yet the question was, who had originally needed this information? And what, exactly, were they intended to do with it? Because they were certainly intended to do *something* with it. The instructions for a '*Nitric Acid/Nitrobenzene ('Hellhoffite') Explosive*' included a footnote, '*Prepare mixture just before use.*' Before use for what?

This much Paul and I quickly established: the document was American. In the places where it helpfully listed '*sources*' for the necessary materials, it referred to '*stores*' rather than shops; plus it spelled sulphur '*sulfur*' and litre '*liter*'. Furthermore, the document was official. We knew that because each page was franked – at the top AND at the bottom – with the thrilling words '*FOR OFFICIAL USE ONLY*'.

So what we had unearthed here, we decided, was some kind of limited-issue, top-secret US Special Forces pamphlet. Paul and I looked at each other with open mouths. This stuff was absolute dynamite. Even better: it was the instructions for how to make absolute dynamite.

We paid the 10p, or whatever the seller was asking for this priceless volume, and walked away, electrified. OK, there were some accompanying feelings of guilt, maybe, at the thought that we had just taken ownership of something that we probably shouldn't have had. But actually, that thought was quite thrilling too. And at least we knew this potentially dangerous instruction manual had fallen into entirely responsible hands.

It was not long after this that we nearly burned down Paul's bedroom.

I used to like going round to Paul's house because his parents (unlike mine) allowed him to have an air rifle, which meant we could while away the hours shooting up plastic bottles in his garden. However, the discovery of the manual had moved our weekend activities on to a whole new plain. We had been using the book to rustle up a little HMTD – hexamethylene triperoxide diamine, to give it its less catchy name. The text had temptingly described this substance as 'a primary explosive', handy for detonators, and possible to assemble from various readily available and ostensibly innocent domestic items, including hair bleach. It was clear to us that if you wrapped tiny quantities of HMTD in little knots of paper, you could make your own bangers – an outcome much to be desired.

By following to the letter the manual's eight-stage mixing and filtration process, Paul and I had ended up with a mound of solid particles which, as instructed, we rinsed with cold water and transferred to a container to dry, the container being, in this case, an old plastic 35-millimetre film canister. Paul then set the canister down gently on his bedside table.

Kaboom! The plastic pot became a fireball and the room abruptly filled with a searing light, hot enough to scorch the poster on the wall above Paul's bed, which was probably of Madonna, given the period we're talking about, though I can't exactly remember. Whoever it was, they didn't come off too well in the brief but properly startling conflagration.

To be fair, the warning had been there in the form of a thick black box at the foot of the relevant page, which stated, '*Caution: Handle dry explosive with great care. Do not scrape or handle it roughly.*' Even so, Paul had hardly touched it, really – just picked it up in its container and set it down perfectly gently maybe with just the tiniest jolt at the end. But that little jolt was enough, clearly. From which we were obliged to conclude that home-made HMTD is hellishly explosive.

The pair of us were in shock, of course. Yet once the initial trauma had worn off and we had worked out that we were a) still alive and b) unlike the poster, surprisingly uncharred at the edges, the moment went from shocking to hilarious. We lay on the floor and wept with laughter.

But of course, the mirth eventually subsided and in due course we settled down to reflect more soberly on what had just happened and to absorb its many important lessons. And at the end of that process of reflection, as you would no doubt expect, we took the mature decision to set the manual and our explosive manufacturing practices aside and never return to them.

Are you kidding? Of course we didn't. We carried on making HMTD with, if anything, even more enthusiasm than before. Indeed, in an entirely separate rogue explosion, this time in the open air, Paul would later manage to embed in his thumb a strip of hot magnesium from my chemistry set which we had misguidedly decided to use as a fuse. Just a flesh wound, luckily. (The incident also left his face blackened and his hair charred, just like in the cartoons.)

However, I did feel chastened enough by what we had been through to take one precaution. Back at home, I transferred my personal HMTD stash from under my bed, where it had been casually residing, to the garden shed, where I placed it, extremely carefully, out of the way on a shelf. That way, if anything was going to be consumed in a random explosion in the middle of the night, it wasn't going to be my bedroom. It was going to be my dad's lawnmower. I felt that was the sensible thing to do.

One time, in a sponsored event for charity, a few of us cadets spent a cold and miserable night on an inflatable dinghy in the Chichester canal, and the following day, in a prearranged show-stopping moment, the RAF Search and Rescue team flew in from Portsmouth in a Sea King helicopter and winched us all out. It was a godawful experience, if I'm being honest – cold and damp. But it obviously bit deep for my friend Paul. He went on to join the RAF and become a Search and Rescue crewman – the guy on the end of the winch, in fact. His nickname there was 'Haz' Hunter, short for 'hazard', so it seems he didn't change.

For all its discomforts, the event left a sizeable impression on me, too. Being plucked up and spun away in the clattering Sea King that day had amply confirmed my early impression that helicopters were where it was at. And so it was that, on days off from manufacturing highly volatile explosives in friends' bedrooms and crashing radio-controlled planes into fields on the West Sussex/

Hampshire borders, I was trying to find out everything I could about the Army Air Corps – the point where, I had discovered, being a soldier and being a helicopter pilot could happily come together.

I found a book in the school library – *Today's Army Air Corps* by Paul Beaver – and devoured it. Somewhere along the line, I also came by a fourteen-page colour recruitment booklet, published by the Army and titled *Officer, Army Air Corps*, which I took home and pored over, drinking in the pictures of Lynx and Gazelle helicopters lined up in the snow, of soldiers in flight suits and headsets concentrating hard on maps – and possibly concentrating even harder on trying not to look like they were posing for the photographs.

Amid the story this booklet told about the potentially rewarding career path that lay ahead of you in the AAC, I was especially drawn to the paragraph headed '*Confidence in the Air*'. '*Your first fixed-wing flight in a Chipmunk will be one of the most memorable experiences of the course,*' the booklet breathed. '*To begin with you'll fly "dual" with your instructor, but as you gain in experience, confidence and skill you should be flying solo in a matter of weeks . . . You then move on to basic helicopter training with some 55 hours of flying Gazelles.*'

Sounded pretty cool to me. Yet I was also already aware that the Army Air Corps was not considered cool by everyone. By comparison with the SAS, say, or the Parachute Regiment, or the Cavalry, or aspiring to fly fast jets, to some it seemed to lack glamour – seemed to be regarded as a bit of a backwater, in fact. The RAF – and, indeed, other parts of the Army – sometimes referred to the Army Air Corps dismissively as 'Teeny Weeny Airways'. (A nickname that vanished overnight when the Army Air Corps acquired the Apache attack helicopter – but we'll get to that later.) None of that put me off, though – not even for a second.

My heart was telling me that, if you got to be in the Army and you got to fly helicopters, then the Army Air Corps was obviously the most perfect amalgam imaginable. And if that was where your heart was, who cared about 'cool'?

It must have been late on in my time at Chichester. They held a big careers convention in the school hall, tables and stalls laid out, festooned with brochures and manned by company reps and careers advisers so that we could all mill around, in the company of our parents, and work out what we wanted to do when school was over. But I already knew. I went straight to the Army desk and told the man sitting behind it that I wanted to join the Army Air Corps.

He gave me a patronising smile and a slow, sage shake of his head.

'You don't want to do that.'

But I really did. At that point in my life, I knew it as certainly as I knew anything.

CHAPTER THREE

The wide world opens up, a voyage to
Middle Wallop, and my wild nights as
Chichester's cocktail king

The 1990 prize-giving at Chichester High School for Boys took place in December in the school hall, under the direction of the headmaster, Mr Austin, and in the presence of distinguished guests: including Professor Alec Eden, director of the Christian Doppler Institute in Salzburg; and Councillor Spawton and Mrs Spawton, the deputy mayor and deputy mayoress of Chichester.

As a member of the outgoing year group, I knew the honour of being summoned to the stage twice on that occasion. The John Wiley Award for Physics? No, not for that. The Ruth Wrigley Prize for Mathematics? No, not for that, either. The Rotary International (District 125) Public Speaking Shield? No, not even for that, actually. The academic decorations, along with many others on the night, went to my more distinguished peers. And this was probably only to be expected after my performance in the exams that previous summer.

For the record, I sat for A levels in Physics, Maths and Chemistry, and when I opened the envelope in August, I found I had got a C, a D and an E.

Now, let's be positive: that's three A levels, which is not to be sniffed at – although Rebecca, my wife, maintains to this day that E didn't actually qualify as a pass at that time. But she is wrong. It did, and it does, and it always will.

Three A levels, then: something to be proud of. Nevertheless, I would have to admit that a C, a D and an E wasn't exactly flying high, grades-wise. And they certainly weren't the kind of grades that won prizes and a handshake from Councillor and Mrs Spawton at the annual Presentation Evening.

My two trips to the stage were: first, as part of a batch of nine of us who were given our hockey colours; and second, on my own, to receive what was described in the programme as a 'Flying Scholarship', but which was really just the school's generous way of acknowledging that, after the summer exams were over, in June and July, I had earned a scholarship from the RAF at Compton Abbas in Dorset, logging thirty hours in a Cessna 150 Aerobat. (I was turning out to be a far more natural pilot of real planes than I was of remote-controlled ones. Which was probably just as well for the residents of Compton Abbas.) That Flying Scholarship wasn't so much a school prize, then: more an out-of-school prize.

Weighing all this up, I don't suppose it occurred to anyone in the school hall that evening – and to me least of all – that some of us would be reassembling on the premises, nineteen years later, to see a building named after me. In fact, if the headmaster had stood up at the end of the ceremony and said, 'We look forward to welcoming you again in July 2009 for the unveiling of the brand new Tim Peake Sports and Conference Centre,' I'm not sure what the reaction would have been. Maybe a bewildered silence would have descended, with the parents and students and Councillor and Mrs Spawton exchanging puzzled looks. Who knows? If the head had timed it right, he might even have got a laugh out of it.

Yet it came to pass. My old school chose to put my name on a building after I was accepted on to the ESA astronaut programme in May 2009. I went back that summer for the formal ceremony to find my physics teacher, Mr Gouldstone, still there. We exchanged some rueful remarks about my performance at A level. He remained, as he had been at the time, forgiving. In 2015, students at the school would gather in that new building, accompanied by a BBC news crew, and watch the television coverage of the rocket launch from Kazakhstan as I left Earth for the International Space Station. It's a surreal thought, even now. Crazy. None of this was in any way prefigured.

Still, one thing I had achieved at school was absolute clarity about what I wanted to be doing once school was over. And what I wanted to be doing was joining the Army Air Corps. I'd had lots of conversations with my Cadet teachers and career advisers about how the process worked. You could go straight in from school as a soldier, or you could join as an officer, via the Royal Military Academy, Sandhurst. The teachers thought I would have a good chance of getting into Sandhurst, so that's what I decided to aim for.

At the same time, for all this certainty of purpose, it wasn't as though my career plan extended off to the furthest horizon. On the contrary, one of the reasons the Army appealed to me at that stage was that it offered what are called 'short service commissions'. In some regiments, these were merely three years long, although in the case of the Army Air Corps the shortest was six. Even so, that didn't feel too threatening, as a block of time. I wouldn't have to feel I was committing myself to an entire life of military service, which would have seemed daunting.

At eighteen, I was thinking I would probably serve my six years, do lots of flying and other interesting stuff and then, somewhere in my mid- to late twenties, come out and get a 'proper' job. I had some fairly clear notions about what that 'proper' job might be. Maybe I could go and live in the Brecon Beacons and run an Outward Bound school. If you had asked me to name my dream occupation outside the military, that's what I would have told you. And I could see how a few years in the Army would amply qualify me for it.

I didn't envisage that, halfway through my short service commission, the Army Air Corps would offer me a regular commission, and that I would end up staying in the military for eighteen years. That wasn't the game plan at this point, any more than 'becoming an astronaut and going to space' was.

For Sandhurst, you needed to have chosen a regiment in advance, to act as your sponsor. And to help you make your choice, and to show you what you were letting yourself in for, the regiments encouraged you to go on familiarisation visits. Although you felt a little under pressure to leave a decent impression of yourself, those visits were good fun because they were as much about the regiment selling itself to you as about you selling yourself to the regiment. They were recruitment opportunities, and 'Join us, and all this could be yours' was the essential message they were trying to impart.

The Army Air Corps familiarisation took place at the Corps' headquarters in the splendidly named village of Middle Wallop in Hampshire (one of 'the Wallops' – Over Wallop, Middle Wallop and Nether Wallop). I got shown around the hangars and whizzed about in a helicopter and generally had a ball. I would have been sold, if I hadn't already been.

I also went on a familiarisation visit to the Royal Engineers at Chatham in Kent, where they took us out on the river in Rigid Raiders, which was an absolute blast. And then they led us to a pit and set us all a task where we were given a small amount of plastic explosive and a detonator and challenged to use our skill and judgement to place the explosive around a mock-up of a concrete building in a way that would destroy the entire structure. And of course, we all got it wrong achieving partial destruction at best (even those misspent hours with Paul Hunter and our US Special Forces guide hadn't prepared me for this), after which they showed us what we should have done to achieve a perfect demolition. All this was an extremely compelling way to spend a day, and I was grateful to the Royal Engineers for laying it on, but I couldn't be swayed. I still wanted to join the Army Air Corps.

In due course, and much more nervously, I went back to Middle Wallop for my formal selection test. As mentioned, I had done my reading of Paul Beaver's *Today's Army Air Corps*. I knew a bit about the Corps' history – how it had functioned in the Second World War in a very different and much broader form which included the Parachute Regiment; how, in its present form, it had come together in 1957 as a merger of the Glider Pilots Regiment and the Royal Artillery's Air Observation Squadrons. I knew that the Army needed to have its own pilots to fly Army-specific missions – reconnaissance, say, or casualty evacuation from the battlefield, or resupply, or aerial observation – and that most of those jobs quickly became work for helicopters.

I also knew that the British Army was, at this time, exploring the possibility of buying an attack helicopter, and was considering the merits of various models. The contenders included the Bell Cobra, the Apache and the Black Hawk (all American models), and the Denel Rooivalk from South Africa. ('Rooivalk' is Afrikaans for red falcon.) I was able to do a bit of research into all of this and then talk about it in my interview.

Superficially that interview was a highly intimidating set-up. Five senior officers, including a brigadier, sat opposite me across the table in the formal setting of the Green Room in the Officers' Mess, where military memorabilia loomed above us on the walls.

At the same time, my years in the Cadets had already exposed me to these kinds of austere military surroundings, so the environment wasn't unfamiliar to me. Plus the brigadier who was in charge, Ed Tait, was a lovely man who took the trouble to put me at my ease and seemed more concerned to find out about me than to try and catch me out or hunt for my weaknesses. And I had gone along pretty thoroughly prepared, which is always to be recommended. It went well and I got accepted.

I then had to go in front of the Army's Regular Commissions Board (now known as the Army Officer Selection Board). That involved going to Westbury in Wiltshire and completing a two-day selection course, featuring some written tests but also with a lot of outdoor challenges, group and individual, designed to highlight things like initiative and leadership potential. For instance, we were presented with a big, cumbersome box of ammunition and instructed to get it across a river using ladders and ropes. Everybody was wearing numbered bibs and the examiners were looking on with clipboards, making notes about who was doing OK, who could lead, who could follow, and how well you could communicate. My voice had long since broken by this point, by the way, so I didn't have any problems with unacceptable squeakiness. At any rate, nobody mentioned it if I did. Again, I passed and, with that on top of my acceptance from the Army Air Corps, I secured a place at Sandhurst for the standard year of officer training, starting the next autumn.

Lots of my peers were going to university to study for three- and four-year degrees, and my sister Fiona was already there, reading French and Russian Studies at Keele. I had a few pangs about not joining them. At the same time, not many of those friends could say what they would be doing beyond their degree courses, and maybe the uncertainty of that wouldn't have suited me. I seemed to like the reassurance of knowing what the next few steps in my

life were going to be, which in this case was Sandhurst, then flying grading, then the Army Pilots' Course, then an attachment.

And then the Brecon Beacons. Maybe.

My first term at Sandhurst would begin in September 1991.
Before that, I had a year to myself. A whole year of freedom! But
I had plans to fill that, too. In the summer, all being well, I would
make the trip of a lifetime to Alaska. But before then (and in
order to help pay for the aforementioned trip of a lifetime), I
would become Chichester's answer to Tom Cruise.

*   *   *

The Nag's Head was on St Pancras in the centre of Chichester,
part of a terraced strip of shops and businesses – a typical English
town pub with a Tudor beam frontage and old, thick oak doors
opening directly on to the pavement. Inside were two bars and a
small dining area and upstairs was a handful of hotel rooms, mostly
booked by tourists or business people but occasionally used by
locals whose consumption downstairs had rendered going home
at the end of the evening either a bad idea or technically
impossible.

During the week 'the Nags', as it was known, saw a slow but
steady stream of custom from the betting shop next door and from
the workers at the Shippam's meat and fish paste factory, just
round the corner on East Wall. At the weekend, however, the
place would be packed out, transforming into one of Chichester's
premier young person's drinking destinations.

Behind the bar, arms aloft, the handsome young Frenchman
Olivier Barbedette would be emptying two bottles of mixer from
an unnecessary height simultaneously into two glasses. Passing
behind him on my way to the till, I would gently touch his back
to let him know that I was there. Fixing the gaze of the wide-eyed

customer waiting in front of him, Olivier would nonchalantly toss the empty bottles over his shoulders, where, apparently through the wonder of telepathy alone, I would be on hand to catch them.

Just one routine from our extensive catalogue of novelty bar moves – 'flairtending', to use the accepted term for the art. Other specialities: glass juggling, bottle spinning, and lobbing stuff the length of the bar. The Tom Cruise movie *Cocktail* was nearly three years old at this point, but here in the Nags it might as well have been this month's hottest release and from Thursday through to Saturday, I was Brian to Olivier's Doug. Like the tag-line on the movie poster said, beside the shot of Cruise behind the bar, '*When he pours, he reigns.*' Same for me and Olivier. We poured, we reigned.

You had to know your mixology, obviously. A Black Russian? Five parts vodka to two parts coffee liqueur, over ice. An Alabama Slammer? Southern Comfort, sloe gin, amaretto and a double slug of orange juice with a cherry for garnish. A Sex on the Beach? Vodka, peach schnapps, orange juice, cranberry juice, shaken with ice and strained into a high glass. Frankly, though, it was all in the style. Even a pint of bitter shandy can become a feature occasion if you pour in the lemonade from high enough.

Cheesy? Maybe. A complete laugh? Definitely. I had needed work in order to get some money together for the summer expedition to Alaska that I was planning. But I hadn't expected to find work as entertaining as this. Barman at the Nags was a magical job to fall into, not least for a teenager fresh out of school. And Olivier was a great partner to find myself working alongside. He was twenty-four, cool, self-assured and very funny – a phenomenally hard worker who knew how to enjoy himself thoroughly, too. I was mightily impressed by him. I'm not sure what had brought him to Chichester but there was quite a strong French crowd in the city and I think a friend had lured him over.

One thing was clear, though: Olivier loved bar work. Indeed, he would eventually go back to Paris and work for a number of years in the Bar Le Breguet in the Marais district. And I would eventually visit him there after graduating from Sandhurst. For now, though, he was devoting his immense energies to creating a party atmosphere in the Nags.

Courtesy of Olivier's grandparents in Normandy, a bottle of home-made Calvados was frequently to be found stashed behind the bar – legendary stuff that would strip paint. We would practise our flair routines in the side alley with empty bottles filled with water, create a merry mess and then have to sweep up the broken glass before opening time. There were, just occasionally, costly mistakes – damage sustained in action. I seem to recall a stray bottle taking out the gin section one night. But it was nothing that couldn't be covered up after a good night's takings – and on those weekends, the takings invariably were good because the place would be jumping.

I didn't have much luck with girls before the Nags. Did I mention anywhere in these pages that I was ginger? And a bit small? I did? OK. Throw in on top of that an all-boy secondary education that seemed to have left me intensely shy around women, and I was effectively stymied, romantically speaking, for a few years there. Claire Smith, my Whitley Close neighbour, and I had been boyfriend and girlfriend for a while in our early teens, but we were more like brother and sister, having known each other since we were two years old, and decided that being close friends worked better.

But now, at eighteen, behind the bar of the Nags and under the expert tutelage of Olivier, I found myself surveying a whole new world of opportunity. There were two women in particular who, at different times, I had managed to get talking to between feats of bottle juggling, and with whom I seemed to be getting on very well, if I said so myself. And my impression seemed to be confirmed when one of those two women came in on her own to see me on my Monday night shift – a quiet evening in the bar, with just me on duty. This was a highly promising development.

It seemed less promising when, moments later, the door opened again and the second of those two women walked in, also on her own. These two didn't know each other, but now the pair of them were sitting side by side at the bar while I fumbled with the drinks and tried to think of somewhere I could take the conversation. I simply had no idea how to play the wretched hand that fate had dealt me. So flustered was I by the whole experience that, closing up at the end of the shift, I left the front door of the bar open. Not just unlocked: wide open. Olivier found it that way in the morning and properly chastised me. Nothing was stolen, thankfully.

When those big weekend nights ran on past closing time and things got a little messy, the Nags proved to be the perfect venue. Instead of making my way unsteadily back to Westbourne, I could crawl upstairs and crash out in one of the vacant hotel rooms. Up until this point, my idea of nights away from home had been family holidays and Cadet trips. Now here I was at the centre of a constant good-time throng, with limitless access to alcohol and a bedroom if I needed it. If the university experience was substantially about partying (and reports coming back from reliable people on the ground seemed to suggest very strongly that it was), then it could be argued that I did my university years in condensed form in those six months of 'flairtending' at the Nag's Head.

Those Tom Cruise bar skills are forever, by the way. Well, kind of. Flash forward to Bosnia during my deployment with the Army Air Corps' 652 Squadron in 1996 – a warm, sunny afternoon on the helipad where I am giving my fellow officer pilot James Gwynn the benefits of my expertise in an impromptu flairtending tutorial.

I am attempting to teach James mastery of the 'bottle cross', where you simultaneously throw bottles to each other from behind your backs and they pass thrillingly close in mid-air. Or, at least, you hope they pass thrillingly close. Our bottles collide halfway. James ends up catching a miraculously intact bottle and I end up getting covered in a shower of broken glass.

Laughter inevitably ensues – and then I look down at my feet. I am standing in a red puddle, and it isn't wine. A shard of broken glass has slashed my right ankle. I am able, at least, to reassure myself pretty quickly that it's a severed vein and not a severed artery: the blood is flowing dark red, as if from a low-pressure fountain, rather than pulsating. Even so, it's more than a scratch. I clamp my hand over the wound and hobble off the helipad to our ops room where I sheepishly ask the duty medic if he would mind turning his hand to a few stitches.

I still have the scar. Let no one say that being Tom Cruise in your spare time doesn't involve sacrifices.

CHAPTER FOUR

*Encounters with whales, a marathon struggle, and
some adventures in primitive lavatory construction*

You probably read in the papers about my selection for the
Operation Raleigh Alaska Expedition, May to August 1991.
It was all over the media at the time. Or, at any rate, it was a story
in the *Chichester Observer*, where this breaking news appeared
under the headline 'TIM GOES ON THE BIG ADVENTURE',
alongside a mug shot of me in a school hockey shirt looking about
eleven. In case you missed it, here are the opening paragraphs:

Westbourne cadet warrant officer Tim Peake has been
selected for Operation Raleigh, the adventure training course
for young people.

Tim (18), from Westbourne, is a senior cadet of the
Chichester High School for Boys combined cadet force.

*Next May, he will be off on a three-month expedition to
Alaska in what should be the experience of a lifetime.*

In an exclusive interview, quoted a little further down in the
piece, I revealed to the *Observer*'s reporter that I had just started
my Duke of Edinburgh Gold Award, of which the expedition
would form a part. I also confessed to enjoying rock climbing,

43

hiking, orienteering and canoeing. It was 'hold the front page' stuff, most certainly. Or, in this case, 'hold page seven'.

Still, it was the first time I had made the papers so, naturally, I was very proud. What's more, the *Observer* was dead right in its assumption: those three months with Operation Raleigh would indeed prove to be the experience of a lifetime. Here I am now, in my forties, a qualified helicopter test pilot with eighteen years in the Army behind me, having travelled to space. Yet I still find myself from time to time thinking back to a trip I made when I was nineteen and reflecting on how much it shaped my attitudes and altered my aspirations. In any number of subsequent situations down the years, I have drawn on things that that expedition showed and taught me. It was, like the headline said, a 'big adventure', and an absolutely formative one.

Operation Raleigh was quite a new organisation at the time. It had been launched in 1984 as a land-based follow-up to Operation Drake which had been arranging seagoing expeditions for young people since the late seventies. A year after I became a 'venturer', they changed the name of the project to Raleigh International, which is what it goes by now, setting up educational voyages to remote places, normally incorporating a scientific or environmental element, or with some aspect of community service.

The proposed 1991 summer trip which had caught my eye was for a trek to south central Alaska. Here the chosen participants would get to do some forestry conservation, some river surveying and some construction work in a lagoon, with lots of camping, kayaking, mountain climbing and some glacier exploration thrown in. It sounded like an ideal way for me to occupy the time between school and Sandhurst. I would need to find £2,500 to fund myself, which wasn't the kind of money my parents were going to be in a position to fork over. But I calculated that six months of steady employment in the run-up ought to cover it – and, indeed, six months of bartending at the Nag's Head eventually did.

First, though, I had to attend Operation Raleigh's selection weekend. I had heard that these could be quite exacting – understandably. Aspects of the trip were likely to be physically arduous and you were going to be a long way from home for quite a while, so the organisers would want to be sure they were taking people who could hack it. But I went along to the Raleigh training and selection centre in East Grinstead feeling pretty sure of myself. I had been through a lot of this kind of thing with the Cadets – had done a decent share of assault courses, night hikes and comfort-free camp-outs. Heck, I had even taken part the year before, with my school, in the 100-mile Canoe Test – an annual endurance event involving four consecutive days of canoeing on a set course, after which it took my arms about a month to recover. I didn't see how the Operation Raleigh people could throw in anything during this selection weekend that was likely to surprise me.

I was about to be surprised.

The first startling development was when, freshly arrived, we were instructed to stand in a line and tip our bags out onto the ground, shaking everything into a pile at our feet. We were then issued with a smallish overnight rucksack and instructed to move five places down the line. We now had twenty seconds to pack the rucksack from the pile in front of us with everything we thought the owner of those things might need for the night orienteering exercise that lay ahead of us.

Oh, and this wasn't a purely academic exercise, by the way. The kit you ended up with when you were eventually reunited with your stuff would be the kit you took away with you for the night.

Now, for me, this was unsettling straight away. I was very meticulous about my kit and extremely organised: I had put a lot of thought into my packing and had brought with me the things I imagined I would need to make a decent go of it. The idea that a) my carefully ordered possessions now lay in a heap on the ground and that b) some complete stranger was about to pick through it all and select which items would be coming with me for the night was instantly disruptive.

But what can you do? I tried to set aside my psychological discomfort and, on the word 'Go!' from the instructor, attempted instead to concentrate on picking the right things from the mound of somebody else's underwear, socks and other belongings in front of me. Meanwhile, along the line, one of my colleagues was busy assembling me a rucksack which (I would later discover) contained no sleeping bag and no swimming trunks. So a cold night lay ahead of me and, what's more, in the early hours of the morning, when we got to the part of the exercise which involved a 500-metre swim in a bitterly cold outdoor pool, I would be doing it in my underpants.

Thanks a bunch, colleague.

The second big surprise came in the form of a challenge which was definitely new to me: a blindfold obstacle course. We were broken into groups of three or four and everybody had to put on a blindfold. Then, while following a rope that wove its way through trees, brambles, mud and water, the leader had to use all their resources to guide the group over a series of increasingly challenging obstacles. Just to add another element, one of the girls in the group that I ended up leading was deaf. Deaf and now blindfolded. One section of the course featured a short but extremely confined and water-filled tunnel. The only way to get through and continue onwards was to duck under the water and swim. Experiences like that stay with you. I think I learned more lessons about leadership and communication and trust-building in the minutes that challenge took to complete than I had learned in my whole life up to that point. All respect to Operation Raleigh for creating a selection process with that much rigour built into it.

Anyway, we made it through the tunnel intact, and I got through the rest of the selection process, and soon afterwards a letter arrived at home confirming that I had been offered one of the seventy places for Alaska. I was thrilled. This would be my first journey outside Europe and my first serious chunk of time away from home. It felt like a big step out into the world.

We flew from London to Minneapolis, then from Minneapolis to Seattle, where we spent the night on the floor in our sleeping bags in a downtown church hall. The following morning we took another flight from Seattle to Anchorage, where two of those typical American yellow school buses were waiting to drive us south for three hours through mountains and glaciers to Moose Pass, 4,500 miles from Westbourne.

Here we were given beds in huts. Or sixty-two of us were. Eight of us had to sleep outside on the ground. I'm not quite sure how I drew the short straw there, but I did. We were so far north that it was still broad daylight at 11.00 p.m. There was also snow on the ground. What with the combination of the excitement, the cold, the damp and the jet lag, I passed a largely sleepless night.

The following evening, four of us would find an equipment hut with some room on the floor and sleep very soundly in there. In the meantime, there was the orientation process to get through. It was only a few hours ago that I had been sitting in a café at Gatwick Airport, and now here I was in a wooden hall, being lectured on the dangers represented by bears, sea lions and moose. Briefly, bears you probably know about. Sea lions are large and unpredictable, with very sharp teeth. And the best advice should a bull moose charge at you is simple: run. I didn't know whether it was a relief, or simply alarming, to learn that when we were out in the wild we would always have a 12-bore pump-action shotgun along with us, just in case.

Of course, those 'big three' weren't the only hazards we would need to be alert for. I picked up a copy of the *Anchorage Times*, the local paper, which offered the following compelling list of Alaska's other attractions:

> While visitors don't have to worry about rattlesnakes, skunks or poison ivy, there are calving glaciers, roaring glacier-fed rivers, avalanches, prickly devil's club and cow's parsnip that can blister the skin, deadly cold ocean waters, killer whales, treacherous mountain edges and hypothermia. And if that isn't enough, Anchorage is surrounded by mudflats that are like quicksand, the state's gigantic mosquitoes have been known to cause allergic reactions and a good number of Alaska clams contain paralytic shellfish poisoning that can lead to paralysis and even death.

A few things to think about there, clearly.

Anyway, orientation continued with a preliminary day-trek out into the bush. Here, pleasingly, I encountered no outsized insects or skin-peeling plants, and certainly no charging moose. Instead our biggest challenge was the fifty-year-old map that we were issued with, which, for obvious reasons, didn't take into consideration alterations to the landscape caused by the Great Alaska Earthquake of 1964 – a 9.2 megathrust which is still the largest earthquake to have afflicted North America and which left all maps to that point struggling to keep up.

Top hiking tip: use a recent map if you can find one.

These initial exercises were a chance to get to know a few of the other venturers. Raleigh try to assemble diverse groups for these trips. Our intake included people from Lytham, from Cardiff, from East Kilbride – from Wellingborough, from Leicester, from Newcastle. We were mostly British, but there was also a sprinkling of people from Australia, someone from Kuala Lumpur, someone else from Trento in Italy. And there were two guys, Iain Bersten and Kevin Hewitt, from the Falkland Islands. They had grown up during the Falklands conflict and it was fascinating to talk to them about that and also about the lifestyle down there, which was completely alien to me. There were some public-school kids in our selection, but there were also kids with inner-city backgrounds and kids who had had a tough time of it growing up. My background was solidly middle class, solidly southern – solidly 'ordinary', as I have said before. This was a major eye-opener for me – my first proper exposure to people from beyond my own social set.

It was fascinating, too, to watch the personalities emerge: to notice the ones who were natural leaders and could instantly take people along with them; the ones who were bossy, yet lazy; the ones who didn't pull their weight; the ones who moaned; the ones who were struggling to adapt, and the ones who took to it as naturally as breathing. Where would these guys have been placing me in those first days? I suspect that some of them would have filed me initially in the category of 'the ones who are small and ginger'. I probably seemed quite quiet, too. I had lost a lot of my shyness during those months chucking bottles around in the Nags, but I still wasn't the kind of person who would immediately dominate a large gathering.

As time wore on, I would gradually begin to speak up more, and put myself forward a bit more firmly. But I had also started to work out that it was often better for me to assert myself by doing things rather than merely vocally. For instance, in the expedition's earliest days, I twice in succession found myself given the unglamorous task of digging the hole for the camp loo. That seemed unfair, and I guess I could have complained about being typecast too quickly as the camp bog-digger. Instead, I went the extra mile with the second loo and fashioned a wooden seat for it. This was wildly appreciated by my camp-mates and fellow squatters and earned me some early respect. What's the old expression? 'When life gives you lemons, make lemonade.' Well, equally, when life gives you camp lavatories to dig, make comfortable ones.

Triumphs in loo-building aside, the expedition got off to a challenging start. My group of twelve was initially put on forestry conservation work. We got a train to Whittier and were then transferred by power boat to Harrison Lagoon, screaming through Prince William Sound and passing sea otters and killer whales. That was a hair-tingling opening to the experience but unfortunately the weather immediately, and literally, dampened everybody's spirits. It rained and rained with no sign of a let-up.

Our task was to construct a gabion weir. This was intended to channel water from a glacial river into a smaller stream which apparently had salmon-spawning potential but which had been drying out in the winter, exposing the eggs and killing the salmon before they could hatch. Gabions are wire baskets that you can place in the water and fill with rocks to create a dam. So the work was essentially a matter of wiring gabions together and pulling rocks out of the river to fill them with. Most of the time we were thigh-deep in freezing-cold river water. And on top of that it was raining.

Wet weather and camping are a notoriously unhappy combination. You would go from damp clothes to damp sleeping bag to damp clothes again. If, after a long day in the river, you could manage overnight to get your wellies down from soaking to merely damp, you were pleased. Then things started to go wrong up at the work site. In an unfortunate double whammy, Kevin from the Falklands put a shovel through his foot and damaged himself, and then, while sitting on the ground recovering, went down with hypothermia and had to be returned to the camp. Over the next couple of days, three more venturers from our group of twelve had to limp away from the river with hypothermia. That was a third of the workforce down.

Just to add a little edge, before bed one night I wandered off to have a pee and the tree I chose had fresh and still-steaming bear droppings at the foot of it. They hadn't been kidding us during the introductory lectures, then.

In the diary which I dutifully kept through the expedition, the entry for 28 May states, somewhat poignantly, 'No cases of hypothermia today.' I was also clearly melancholy enough to list the things I missed: 'Bath, bed, airing cupboard with warm dry clothes, CD player, kettle, sofa, house, home . . .'

On the plus side, I had learned how to cook damper, the Australian bush food. Recipe: mix some flour with some water and some salt to form a dough. Slap the dough on the end of a stick and then shove it in the camp fire for ten minutes or so. Remove, slather with jam and consume. Delicious. Or certainly if you've been up to your thighs in glacial water all day.

After ten days, as soon as the river task was completed, the rain finally let up. But by then we were off by boat up the Coghill River, travelling into dense woodland where the hummingbirds were delightful and the mosquitoes less so. Our task was to pitch camp and then help in the construction of a long boardwalk path with steps and bridging, to link up the river with a public cabin at Coghill Lake, thereby opening out the area to hikers and kayakers.

Again the work was tough and the conditions were challenging, and all of us struggled from time to time, but the surroundings were always there to lift our spirits. I remember once walking a short way out of camp to a viewpoint and seeing the cloud and mist form over the mountains, with the water absolutely still in front of me. It was a primal scene – almost like being catapulted back through billions of years to when the Earth was being formed. You didn't get that sensation very often in Chichester.

Or there was the time when we were all sitting around the campsite one evening when there was a great rush of air from the water and a huge humpback whale surfaced directly in front of us, almost as if posing for a photo. Everybody dropped their mugs, radios, diaries or whatever they were holding and rushed to the beachhead to watch the whale finish posing, duck back under and disappear off in the direction of Whittier, bound for its next photo shoot.

I was sharing a tent with a dry, quietly spoken guy from Devon called Jeremy, with whom I got on extremely well. One day the pair of us decided to head off and look for gold. This wasn't an idle fantasy: the Alaskan ground was rich in minerals. Silver we had already come across, and we knew there was gold out there. We took a pepper-spray canister with us, in case we happened on a wandering bear – or rather, in case a wandering bear happened on us. Even then the supervisor, Mark, stopped us as we were leaving the camp and insisted that we take the shotgun. So suddenly I'm off in the mountains with a shotgun, panning the river for gold, like some kind of nineteenth-century pioneer. We did find some gold in the water, by the way – but only faint dustings of it in negligible quantities. Still a thrill, though.

However, the activity on the trip that I was most looking forward to was the kayaking. Our group leader, a great ex-Para from Worcestershire called Mark Bridgeman, described this three-week portion of the expedition as 'the vacation part'. Well, eventually maybe, but not at first. Close to the coastline, we exhausted ourselves in the first couple of days trying to kayak through 4-feet tall waves into strong winds. The weather and the water soon settled, though. The water remained treacherously cold – survival time in it would be about four minutes, so you needed to be pretty confident that you had your capsize routine down – but we were bathed in virtually unbroken warm sunshine of the kind in which you could lie out, under searing blue skies. Our lives at this point became rather wonderfully laid-back and dreamlike. In my diary, I would eventually rate these three weeks as '*the best time of my life*', further describing myself as '*totally at peace with the world*', and adding, '*Tonight there is absolutely nothing that I yearn for.*' And even though I had consumed my share of a bottle of Southern Comfort round the campfire directly before writing those late-night words, I stand by them and the sense of teenage contentment that they were trying to convey.

One of our tasks was to map with GPS some of the coastline of Prince William Sound which had previously only been charted by aircraft. We had primitive Magellan GPS devices, which were the size of a house brick, and we mapped the coastline as we travelled. Most of the time, though, we were free to hop from island to island. Humpback whales, sea lions and otters would be popping out next to you, eagles would be soaring overhead and all around you was this classically beautiful backdrop of Alaskan mountains. And then, at the end of the day, you would pull the kayaks onto a beach, drag some dry wood together, get a bonfire going and cook up the night's meal. It was bliss.

One time we decided to take advantage of the clear skies and paddle through the night to our destination of Olsen Island. It was nearly midsummer, and hardly dark at all at night, but at one point a huge moon rose to our left while an orange sunset played out to our right and Jupiter shone like a diamond in between them. We rafted our kayaks together, tied a tarpaulin around the paddles to create a sail, and for the next half-hour used the strong backing wind to blow us forwards. When the wind dropped, we went back to paddling, until the sun started to rise again, causing a lilac-coloured mist to come pouring down the hillsides and across the water. We pulled up, made a breakfast of sausage and beans and then fell asleep in our tents for a few hours. Only when we woke up did we realise that we had completely overshot Olsen Island and travelled on another five miles to Little Axle Island. Overawed, we had squeezed two days' worth of kayaking into one night.

The one blip during this idyllic phase of the adventure was the cock-up we had made with our provisions for the trip. Halfway through the three weeks we did an inventory check and discovered that our food stocks were shockingly low. We worked out that we must have loaded up at the beginning of the trip with a third less food than we would need – yet we had been merrily eating for the first ten days as if we had three weeks' worth on board. This meant that strict rationing had to come in for the second half of the trip. We started dragging fishing hooks with bits of tin foil behind the kayaks in an attempt to lure salmon – which occasionally worked, in fact. Also, every now and again we encountered a passing trawler and would shout up and beg the crew to throw us down any food they could spare. That also worked pretty well and the odd can of beans, loaf of bread or, better still, freshly caught fish would supplement the evening's rations.

Nevertheless, the meals did get a bit skimpy. We were never in any danger of starving, but the general hunger levels rose – and with them the general irritability in the group. Interestingly, some of the people who had been the life and soul of the party when things were going smoothly now became absolute bears and really struggled with the situation. At the peak of the rattiness, one of the larger guys among us suggested that rationing ought to take into account individual body size. That didn't go down terribly well. It was revealing to see people's reactions when things got a little uncomfortable. People often ask me about the soft skills you develop for spaceflight and the psychological preparation you do, and I realise that, for me, in some ways it started right back in Alaska. Situations would develop later and I was able to draw on some of these experiences – to think, OK, I've been here before. I know why this person is getting upset and why their personality suddenly seems to have shifted.

Again, though, amid all these potential irritations, the natural surroundings were there to override any discomfort by taking you out of yourself completely. Like, for instance, the hour we spent picking our way through icebergs in the brilliant white light of Colombia Glacier. Or like the moment on Midsummer's Day, just after lunch, when Mark plunged into the water off a rock and got out to dive again. I was just taking my boots off to join him when two 12-foot killer whales swam right into our diving pool. They must have been attracted by Mark's swimming. They rolled around in the water in front of us for about a minute, putting on a show. I could have reached out and touched them from where I was standing. And then they carried on around the corner and were gone.

*   *   *

We got back to Moose Pass camp from the kayaking trip on the eve of 4 July to find everyone making plans to go to the Independence Day celebrations in the nearby town of Seward. The festivities there would include the Mount Marathon. First staged in 1915, the Mount is said to be America's second oldest public race after the Boston Marathon. People travel hundreds of miles from other states to take part, there's a fair with stalls and a parade, and the whole town seems to turn out to watch. But it's not a marathon in the Olympic sense: it's a 3.5-mile run, straight up a 3,022-foot peak and straight down again.

So you've got a lung-busting ascent, followed by an uncomfortably steep, 40-to-50-degree return, slipping and sliding on the mountain shale – 'nature's razor blades', as one commentator has described it. By the end of the race the medical tent right by the finish line is typically doing a roaring trade in Elastoplast and bandaging for the unlucky contestants who have managed to strip the skin from their legs and arms. The average competitor is looking at a stint of high-risk agony lasting about an hour – assuming you stay upright.

I loved the sound of it. A few of the Raleigh people had managed to sign up in advance to take part. I was really keen to join them but by this point the deadline for entry was well past. However, one of the staff members, Joe Cornish from London, had registered but dropped out. It seemed like I could have his race number and his electronic timing tag and nobody would know any different.

So the next morning I found myself getting up with the dozen or so other runners to feast on a breakfast of porridge and pancakes, followed by a plate of fried-up pasta – carb-loading, Moose Pass-style. We then drove into Seward, walked around the stalls and watched the parade of floats until the races started. Further fortified by coffee and a Mars bar, I made my way to the start line. Ahead of us, the mountain rose up alarmingly sharply. You had to guess where the top of it was because it was shrouded in cloud. I began to wonder whether this was such a good idea after all.

No time to turn back, though. The klaxon sounded and we were away. On the flat road segment, leading out of downtown Seward, I came from the back to make up some ground on the rest of the field, and I was still passing people as we reached the mountain and began the ascent. Soon after that, however, the pace really began to tell. As the greenery off to the side became rocks and we passed up into the mist, my legs began to complain almost audibly. By the time I reached the summit, with the air noticeably thinning, my lungs were practically sending up distress flares.

Still – downhill all the way from here, no? Well, in a sense. But containing yourself on the downward slope, tensing your muscles to prevent yourself going over on the shale, turned out to be even more exhausting than forcing yourself upwards – especially when you had already exhausted your muscles going a mile and a half in the opposite direction. I dug in my heels and staggered down, my calves thrumming, and, despite a couple of slips and a brief passage where I seemed to lose the course altogether, somehow got to the cheering crowds at the finish line with practically all of my skin still intact.

So that's how I made it into the sports section of the *Anchorage Times* – my second-ever appearance in the papers. OK, the bulk of the report is taken up with the victory of Bill Spencer, a former Olympic cross-country skier, who won the Mount Marathon for the second time with a pretty awesome time of 44 minutes and 15 seconds. And, yes, greater space on the page is given to tales from the casualty tent. (Fifty runners apparently required treatment for their wounds that day, either as a consequence of falling on the shale or getting hit by dislodged rocks – there's a reason it's called the toughest 5k on the planet.)

But scan the results list at the foot of page eight, and there I am: Joe Cornish, in 60th place out of 135 with a time of 1:03:15.

Not too shabby. That placed me fourth out of the dozen or so Raleigh guys who had taken part, most of whom had been in training for a while, and none of whom had competed with a day's notice after just spending three weeks sitting in a kayak.

My reputation among my fellow venturers rose exponentially on the back of this performance. They were nearly all in town to see it, and people were coming up to congratulate me warmly afterards. Stardom! At the time, though, I wasn't sure what felt more satisfying: the acclaim or the hot shower that was made available to the runners afterwards, and the all-you-can-eat post-race barbecue in Seward which, after the rationing problems on the previous trip, took on the quality of a dream come true.

Full of smoked meat and bottled beer, four of us eventually hitched the 25 miles from Seward back to the camp at Moose Pass. Attracting lifts proved initially difficult, even when we went to the trouble of forming a three-man human pyramid at the roadside or limping along in an exaggerated manner. Actually, maybe it proved difficult *because* we did those things. Anyway, we owed the first 5 miles to a madman in a customised 3.5-litre Ford Bronco truck, who drove at about 100 mph with the four of us in the back, clinging on for our lives. The rest of the journey happened more or less inside the legal limits, with two women in a pickup truck.

Back at the camp there were 4th of July fireworks on the beach and more beers. It seemed as good a time as any to break out a gift that had just arrived in the post from my French bartending friend, Olivier: a small shipment of his grandparents' lethal, paint-stripping brandy. Life seemed very good and very full.

* * *

News travelled slowly. The first I knew about the accident in the glacier was three days after it happened. A few of us had walked from the Moose Pass camp into Seward to do some shopping. We were in the liquor store when we bumped into some people from one of the other groups, on their way back from trekking, and they told us what had happened.

A group of four venturers – Nicki, Jackie, Gavin and Ying Ju – and a staff member, Helen, all roped up and carrying 50-pound packs, had slipped more than 200 feet down Eklutna Glacier, a 40-degree slope, and hit rocks at the bottom. Gavin had sustained a cut to his head which needed stitching. Nicki had fractured her pelvis. Ying Ju had damaged her hip. Jackie was the worst hit. She was knocked unconscious with a triple lower leg fracture and a dislocated hip. Helen, the staff member, was a qualified nurse and she administered some first aid at the scene but she, too, was injured, with a broken collarbone.

Meanwhile Mark, who was overseeing the group, and a couple of other venturers who weren't pulled down, hiked 15 miles to the nearest campsite to get help. Helicopters from the Alaska State Troopers and Providence Hospital in Anchorage were raised and all five of the fallers were taken to hospital, after a tricky rescue by searchlight. All this, obviously, had taken some time. The accident happened at around 3.00 in the afternoon. It was 1.30 a.m. before the airlift was completed.

All of this was really upsetting to hear. Apparently Ying Ju had slipped and pulled the others after her. Everyone had tried to use their axes to brake the fall, following the procedure they had been taught, but they were in soft snow and the blades wouldn't bite. They had clearly had a very lucky escape, all things considered. Gavin and Helen were discharged from hospital almost immediately and returned to camp looking bruised and battered but otherwise fine. But the other three remained under supervision so we set up a trip to Anchorage the next day to visit them. Nicki and Ying Ju seemed to be doing OK, and it was obvious that they would be out within a day or two. Jackie, however, we weren't even allowed to see. She was no longer unconscious but she was still in the intensive care unit and she was going to need surgery on her broken leg. Her parents had been informed and they were flying out to join her.

We were very subdued, coming back from that visit. The incident firmly brought home the reality of the risks we were all taking out here. At the same time, we were young and we had the blitheness that comes with youth. I think all of us would have confessed that, even as we shivered to think about what might have happened to Jackie and the others in Eklutna Glacier, at some level we were worrying about what this incident might mean for our own glacier trip, which was imminent. Would it now be cancelled? I know I would have felt bitterly disappointed if that had happened, even after being shown graphically how badly things out there in the mountains could go wrong.

In fact, to my relief, the trip went ahead, only slightly amended. There would be no traversing. Instead we would mainly stay at the base camp and set out from there each day to climb peaks. We packed light and left, a couple of days later, for a drive in the minibus and then an 8-mile trek, including a two-hour mountain climb, to Mint Glacier, at about 5,000 feet. The hut that would be our home for the next ten days was within a walk of the glacier and surrounded by twelve mountain peaks, each around 7,000 feet – a stunning location to wake up in every morning and to play around in every day.

We had a superb teacher in Ron Enzler from New Zealand. He was young, extremely talented and just a touch crazy. Ron taught us to abseil and to use an ice axe and crampons – to the point where we were capable of doing a 650-foot climb up a 65-degree ice wall, the glacier intimidatingly far away beneath our feet, and with only a Prusik knot to hold us up. He also taught us to do crevasse rescue, and then happily threw himself down a 30-foot crevasse, trusting us to be able to use the techniques he had shown us to haul him back up again and save his skin.

Again, the breath-taking sights came in droves. I got up at 5.00 one morning to climb a ridge and watch the sun rise over Mount McKinley. I hiked across Bomber Glacier where the wreckage of an American TB-29 Superfortress, which crashed during a training mission in 1957 killing six of its ten crew, still lies strewn across the ice – an eerie vision. I climbed over waterfalls, swam (very briefly) in mountain lakes and drank hot chocolate late at night in very overcrowded tents.

And before I knew it those ten days were almost through. The following morning we would be driven back to Moose Pass to start packing up for the journey home. And we would stop in at the hospital in Anchorage to visit Jackie again, whose parents had flown out to join her and who was now thankfully on the way to recovering. But before that there was time for one last 53-metre rock climb, one last abseil, and then one last night in the local Roadhouse, 10 miles down the road, eating taco and French fries and drinking beer, and already becoming mournful at the thought that this brilliant time was all but over.

At Moose Pass, the mood was bitter-sweet. There was excitement about returning home, but also lots of melancholy at the prospect of breaking up the group and much sad talk about 'returning to civilisation'. We held a last-night party with a disco until 3.00 a.m. In the morning we staggered out to pose for an expedition photo by the lake in the rain. I didn't want to leave the people I had met, and I didn't want to leave Alaska. I felt like I wanted to stay there forever.

Still, home had things to be said for it, too. My dad came to collect me from Gatwick. Over the phone I had given him a list, which I had composed in advance with great thought at the back of my diary, of requests for things to bring to the airport. I guess the list gives an insight into what from my home life I had been missing most badly, and was most urgently looking forward to being reunited with, so I'll reproduce it here. Those things were:

1 pint fresh full fat milk
2 Scotch eggs
1 fresh cream bun (anything with chocolate/jam/cream)
4 sandwiches on brown bread, thickly buttered (suggested fillings: grated cheese, scrambled egg, sliced ham and pickle)
1 Snickers
1 Twix
1 banana

It was quite nice to see my parents, too, by the way.

So, yes, it stays with me, that Alaska trip. This is why, much later, when I was with the European Space Agency and planning my Space Station mission and considering what kinds of outreach work I could usefully do from the platform that it was going to provide me with, I decided to set my emphasis as much as possible on engaging with young people – with STEM, with the Scouts, with the Prince's Trust. I knew what it meant to me, as a young person, to be given an opportunity that ended up radically altering my horizons. I also knew that it doesn't necessarily take much at that point in your life to leave a big impression – that it can be a nudge from somewhere, a passing suggestion, or somebody showing you something that you hadn't considered and thereby opening a door that you end up walking right through. Who knows where the inspiration that will carry someone onward for the rest of their life might come from? For me it was that Operation Raleigh selection weekend, and the adventure that followed it. For another young person it could be something else completely. But all they need are ideas and opportunity.

CHAPTER FIVE

*Ironing boards on parade, clothes out the window,
and the less than gentle art of bulling*

After I returned from Alaska, I had a few weeks knocking around at home and feeling flat. That huge experience had come to an end, and saying goodbye to everyone and leaving it all behind proved hard. But then almost immediately my enlistment date at Sandhurst was looming and I had to start thinking about that. The closer it came, the more daunting it felt – a pivotal moment in my life, clearly. When I was all packed up, I lay in bed the night before, in the only house I had ever lived in, thinking, 'Blimey, tomorrow I leave home.'

The following day, as my parents drove me the 60 miles or so to Camberley, the car was very quiet. I was feeling apprehensive, without particularly wanting to admit it, and it was clearly a big moment for my parents as well – a symbolic severing of the cord as their son went off into the big bad world. Now that I have children of my own, I can envisage a similar day eventually coming in their lives and I can imagine how emotional it's going to feel. I think my mum, in particular, was putting a brave face on it.

Again, though, I can only reflect with gratitude on the way my parents fully supported me in this career plan that I had cooked up for myself. They weren't military people, they weren't from military families. (Both my grandfathers fought in the Second

World War, but that was conscription, which was different.) The world of the military was wholly alien to them. Yet here was their son who had somehow developed this overwhelming passion for Army life, and who was determined to take that direction over university or anything else on offer to him. They must have had concerns. Later, when I ended up on active duty in Northern Ireland, they *definitely* had concerns. Yet they could see that this was where my heart lay and they were willing to set their reservations aside and encourage me all the way. Being nineteen, I didn't give them enough respect for that at the time, but I certainly do now.

So there was a fair amount of suppressed anxiety in the air as we turned off the A30 and through the gates of the Royal Military Academy that Sunday morning in September 1991. However, the sight of the ironing boards lightened the mood a little. Sandhurst feels pretty grand on arrival, what with its 700 acres of parkland and woods and its black ornamental cannons, as well as the famous long white Georgian building known as Old College, whose steps and Corinthian columns dominate the parade square. But the grandeur is somewhat undercut when what also greets you is the sight of the new intake of officer cadets struggling in from the car park with their ironing boards, a piece of kit that the Academy stipulates you bring.

At this potentially sensitive point in the day, the Academy played a very smooth PR game. Upon arrival, my parents were immediately whisked away with all the others into a swish anteroom for coffee and biscuits. They would eventually be given a leisurely show-round, but only after a nerve-settling introductory talk from a charming sergeant major during which they were jovially informed, 'The next time you see your offspring, they are going to fall through the front door hollow-eyed and desperate for sleep.' (Completely true, as it happened.) Nevertheless they were amply reassured that this was nothing to be worried about, and all part of the wonderful, opportunity-filled adventure that their children were now fortunate enough to be embarking upon, lucky old them.

I, meanwhile, along with my fellow officer cadets, was issued with a red tracksuit, told to get upstairs and change, and to be back on the parade ground and lined up in thirty minutes. No coffee and biscuits for us, then, and no soft introduction. We were straight into it. There was a little window later on in which we were able to say goodbye to our expertly mollified parents. But with that they disappeared down the drive and our lives abruptly became hell for five weeks.

The RMAS has been training British Army officers since 1812. Before then, by and large, the way you rose to high military rank was by paying for it. But perhaps inevitably, that hadn't been working out too well in terms of getting the right people for the job, so the decision was taken to create a special school for officers instead. Nowadays Sandhurst reckons on turning a cadet into an officer in forty-four weeks – provided, at least, that the cadet manages to survive the first five.

I shuffled off clutching my ironing board and, now, my tracksuit to a red-brick Victorian building called New College. The building is outwardly beautiful but my bedroom on the first floor was a slim, high-ceilinged rectangle, drab and un-cosy to a fault, starkly furnished with a bed, thin curtains, a wardrobe with some open shelves, a sink and a mirror. The loos and the showers were in a communal block at the end of the hall. You weren't allowed to put posters on the bedroom walls. That was a privilege reserved for later. I had also brought a kettle with me and a CD player, but these, too, were taken from me on arrival and locked away. They would be returned at the end of the first five weeks, assuming I was still there.

Later I would also be allowed a duvet, but for now the bed was laid out with sheets and Army blankets, and one of the first things you were shown was how to make up that bed to the required standard. You had to tuck your sheet under your mattress using razor-sharp 'hospital corners', folding the excess neatly underneath. (I had a head start: my mum, the midwife, had taught me how to do this already.) Excess pillowcase material also needed to be tucked under, and the closed end of the pillowcase had always to be the one closest to the door. The blankets were only to cover the bottom half of the bed and had to be wrapped under the mattress tightly. The space of exposed bedsheet between the pillow and the top edge of the blankets had to be exactly the length of a sheet of A4 paper.

There were similarly precise rules about placement for every item of your clothing. Your shirts needed to be crisply ironed, it went without saying, but they also had to hang facing the same way. Your T-shirts were to be folded in such a way as to occupy the area of a piece of A4, and they had to be perfectly stacked on top of each other on the specified shelf, just as your socks had to be perfectly rolled together and arranged. Your toothpaste, toothbrush and deodorant needed to sit equidistant along your folded face flannel.

And then there were your drill boots. These had to be kept polished, obviously, but to say that is barely even to hint at the hellish labours that their upkeep was about to condemn me to. For at least the first three weeks of my life at Sandhurst, I spent every evening, along with everybody else, in the platoon room, sitting at our individual tables 'bulling' our boots. When the boots were first issued to you, the new black leather had a matte finish with something of the dimpled texture of a golf ball about it. Before you could start polishing, it was your job to heat a spoon over a candle and work at that leather until it was flat, and properly able to take the polish. You would press out the dimples with the back of the spoon, the delicate point being that the spoon had to be hot enough to alter the leather, but not so hot that it burned it, because if you scorched the leather anywhere, you would have a patch of boot that would never properly shine.

I took it slowly and steadily and avoided burning anything, but nevertheless I was never 100 per cent happy with those boots. I would go at them hard with the polish and the water and the beeswax but I could never get the shine on them that I thought they ought to have. It was a source of constant frustration to me – this endless task that I could never complete to my satisfaction. Some people seemed to find bulling therapeutic, but for me it was a form of torture.

Those drill boots were also, incidentally, quite difficult to wear. In addition to the thirteen metal studs hammered into their soles, they had clumps of metal at the heel and the toe, almost as thick as horseshoes, so that they clacked loudly on the ground. Until you got used to them, they slid around like ice skates. Those early days sometimes took on a slapstick quality, with cadets randomly going over or skidding in and out of view as they accustomed themselves to the iron on their feet.

Your kit would need to be in order for the morning room parade. At 5.55 a.m., the officer cadets had to be up and dressed in their tracksuits, standing to attention outside their rooms behind the black painted line which ran along the floor of the corridor. At 6.00 a.m. you sang the national anthem. Then, while you continued to stand to attention, eyes front, the colour sergeant passed from room to room on his tour of inspection.

Now, our colour sergeant in One Platoon, Alamein Company, was a huge Welsh Guardsman, Colour Sergeant Roberts. I've been back to Sandhurst subsequently and had a beer and a chat with him and I know him to be a fantastic guy. All of the colour sergeants who get to do that job are at the absolute top of their game. They have to be, because it takes a very special senior NCO to take on the responsibility of turning cadets into officers.

However, that doesn't mean that, to a fresh-faced officer cadet still finding his feet at the Academy, Colour Sergeant Roberts wasn't a powerfully intimidating figure. The rank structure means that you have a platoon commander, too, who will be a captain and therefore senior – Captain Breen, in our case. But you might only see him once a day, for a morning lesson or an evening lecture. The colour sergeant is the real presence in your life, particularly in those initial five weeks. Accordingly, Colour Sergeant Roberts was God in our world – and not necessarily always merciful.

He was known as 'Big 80'. Or, at least, he was known as 'Big 80' to his friends around the place. We officer cadets knew him exclusively as 'Colour Sergeant', and for us he only became 'Big 80' when he left the room. The '80' referred to the last two digits of his military number, and helped to distinguish him from the large number of other people in the Welsh Guards called Roberts. The 'Big' referred to the fact that he was six foot five and built like an upright fridge-freezer. Such was the man who appeared, stern-faced, in our corridor at 6.00 a.m. for room parade.

You would be standing at your door hearing him further down the way, inspecting your neighbours, bawling insults with a broad Welsh accent about how they were 'idle bastards' who 'can't follow simple instructions', and trashing their rooms – ripping the carefully made bed apart, throwing the patiently ironed T-shirts across the floor. And you would be wincing in advance at the horrible inevitability of it all, this inexorable force coming gradually towards you, like an advancing storm cloud.

Your room could be perfect, or as close to it as was humanly possible, but he would still find something – a marginally misaligned toothbrush, maybe, or a minimally skewed T-shirt, or even a thin layer of dust on the top of the door. For me, in the first three weeks, it was normally the drill boots. Behind me Colour Sergeant Roberts would explode with contempt and then I would hear the sash window go up and the boots that I had spent hours the previous evening buffing to the best sheen I could manage would be slung out to crack on the parade ground, one storey below. Often other items of clothing would float down to join them and then, after inspection, you would have to run downstairs with any other cadets who had suffered a similar fate, and gather your belongings off the ground – assuming you could find them, jumbled up with everyone else's stuff. And in the process of doing that, you might again find yourself face to face with Colour Sergeant Roberts.

'Why have you made a horrible mess of my parade ground, Peake? Restriction of privileges!'

New cadets didn't have many privileges to restrict, of course, so 'ROPs', as we called it, would normally mean extra cleaning duties – mopping floors, scrubbing toilets. And you would be obliged to do those additional chores in the evening, further limiting the time you had to get your kit and your room ready for the next day.

On the occasions when, by some miracle, your room passed muster, there would be no warm words of praise. Colour Sergeant Roberts would simply move on silently to the next room. Perfection in this area, you realised, wasn't something to be celebrated. It was something to be expected.

Little wonder that people came up with ruses. Some cadets, having got their beds made to a condition they were happy with, would elect not to mess them up by sleeping in them. They would use the sleeping bag and the roll mat that we were given for exercises, and sleep on the floor instead. That tactic could buy you some peace of mind and a few minutes' extra kip in the morning. But, of course, Colour Sergeant Roberts was wise to all the tricks and would occasionally perform a spot-check at around 1.00 a.m. to be sure that everybody was where they were meant to be in relation to their mattress. Found cheating on the bed inspection? Restriction of privileges, obviously.

Some of the more experienced and therefore slightly less nervous among us – the 'olders and bolders' – would wait until the colour sergeant was in someone else's room and then dive back into their own room to make last-second adjustments. That took nerves of brass, though. You needed to time it right in order to be snapped to attention in the corridor again when he emerged. Again, extra duties and the most almighty verbal dressing-down awaited you if you were caught trying to buck the system.

The whole thing was a comedy sketch, really, and would have been pretty funny if you hadn't been at the centre of it. You could protest and answer back, but you knew there was no winning that war, and those who had a go didn't last long. In fact, it never even occurred to me to try. In that respect, I was perhaps at the perfect age and stage in my life for all of this. My platoon was quite a mixed age group – from eighteen up to about twenty-five; from people fresh out of school to graduates and others who had already served a few years in the Army. The older cadets might have been tempted to think, 'Hang on, why am I standing for this?' If you had seen a bit of life, worked out how to play the system a little, you might have been a harder nut to crack, from Sandhurst's point of view. But I was nineteen, just a year out of school, never left home, with one pub job and an Operation Raleigh trip behind me. I was malleable. I was going to do what I was told and, moreover, say, 'Thank you, Colour Sergeant.'

Incidentally, our group wasn't just mixed in terms of age. It's a misconception about Sandhurst that it's all posh Cavalry types from privileged silver-spoon backgrounds. The social range is far broader than that. In my year there were a couple of guys – John Fisher and Max Francis-Jones – who went on into Cavalry regiments. They became good friends of mine, as did Leon 'Tommo' Thompson, who ended up in the Princess of Wales's Royal Regiment and whose room was opposite mine. But there was also Andy Inman, a Jersey policeman, a little older, who had decided on a change of career. (He, like me, went on into the Army Air Corps.) And there were Dave Elms and Al Rogers, again a little older, both of whom had gone straight into the Army from school and served as corporals but had been encouraged to try out as officers. Those two were the steady hands in the platoon.

And then there were the cadets who had come straight from A levels at Welbeck College, the Army Sixth Form college, and international cadets such as Udaya Thapa, who was a young Gurkha from Nepal. I would say 90 per cent of the intake were from relatively ordinary backgrounds like me. Of course, we were all men: there were no mixed companies at that time. These days, sensibly, there are. (My wife Rebecca went through Sandhurst in a mixed company.) In 1991/92, there were women going through training, but they were in a completely separate block and we barely glimpsed them at any point in our forty-four weeks.

Not that we had much time to look up. At first our days were a spin of relentless activity. At the weekend, you might get a lie-in – which would mean not getting woken until 7.30 a.m. And then, more likely than not, you would have to jump out of bed for a boot inspection. Other mornings might find you out on a training run by 5.00 a.m. There were sessions on the parade ground, sessions in the Montgomery gymnasium, sessions on the assault course on Barossa, the vast woodland training area that sits through a gate at the back of the compound. Your timetable was so tight that meals had to be consumed on the run. There just wasn't time to do anything but rush in, stuff your face as quickly as you could with sausage and beans and then rush out again. Otherwise you would be late for your next appointment, and to be late was arguably the worst offence, even worse than ducking your bed-making duties, because it showed lack of discipline, disrespect for others, disorganisation ... the full menu of personal weaknesses which the Academy was determined to eradicate.

By the middle of the first week my legs were aching from marching at 140 paces a minute. We were learning to march as a platoon, and learning drill. We were doing weapons training. We were learning military tactics. A lot of this was practical, but some of it was in the form of classes and lectures in Churchill Hall, known as 'the concrete sleeping bag' on account of the extremely high chance that your tired limbs would overtake you in the warm air coming from the radiators and you would drop off mid-lesson. During a lecture in week five, one officer cadet fell off his chair and rolled gently down the aisle for a few feet, sound asleep, to the great amusement of the room – those who were still awake, at any rate.

The idea was that you had barely any free time – no life to speak of. The evenings would be filled with 'personal administration' (sorting your kit out), and when you were finished doing that, which might be 11.00 p.m. if you were lucky, you were so exhausted that you just crashed, trying to cram in as much sleep as you could before the next dawn call and the next active day. It was an exercise in sleep deprivation as much as anything else. And there were moments, during that phase of absolute exhaustion and unceasing demand, when you were at the end of your tether, too tired to get things right, too tired to organise the thoughts in your head. But you just had to keep telling yourself that this was the plan: this was exactly how they wanted you to feel, this was how you were *supposed* to feel, and if you just took it one step at a time and saw it through, things would eventually get better.

A few people did fall by the wayside. But this also was the point of the exercise. If there was any doubt at all about whether you had chosen the right career path, those five weeks would show it up. And conversely, if you came through those first weeks it was practically incontrovertible proof that you had made the right choice.

There was an important group-element to it all, as well. It was hard on you as an individual. Yet, at the same time, you knew it wasn't just about you: the whole platoon was going through this at the same time. There was something extremely bonding about that. The fact that we negotiated those first five weeks together created a unity that stayed with us throughout the whole of our time at Sandhurst.

When the five weeks were up, it was like a mini-graduation – a real moment of pride that was shared, you could sense, by the staff, even the reliably unsentimental Colour Sergeant Roberts. At this point – and at this point only – we were presented with our uniform, the thinking being that now, finally, we had earned the right to wear it. It would be very easy for Sandhurst to dish out the uniforms on arrival. But no. You don't get to dress like a soldier until you've properly shown you can act like one.

At this point, too, you got a weekend off – which is when I returned to my parents in Westbourne and collapsed through the front door, hollow-eyed, exactly as predicted.

Back in my New College room that Sunday evening, well rested after some time at home, I put up my posters, installed my kettle and listened to my CD player. I had a jar of instant coffee to call my own, and a tub of Marvel powdered milk. All in all, my world seemed to have taken on a whole new glossy sheen.

* * *

Needless to say, once the Academy had allowed the brakes to come off a little at the end of week five, the first thing they did in week six was to slam them straight back on again. It was right at the beginning of that week that we boarded coaches for Wales and set off to 'Exercise Long Reach' in the Brecon Beacons, an exercise that made all the others we had done on Barossa to this point look like walks in the park.

'Exercise Long Reach' was a cross-country orienteering exercise, essentially, but it was the toughest one of those that I had ever done. We had to navigate our way in teams of four around fifteen checkpoints, competing to be first team back. It involved about 65 kilometres of walking, night and day, with 60 pounds of kit on your back, stomping through bogs, streams and fields, and it was where you found out exactly how much endurance you had in your tank. The march element had to be completed in thirty-two hours, and if I got as much as a couple of hours' sleep in all that time, I would be surprised. Still, our group of four, with Dave Elms leading the way, won it. Our prize was an RMAS hip flask for each of us, which would come in very handy on future exercises.

It might seem unlikely, but I was starting to enjoy myself now that the pressures of the initial 'make or break' phase had lifted. The challenges presented by the exercises were still daunting, but the satisfaction of completing them and moving forward was immense. I had settled into the company hockey team, who played every week. We had even had our first formal Company Dinner night, for which everybody donned their high-collared No. 1 dress for the first time. Inevitably, those new, starchy items began to rub within about ten minutes and by the end of the evening we all had red-raw rings around our necks and were determinedly staring straight ahead to avoid turning our heads and adding to the chafing. As Christmas neared, I was knackered – but obviously not *that* knackered. During the three-week break, you could select an Army-organised activity to do. I went off and did Nordic skiing for a week – my first time on skis, which I absolutely loved.

In the New Year, in term two, our privileges were further expanded. I replaced my scratchy army blankets with a far more comfortable cotton duvet – luxury unbounded. But that didn't mean the work softened up in any way. Almost as a kind of welcome back, we were taken out onto Salisbury Plain in the bleakness of January for 'Exercise First Encounter'. We were working our way through the stages of trench digging and by this point we were at the level known as 'stage three with wriggly tin'. A stage three trench is two firing positions with a sleeping area that links them in a tunnel 6 feet down, all lined with wriggly tin, which is Army-speak for corrugated iron.

At 2.00 p.m. on a grey Wednesday afternoon, in the middle of Salisbury Plain with the ground dismayingly cold and hard, we were given the simple instruction, 'Start digging.' And off we went, working away with picks and shovels, digging and digging and digging. It was hot work: we were in T-shirts and combat trousers for the most part. Indeed, we were so hot with the effort that none of us even noticed the temperature drop to minus 8° C – the coldest night of the year. The full harshness of the climate only became apparent in the morning when, with the job finally done and all of us gasping for refreshment, we went to brew ourselves a cup of tea and couldn't because the water in our bottles had entirely frozen.

The rest of that day offered another twelve-hour shift with the pick and shovel, now creating a series of four-man battle trenches. Then through Thursday night we had to stand guard over our work.

That time the cold really did bite – to the extent where, despite my bruised hands, I almost longed to be doing some more digging.

The trenches we had made were meant to be able to withstand the approach of advancing armoured vehicles and infantry so, on the Friday, to conclude the exercise, soldiers from the 17th/21st Lancers and the Light Infantry obligingly charged all over our work, bringing some Challenger tanks with them to see how well our endeavours with the wriggly tin stood up. They didn't position us in the trenches for this test, I hasten to add, having very little faith at that point in our construction abilities. And just as well, really. The Challengers did some damage to our work that it was painful to witness. All that remained was for us to get the shovels out again and fill in all the holes we had made, completing the work that the Challengers had more than started. Another classic Sandhurst exercise completed.

Soon after this it was time for the dreaded Log Race, in which teams of officer cadets carry a cumbersome lump of timber round the Academy's grounds. That's a Sandhurst staple. Indeed, whenever I saw photos of Sandhurst before I went there, they always seemed to show a group of men in blue rugby tops, lightweight trousers and boots, carrying a telegraph pole through some woods, and below would always be the caption, 'At Sandhurst you will become fitter than ever before.'

Well, that was true. I was fit when I arrived at Sandhurst, but I was fit in the way that runners are. I wasn't 'upper body strength' fit and that's what Sandhurst very quickly builds – not least by getting you to cart a giant tree trunk through the undergrowth in a timed challenge. (For the record, One Platoon didn't do terribly well in the Log Race. We finished a lowly fifth, partly because so many members of our group had collapsed vomiting into the bushes by the time the finish line came into view. It wasn't for want of effort, clearly.)

Throughout our time there would be competitions on the assault course on Barossa – against each other and sometimes against visiting units or colleges. The Royal Naval College at Dartmouth would come up, or the Royal Air Force at Cranwell, and once a year the US Army would send a team from West Point. A lot of prestige was attached to winning those competitions, which, I have to say, the Academy mostly did. One of the big moments on the assault course was when you would have to get your team of eight over a 12-foot wall. While you still had enough people to stand at the bottom and haul others up, it was easy enough, but you had to be organised about getting the last three over. The best tactic was to get the two tallest people with the longest arms to lie on top of the wall and stretch down; then the last man, who for preference would be the smallest and lightest in the group, could run up, hit the wall, jump and then with any luck be dragged up.

Guess who usually had to be the last man.

By now I was more muscled than I had been, but I was still relatively small and light, so I would be the final climber – the one springing and hoping to be grabbed by his teammates' hands. Meanwhile most of the rest of your team, having hit the other side, had already run ahead, so you were then sprinting to play catch-up. But all of this was revolutionary, in terms of my fitness. I became much stronger – indeed, 'fitter than ever before', just as promised in the brochures.

Of course, even though some of the stringency of term one had now relaxed, discipline remained paramount. I recall being summoned to Colour Sergeant Roberts's room with Dave Elms at some point during the second term, and given the hairdryer treatment – told that we were a disgrace and that we needed to pull ourselves together and also that he was now entirely sick of us and we should get out of his sight. I don't think Dave or I ever worked out what, exactly, we had done to merit this abuse. Most likely it was simply our turn. But we stood there and took it anyway.

And, obviously, now that we had privileges, there were privileges which could be taken away from us, so the stakes on failing to come up to the mark felt even higher. ROPs could be extended in extreme cases to the cancellation of a weekend off. Because you cherished the thought of those breaks with every one of your aching limbs, that would be an appalling blow to be dealt. I was careful to ensure that it never happened to me, but it happened to a friend of mine and I can still recall returning to Camberley that Sunday night, after I had enjoyed a leisurely recuperation at home, and finding him in his room, recovering from two extra days of duties and wearing the expression of a man whose soul had been crushed with a hammer.

Term two flew by and then we were up to the Easter break. For those weeks off, you were encouraged to organise your own adventure training with a few of your fellow officer cadets and get the Academy's approval for it. Legends abound at Sandhurst of the groups who have sold some bold idea for broadening their horizons to their colour sergeant and platoon commander, and then essentially gone somewhere and lain on a beach for a week – before quickly racing off on the last day to get the snaps they needed to suggest that they had been doing something far more arduous or cultural. When it was our turn to think up something, I was fortunate to be friends with Udaya Thapa, who suggested that some of us should go out to Nepal with him and trek around the Annapurna Circuit, a famous mountain hike of around 145 miles in length, through stunning Himalayan scenery. Our platoon commander seemed to think this was an entirely acceptable plan, and off the five of us flew to Kathmandu, with the Academy's blessing.

Udaya spent the first four days of the trip showing us around the city and many of its bars. Then we finally tore ourselves away and took a night bus that twisted precariously round mountain hairpins to the town of Pokhara, a departure point for the Annapurna ring. Once there, however, Udaya suggested that maybe we didn't want to go to the trouble of walking all the way, and that perhaps it would be generally easier on our feet if we flew some of it. We considered this proposal for all of five seconds and then hopped on an overworked de Havilland Twin Otter that took us to Jomsom, a big centre for trekkers high in the mountains, which at a stroke got rid of a few thousand feet of elevation. And then we did a relatively simple walk back down the other side – taking lots of pictures to show on our return, of course. Still, even though we had shirked some of the hard miles, it was a fantastic trip. Being on an adventure with great mates and exploring a remote part of the world – I felt like I was living the dream.

Back at the Academy for my third and final term, I was entitled to call myself a junior under officer, the title given to the highest-ranking officer cadet in his company. It also meant I was one of the two candidates vying for the RMAS Sword of Honour, awarded by the commandant to the year's best cadet. It went to the other guy in the end. Still, to have made junior under officer and to have even been in the running for the Sword was quite an achievement, really, for someone who had come in as one of the youngest and greenest cadets.

In that last term, almost all weekends were free and the possibility of drifting at will into Camberley to go to McDonald's, which would have been the stuff of crazy fantasies at the start of term one, was now perfectly real. Although harsh physical training continued, and included one particularly arduous trip to Norfolk, the emphasis at this stage was much more on studies and classroom work – battlefield tactics, military history, military law. Ahead of us lay the prospect of our final exercise which, like all exercises, would be physically gruelling, but even this, in the spirit of end-of-term frivolity, was going to take place in the South of France. The thought of that treat drew me on through the various written exams – all the way to week thirty-eight, indeed, when it was casually announced that France was off (something to do with restriction of the training areas, apparently) and the exercise would now be staged, not under the warm Mediterranean sun, but up in Scotland, in the company of the midges on the Isle of Jura. One last wicked joke at our expense.

Still, it was a memorable trip even so. The SBS(R) – the Special Boat Service (Reserve) – were helping out and provided Rigid Raiders to ferry us from the mainland to the island and to help us execute mock beach landings and beach assaults. When the exercise on Jura was finished we were despatched to let off steam on three days of Scottish adventure training. There were a lot of high spirits in the two SBS Rigid Raiders that took us back to the mainland that day and the boats were thrashing across the water and crossing in front of each other. In the midst of this larking, one of the boats caught the other's wake at the wrong angle and flipped, landing upside down on the water. Thankfully, everybody on board was fine, but ten SA80 rifles which the boat had been carrying sank to the bottom. Naturally the British Army is keen to account for the whereabouts of its weaponry at all times, so these rifles couldn't simply be abandoned. The Royal Engineers were called in and a massive salvage operation had to be staged before those guns could be given up as a lost cause. By then the story had attracted the attention of the press.

We headed on, a little chastened, to Cultybraggan Camp, a big spread of Nissen huts and a former PoW camp, repurposed as an army training centre, where we were granted our promised three days of outdoor activity. Many years later, when I made what was only my second trip to Scotland at that time, Rebecca took me to one of her favourite local parts of rural Perthshire. As we stood on top of Lednock Dam, just a few miles north of her parents' village in Comrie, and looked out across beautiful Glen Lednock, I became eerily aware of having once abseiled down the dam's sheer concrete. It was, indeed, the scene of that final Sandhurst adventure training. Of all the dams in all the world.

I would find it hard to list definitively all of the things that I owe to those forty-four weeks at RMAS, exhausting and infuriating though that period frequently was. In some respects the Academy's infantry officer course could be regarded as narrow. It's not like West Point, the US Military Academy, where you do a four-year course with a broad academic curriculum and come out with a degree on top of your military commission. There is no recognised educational qualification at the end of Sandhurst. It's a degree, very purely, in military leadership.

But then I think of the valuable assets that it so efficiently bred in you: courage, discipline, integrity, loyalty, respect for others, commitment. It instilled self-belief, giving you a confidence about problem-solving and decision-making under pressure in complicated and dangerous environments. It gave you the confidence, also, to be true to yourself and therefore authentic around other people.

And it taught me about resilience, too – about controlling your morale, learning to choose your mood. Take the room parades, for instance: it wasn't simply about getting an earful, it was about getting an earful when you least felt like it – at 6.00 a.m. in the morning after very little sleep. So how were you going to respond? How much self-control could you tap into? We were constantly trained to seek what is called at Sandhurst a 'Condor moment' – named after the old cigar advertising campaign in which, typically, someone having a tough time would light a cigar and retreat inwardly to a calmer place. The idea, as taught at Sandhurst, was to reach that calmer place without needing the cigar.

And it showed me the value of practising things over and over until they become routine, in the knowledge that routine is what you fall back on when the pressure bites. It showed me, also, the importance of learning from mistakes. When you were being taught how to do a section attack, for example, you were deliberately overloaded with information so that something would be bound to go wrong. What would be the point of the exercise if you simply sailed through? The screw-up was potentially where you did your best learning. The same thinking applied very strongly later in my life with astronaut training. In the simulators, when we were practising for launch or docking or any of the more complex flight procedures, the instructors would invariably throw in an emergency at some stage, or lob you something unexpected to deal with, because otherwise it was a wasted training opportunity.

As a consequence, when it came to the real thing in 2015, and our rocket launched and we eventually reached orbit, I was aware of feeling almost surprised that we had made it into space and nothing had gone wrong. It was like something was missing. What? No fire? No depressurisation? Our life support systems are all good? Amazing. A free pass! That was the extent to which we rehearsed for emergencies.

Graduation day at Sandhurst was warm and sunny. My parents and grandparents came and were there to see me, in my dark blue and red uniform and peaked cap, perform the symbolic formal march off the parade square, following the adjutant who rides a white horse up the steps to Old College. Then there was lunch and tea and eventually the older members of the family withdrew, the young friends stayed on and the scene was set for the Commissioning Ball which, on a tide of food and alcohol, and also music and bumper-car rides, would roll through until dawn.

I invited my sister and a couple of her friends from university, and my pal John Platt who was at Dartmouth Naval College, on his way to becoming a Navy helicopter pilot. He and I had first met when we both did the RAF flying scholarship at Compton Abbas and were thrown together in digs in Shaftesbury for a few weeks – the beginning of our firm friendship.

As tradition dictated, for the ball I wore my Army Air Corps officer's mess dress uniform with my new second lieutenant pips sewn onto each shoulder but covered over with ribbon. Then on the stroke of midnight, in a ceremony repeated all over the room, I stood up and two of my sister's friends, Jane Selby and Sarah Kennedy, one for each shoulder, ripped the ribbons away and kissed me on the cheek. And now I was an Army officer.

CHAPTER SIX

On tour with the Falling Plates, airborne in a
Chipmunk, and the night of the sunken Land Rovers

It's the middle of the night in October 1992. I am twenty years old and only a handful of weeks into my first tour of duty, and I am standing, covered in mud, by a bog in County Fermanagh, wondering whether a tea break has just cost me my career in the Army.

It had seemed like such a good idea at the time – the sensible thing to do, even. I was in command of a twelve-man unit, nearly half of my platoon from the Royal Green Jackets. We were out in two armoured military Land Rovers performing what's known as a 'snatch patrol', where you show up out of the blue and keep watch on an area for a few hours.

Now, in Northern Ireland, in the time of the Troubles, word could be relied upon to go around very quickly about where the British Army's peacekeeping forces were and what they were up to. Accordingly, it made sense, when patrolling, to mix it up a bit – to try not to move in repeated patterns. That way, people hostile to your presence would have less opportunity to anticipate your movements and get organised to have a go at you – to lay the command-wire for a roadside explosive device, say, or put a sniper in place.

Being unpredictable could be hard, though, especially if you were patrolling a small village area, as we were that day – and even more so if you were doing it for a long stretch. Ours was a twelve-hour shift and it was rapidly apparent to me that no amount of ingenuity was going to enable us to keep covering our allotted patch in new ways. As the shift wore on and we kept turning slowly down the same roads, I was getting an increasingly bad feeling about our situation – the feeling that we were potentially setting ourselves up for trouble. The hairs on the back of my neck were starting to go up.

Fortunately, I had earlier had a conversation about exactly this kind of moment with one of my fellow and more experienced Green Jacket platoon commanders, Lieutenant Mark 'Smiley' Riley. Smiley's solution? Knock off for an hour or two. Head right out of the area and find somewhere to take a break. Then go back in. Bingo: pattern disrupted. And you get to have a brew into the bargain.

With the afternoon wearing on and my anxiety levels rising, it seemed like the perfect time to employ the Riley method. I instructed the men to move out and we left the village behind us and drove the two Land Rovers up into the hills above Fermanagh before turning off up a narrow track and stopping, well out of sight, on the edge of some woods. Ahead of us, the other Land Rover carried on further up the track so that we would be separated, just in case anyone, even out here, chose us for a target. The sun would soon be going down and I figured we could make tea, sit around for a bit, watch the sun set, and then go back fresh for the night patrol.

Tea went down well. In the Army, it usually does. As the light began to fade the six of us in my group packed up and climbed back in the Land Rover. The driver started the engine and swung the vehicle round to head down to the road again.

This was when the first less than optimal thing happened.

In the process of turning, the Land Rover put two wheels on the verge beside the track; those wheels spun hopelessly for a few moments, and the vehicle came to a shuddering halt. We all climbed out to look. The ground beside the track was now revealed to us to be a good old Irish bog, as thick and sticky as treacle. Not only was the 3-tonne Land Rover stuck deep, half in, half out, it was now tilting so far over that it was in imminent danger of falling onto its side.

At this point, you could only be grateful for the tactic of travelling in pairs. At least we had a second Land Rover with us that would be able to come back down and pull us out. Still, the first thing we needed to do was shore up the stricken vehicle and prevent it from toppling over. Casting about the place, we found some discarded fence posts and wedged them against the side of the Land Rover to hold it in place. Then I set off up the track to fetch the other group and bring them down to help us.

The scene that greeted me further up the track was not entirely encouraging. The men were standing in a cluster, looking on at something. As I arrived, someone said, 'We've got a problem here, boss.' In front of them, I could now see, was Land Rover number two. Its driver had taken the vehicle off the path altogether and it was now squatting in the bog on its flat armourplated floor, not just two of its wheels but all four of them sunk deep and spinning pointlessly. It was more marooned even than ours was.

That was the second less than optimal thing.

It was clear that, of the two bog-bound vehicles that were now under my command, the one that was merely halfway into the bog was more open to retrieval than the one that was completely in. So I led the guys back down the track and we all pitched in on stuck vehicle number one. It would only be a matter of time, wouldn't it, before the combined efforts of a dozen soldiers freed this partially sunk Land Rover? In fact, no, it would not. We tried everything, from every angle. The Land Rover only got deeper and we only got muddier.

Our efforts continued for some time. Night had long since fallen. At regular intervals, I would radio back to headquarters, as expected, and report that all was well and that our patrol was proceeding safely and without incident. Which was substantially true, of course, as long as you ignored the fact that neither of our vehicles was actually on patrol and both of them were currently up to their axles in a bog.

Eventually, I stepped back and weighed my options. It seemed to me as I stood there in the dark that, as the officer in charge of this mission, I was even deeper in the crap than the Land Rovers were. And with every duplicitous call to headquarters I was probably digging myself in even further. I really only had one plausible course of action: to get on the radio to headquarters now, explain the situation, admit that I had screwed up, and ask them as beseechingly as possible to send a Chinook helicopter in the morning. A Chinook would haul both our vehicles out of the mud in no time. Then we could all return to base and I could answer to my seniors for having come off patrol and sunk not one but two military vehicles, thereby necessitating a costly and man-power-intensive salvage operation.

Would my career as an Army officer survive a humiliation on that scale at this tender, early stage in its life? I really wasn't sure that it would. In fact I was pretty convinced my career would be over just as soon as we got back. But given the position we were in, what else could I do?

It was at this point that I was grateful for the input of two lance corporals in my unit. Have good people around you and, more importantly, listen to them: those seem to be the two lessons in leadership here. When I said I thought it was time to cut our losses and confess what was going on, my lance corporals urged me not to. Their attitude was: our shift still has several hours to run. Sure, we might have to give up in the end. But let's at least use the time we've got and see if we can get out of this somehow.

I still wasn't sure we would succeed. But if it bought me a stay of execution for a few more hours, then why not? So on we laboured. Sure enough, we still weren't really getting anywhere. It was around 1.00 in the morning when I decided a small party of us should walk until we found a farmhouse and seek help there. That wasn't a decision I took lightly. This was neither a place, nor a period in history, where a knock on the door in the middle of the night from the British Army could be relied upon to be received enthusiastically. But we had little choice.

The farmer didn't seem too delighted about being woken up, but he didn't slam the door in our face (or worse) either and, with almost no persuading, he kindly agreed to get his tractor out. I suspect, as much as anything, he was eager to get us off his land.

Anyway, the farmer and his tractor managed to pull out the first Land Rover and we used the first Land Rover to haul out the second. Then we drove back to base, arriving on schedule, albeit with two suspiciously mud-caked vehicles, one of which was going to need a repair to a still-smoking, practically burned-out clutch. I put in my patrol report as if nothing unusual had happened and returned to quarters, where I found Smiley Riley and told him how well his tea-break plan had worked out for me. He seemed to find the whole thing hysterical. I would eventually start to be amused by what had happened, too, but only after a few more days had passed and I was completely sure that I had got away with it.

*   *   *

I had applied for and been accepted into the Army Air Corps, so what was I doing in Northern Ireland with the First Battalion of the Royal Green Jackets? Here's how it worked: as a young Army Air Corps officer, you were going to spend a large chunk of your early years commanding little more than your aircraft around the skies. So, in order to gain some real leadership experience, within weeks of being commissioned and leaving Sandhurst, you would be sent on attachment to an armoured or infantry regiment. It was a standard step in the process and yet one which almost tripped me up completely.

Midway through term two at Sandhurst, all officer cadets have an interview with their sponsoring unit – in my case the Army Air Corps. It's really a check to see that they still like you and that you still like them. Things at the time seemed to be going pretty well: I was one of the three cadets in my platoon who had been chosen as cadet corporals; I'd had good reports from the colour sergeant; I was on my way to becoming one of the intake's two junior under officers. I wasn't really expecting a savaging on this occasion. And, indeed, most of the interview seemed to go pretty well.

Then a rather severe-looking colonel across the table from me asked me what my plans were for my attachment.

I said I was thinking of going Infantry, in order to get a feel for life in the wider Army, and that I also thought an operational tour in Northern Ireland would benefit me.

The colonel listened and then asked, 'Well, who have you approached about doing this?'

The question rather threw me. I had no idea I was meant to be taking care of that. Indeed, I thought my attachment was something that would be sorted out between the Army Air Corps and myself later, assuming I got commissioned.

'I haven't approached anyone, sir,' I replied.

A chill descended on the room.

'What?' said the colonel. 'You didn't think about making contact with anyone?'

Again, I had to admit that I hadn't.

'Well, that doesn't show much in the way of initiative, does it?' said the colonel.

I left the room feeling scalded, and also slightly panicked. How had I misread the system? Determined to put things right, I went straight to my Sandhurst company commander and said, 'Sir, I need to sort myself out a posting to Northern Ireland.' The commander seemed a little surprised, but when I explained that I had just been roasted in an interview for failing to get this done he agreed to help me. He was a Light Infantry man and he said he knew the commanding officer of the Royal Green Jackets, who were going to be in Northern Ireland at the right time for me, so he would see what he could do. Pretty soon afterwards, between the two of us, we had managed to pull the right strings and get the Green Jackets to accept me within the necessary time frame.

I assumed that was the end of it. But not at all. Further down the line, when the Army Air Corps posting branch got wind of this arrangement, they went ballistic. Who did this upstart cadet think he was, organising his own posting? What business was it of his? 'It's up to us to tell him where to go, it's not up to him to tell us where he's going,' was the gist of their response. Apparently, on the back of this perceived presumptuousness, they were on the verge of reviewing my entire application. My colour sergeant later confided in me about just how badly this mis-step had gone down. 'You have no idea how close you were to not being selected into the Army Air Corps,' he said.

That, of course, would have shredded my entire game plan. I would have been devastated. A near thing, then. And all because some colonel had possibly got out of the wrong side of his bed on the morning of my Sandhurst mid-term review.

Anyway, once the dust settled, the Air Corps posting branch seemed to take the view that, as this attachment of mine was already organised, they might as well save themselves some work and leave it in place. I joined up with the Royal Green Jackets in Omagh in October 1992.

There was another hurdle to jump, though, before I could depart for Northern Ireland. At this point, the Army Air Corps still hadn't seen me fly. Neither they, nor I, yet knew whether or not I could fly an aircraft to the required standard. Some people reach this stage of the acceptance process without ever having sat in an aircraft. I, at least, had that RAF flying scholarship under my belt from my last year at school. But that wasn't an acceptable qualification on its own. As far as the AAC were concerned, I still had it all to prove.

So, in the summer after I left Sandhurst, I reported to Middle Wallop for flying grading. This four-week course involved being put through your paces for about thirteen hours in the air with an instructor, plus some additional officer training on the ground. Pass at the end of it and I could finally consider myself an Army Air Corps officer. Fail and ... well, the Army would have found me something else to do, no doubt, probably in the Infantry, and I would definitely have made the most of it. But I would have forever had thoughts about what might have been. My heart wasn't in anything that didn't involve flying.

Surrounded by gently hilly Hampshire countryside, Middle Wallop could proudly claim to be the largest grass airfield in Europe. It had been the home of the Army Air Corps since 1957 but before that it was the airfield from which the RAF flew Supermarine Spitfires and Hurricane Mk.1s into combat in the Battle of Britain, and that history could still be felt among its buildings and tarmac strips and hangars. I was certainly feeling it strongly on the first day, pulling on my flight suit and walking out to the aircraft lined up on the apron for our training, which were Chipmunks – compact tail-draggers with a tandem seat arrangement, one in front of the other, that have something of the classic character of the Spitfire about them.

The Chipmunk is a lovely plane to fly ... once you're in the air. It's the taking off and the landing that are difficult. Normally you wouldn't learn in a tail-dragger. You would probably use a basic type of Grob or some kind of Cessna, like the ones we flew on the RAF course – nose-wheel aircraft. Those are much more forgiving on the ground. You can taxi them like a car, trundle them down the runway full throttle and then pull them up relatively straightforwardly.

A tail-dragger, by contrast, throws you some extra elements to deal with. As you increase power, the airflow from the propeller corkscrews around the fuselage and hits the tail fin, which causes the plane to yaw (usually to the left, since most single-engine props turn clockwise viewed from the pilot seat). The torque from the engine also causes the left wheel to dig in a bit more, increasing that tendency to yaw left. In a tail-dragger, the aircraft's centre of gravity lies behind the main wheels, so when hurtling down the runway, the plane will gladly flip around given half the chance. So you've got to be ready to counter that yawing movement with some pedal or else risk doing a ground loop, which is where you look out the side window and see the back end of your own aircraft trying to overtake you. Which is not ideal.

On top of that, as you build up speed, you've got to get the tail off the ground before you lift the aircraft up, but once you've got the tail up the aircraft becomes even more unstable as you fight to control both pitch and yaw. All in all, there's quite a lot for the novice to be thinking about as they hurtle across the grass for the first time, bound, they hope, for the sky.

Then there's landing. With a nose-wheel aircraft you can more or less plonk it down on the runway and it will be fairly forgiving. The tail-dragger, on the other hand, needs you to aim for a perfect three-point landing – all three wheels together. And in order to achieve that you've got to 'flare' the aircraft as you come in – pitching up to get it into exactly the right attitude. This, too, takes some practice. As soon as you pitch up, you can't see anything except sky over the long nose of the aircraft, so you have to land looking sideways. Most students initially come in too flat, which means that the two main wheels hit the runway on their own, causing the plane to bounce straight back into the air. Again, not ideal. In the first days of flying grading, one grew used to seeing Chipmunks kangarooing off the airfield to spectacular heights of anything up to 20 feet, whereupon the instructor would be obliged to seize control and go full throttle back into the sky for a go-around in order to prevent the plane from crunching down ruinously on its undercarriage.

These days the Army Air Corps uses a Grob 115E for flying grading, which is far more accommodating. Back in the 1990s, the attitude seemed to be, 'Well, if you can fly a Chipmunk, you're probably going to be OK.' I certainly found it fun, especially once I got the landing sorted, though the pleasure of the flying had to compete with the constant pressure of the course. The question wasn't really whether you could become a pilot; most people could become a pilot if they had the resources and put in the hours. The question was whether you could become a pilot in a short space of time. The lessons came thick and fast, in a condensed chunk. Moreover, at no point were we given any feedback on how we were doing. That threw me off-kilter a bit. You would go up with your instructor, get asked to do certain things, do them to the best of your ability and then come back down to silence. The next day you would carry on and do the same. I was enjoying myself, but I was also mildly uneasy the whole time, wondering how I was actually getting on.

Accordingly, at the end of the four weeks, I still had cause to feel anxious when I went and stood with the others in the corridor outside the chief flying instructor's office and waited to be summoned in, one by one, to hear the final verdict. Andy Inman and Jerry Bird, two contemporaries of mine from Sandhurst, were there. Jerry went in first. Moments later, he came out, shook his head and said, 'I'm a no.' Andy and I thought he was joking, but he wasn't.

If I had been anxious before, I was even more so now. I went into that room thinking, 'Is this it?'

'I'm pleased to tell you ...' the chief flying instructor began.

Thank God for that. It would have been a huge blow. Flying was everything to me. And now I would be able to concentrate on it. Or at least, I'd be able to concentrate on it as soon as I got back from Northern Ireland.

* * *

You could spend a lifetime studying the complexities of the political situation in Northern Ireland during the Troubles and still not feel you had mastered them. I had a week-long preparation course at the Ballykinler Training Centre in County Down. Better than nothing, of course. It was an opportunity to learn a bit about what to expect from the operational environment, and to understand at least the outline of the British Army's role in Operation Banner, which was essentially about two things: reassurance and deterrence. We also had some of the more specialist pieces of kit explained to us, things that we hadn't come across at

Sandhurst, such as the radio jamming systems designed to foil remotely controlled improvised explosive devices. The rest I would just have to pick up as I went along.

In the autumn of 1992, the situation on the ground in Northern Ireland certainly wasn't as volatile as it had been in the 1970s. But that didn't mean all was quiet. In the months prior to my arrival, terrorists had blown up a van carrying fourteen maintenance personnel who were going home after doing contract work at the Omagh barracks where I would be based with the Royal Green Jackets. Eight of the workers were killed and the rest were badly wounded. On another day a petrol bomb had been flung at a group of Green Jacket Riflemen, who escaped injury, and a Mk 15 grenade had landed at the feet of a Rifleman but failed to explode. And on another day again a group of Riflemen, on a tea break between patrols, were kicking an old football about on the forecourt of an RUC station when the ball split to reveal that it was hosting 4 pounds of Semtex hooked up to a device that somehow hadn't gone off.

Terrorists had also ambushed a Green Jacket foot patrol, a bullet puncturing a corporal's combat jacket (but not, fortunately, the corporal), and had then fled in a stolen car. This led to an exchange of fire in which all of the car's tyres were shot out, yet without bringing it to a halt. The car veered into a housing estate where its occupants jumped out and ran off. One of the terrorists was later found, during a door-to-door search, hiding out in someone's house, badly wounded and concealed inside the frame of a divan bed. Meanwhile, near the barracks in Clady where I would periodically be quartered, an explosive device hidden in a gas cylinder had been placed among other cylinders adjacent to a petrol station but, again, it had failed to go off. At various times, stones, bottles and one Mk 15 grenade had been thrown over the walls and into the family quarters. This, it seemed, was routine business for the Green Jackets and was the kind of action I should prepare myself to expect.

The Royal Green Jackets had a distinguished history in Northern Ireland. Since the commencement of Operation Banner in 1969, there had been only four years when there was no Green Jacket presence in the country. The regiment could boast 102 Mentions in Despatches, a DSO, six MCs, two DCMs and eight Military Medals. It had also lost thirty-two of its soldiers in action and a further seven bandsmen who were murdered by an IRA bomb in Regent's Park in London in July 1982. In Army circles, this aspect of the regiment's record had earned it the typically dry nickname, 'the Falling Plates'.

'Who is your attachment with?' people would ask.

'The Green Jackets,' I would say.

'Ah, the Falling Plates,' they would reply.

It wasn't reassuring.

I would also, at the age of just twenty, be leading men for the first time – and very quickly, as it turned out. I arrived and, literally forty-five minutes later, after the briefest of introductions, I was out on the street functioning as a platoon commander. By then I had already been caught out in a classic dupe. Arriving as a second lieutenant, I was, naturally, very keen to see who my sergeant and corporals were. These, after all, were the older and more experienced guys who were going to be providing me with critical support, especially during the induction phase. I knew how much, as a novice, I was going to be leaning on them.

You can imagine how unnerved I was, then, at that initial meeting, to be introduced to Sergeant Deaville and to find myself shaking hands with someone who looked barely older than I was. My heart sank. Just when I needed someone with a few miles on the clock ...

They must have kept up the pretence for about half an hour before finally revealing to me that they had dressed up one of the younger-looking members of the platoon in sergeant's stripes for the occasion. The real Sergeant Deaville was over in the corner, nonchalantly chewing gum and pretending to be some old-hand Rifleman. I was massively relieved though, of course, I tried not to show it.

The thirty Riflemen now under my command were mostly rowdy lads from big-city backgrounds – London, Liverpool, Birmingham. They were streetwise and hardened in ways that I definitely wasn't; fantastic soldiers who absolutely knew their stuff, but were also ready to push the limits in their downtime. And they were perfectly ready to push the limits with me, a kid with a young face, just out of Sandhurst. I was going to have to work hard to earn their cooperation. After all, I wasn't even one of them: I wasn't a Green Jacket. I was a temp – an interloper who would be off flying helicopters as soon as he left here, which they well knew. And yet it was my role to be boss and straighten them out when they stepped over the line.

Those first weeks in Northern Ireland were a steep learning curve for me in terms of managing people, and especially in terms of learning how to deal with misdemeanours. Disciplining soldiers wasn't really a topic that was practised at Sandhurst. Looking back, I realise that I had more training in areas like team management and conflict resolution as an astronaut than I did as an army officer. We used role play occasionally during astronaut training, working through some of the tensions that were likely to arise in a close-knit group, whereas at Sandhurst, when I was there, role play wasn't used at all and would probably have been sniffed at as unsoldierly. Yet, straight away, here I was with thirty loud, self-possessed young men to control, and discovering that rank structure will only get you so far.

The only way I could make it work was by gaining their respect, and the quickest way to gain their respect, I thought, was to be as professional as possible – show myself to be competent and somebody in whom they would be prepared to put their trust.

Within days of my arrival, tragedy struck. The battalion suffered the loss in an accident of Rifleman Richard Davey. He had only joined the company a few weeks earlier and his death sent a shock wave right through the platoon. As we rallied, it bound us together, too, which is perhaps the only consolation you can seek at such times. I was responsible for going to Rifleman Davey's room and overseeing the packing up of his belongings, prior to them being sent home. It's poignant enough to contemplate the possessions of someone who has just died, but to know that these things were going back to a young man's grieving parents was unutterably saddening. I've carried the memory of that feeling with me ever since.

Until this point, Northern Ireland had existed almost entirely for me as stories in the news. And now here I was, starkly in the middle of it – and one of its targets, in fact. I would get to know the sensation that all patrolling soldiers in Northern Ireland experienced, when the barrack gates opened in the morning and you would drive out or dash out on foot and know that from this point on you were vulnerable. You would see the murals, you would sense very clearly the intensity of hatred towards you. When you were stopping people at checkpoints, interrupting their daily lives, it would be quite obvious that some of them would like nothing better than to take a pop at you. It was eye-opening and difficult to handle – not least when you didn't, and couldn't, culturally reciprocate that hatred, and when the position you were required to take was in the middle of it all.

Our task was mainly patrolling. We had a strong intelligence unit that briefed us on who the suspected IRA terrorists were, which people were under observation, where weapons and ammunition might be getting stored, which vehicles might be being used to move weapons around. And that intelligence underpinned the patrolling that you did on the streets and elsewhere and the vehicle checks that you might do to try and anticipate the movement of arms, ammunition and explosives. Partly you were there, also, to help the Royal Ulster Constabulary do their job. The RUC were probably at the time the IRA's number one target. The British Army would have been a close number two, but if the choice was between attacking a police station and a military barracks, the police station would be more likely to get hit. So your work was to assist and protect the RUC.

I did most of my urban patrolling in Strabane, which sat right on the border with the Republic, suffered terrible economic deprivation and soaring unemployment levels and for a while knew the distinction of being the most bombed town in Northern Ireland. Patrol in Strabane was always edgy, as urban patrolling tends to be. The enclosed settings meant you were more reactive than active, waiting for something to happen to you rather than getting ready to make something happen. It's an enervating position to be in, in the long term. And there were so many opportunities for snipers to get a look at you. At this point, we knew the IRA had access to some really big .50-calibre sniper rifles, so that made you uneasy. And there were so many places in Strabane in which they could plant devices – abandoned buildings and doorways. You couldn't possibly check everywhere. Obviously you worked to minimise the risk to yourself but you knew that your survival in those circumstances was as much about being lucky as it was about being professional.

Strabane was also where we got called out to support the RUC during some rioting. I have no idea why it suddenly kicked off in the town that particular evening, but kick off it did, and we donned the full riot gear with the shields and riot guns. We didn't end up firing any baton rounds, but I ducked a few bottles and stones until it blew over and, the whole time, given where you were, you couldn't avoid thinking, 'How badly could this escalate?'

Other times it was our job to operate vehicle checkpoints. Sometimes these would be spot checkpoints, which we would drive out and set up in the armoured Land Rovers – the Snatch Land Rovers, as they were known. Or an RAF Wessex helicopter might pick a team of us up and then drop us somewhere near the road so that we arrived in the area completely out of the blue. After a while the helicopter would return and airlift you to somewhere else. Like with your movements on a city patrol, it was all about keeping it fluid and unpredictable.

And at other times again, you would be manning the permanent vehicle checkpoints at the border. There was nothing particularly fulfilling about that work. It was just something that had to be done. Four miles out of Strabane stood the PVCP at Clady, on the border between County Tyrone in the North and County Donegal in the South. A fortnight of duty there could tax the patience quite heavily. A blank-looking, green-painted fortified base with tall walls and a watchtower, Clady was known as 'Rorke's Drift with radio'. It wasn't an entirely cheerless place: they kept a supply of confetti on hand to throw over newly-wed couples returning from the South. Nevertheless I don't suppose too many people wept when it was dismantled in 2010. It was my dubious pleasure to spend Christmas at Clady. We strung up some tinsel to make it look festive, but I'm not sure how successful that was. It just looked like Clady with tinsel.

And then there were rural patrols, known as 'cuds patrols'. Those were probably the ones I enjoyed most because at least you were out in the fresh air. Also the countryside was a more workable military environment: you could be clever about the terrain, use your observation techniques, spot the danger areas and stay out of them. That didn't mean the cuds were safe, though. It was common IRA practice to loose off a few shots across the border from the South and then high-tail it. The one time a unit of men under my command came under fire was during a countryside patrol. One of my lance corporals and his section had shots fired at them from across the border. They saw the firing point and returned fire. My section was 200 metres away, on the opposite side of a wood line, but we heard the gunfire, got the radio call, 'Contact, make ready', and ran round to join them, weapons cocked, adrenaline up. The gunman had given up and run for it.

Another day, I was with a team doing a spot vehicle checkpoint, and with two other patrol groups flanking us in the hills on either side. In one of those teams a Rifleman inadvertently kicked a wire underfoot – only to discover that he had just pulled the detonator out of an explosive device planted by the side of the road. The device had failed to go off, which was just as well, given that it was meant to ignite a barrel full of petrol, explosives and nails. That was a truly close shave, but it was not uncommon, while on patrol, to find signs that someone had obviously thought about trying something but had run out of time or had got spooked and left. They were little reminders that you weren't out there on your own.

For the most part, you would rotate these patrol duties in fortnight-long segments, normally deploying to a temporary base where home was a CORIMEC prefab building, protected by gabion baskets filled with rubble, and with concrete over its roof to protect against mortar fire. There you would most likely be given a windowless cell with little in it besides a metal-framed bed onto which you could throw down your sleeping bag. Between times you would return to the greater luxuries of Lisanelly Barracks in Omagh, which at least offered the comforts of an officers' mess.

Returning to Lisanelly for downtime became something of a treat. Which is why it was annoying to be woken by a bang on my door at 3.00 a.m. one weekend and be instructed to get dressed and get my platoon together for an emergency search operation.

'What are we looking for?' I asked.

It was an HK53 assault rifle. Apparently an Army Air Corps Gazelle helicopter had lifted up out of our barracks earlier that day bound for Aldergrove. It was only when the pilot reached his destination that he remembered he had hung his rifle over the searchlight boom during his pre-flight check and had forgotten to take it into the cockpit with him before leaving. I can only imagine the horrendous sinking feeling that must have gone through his stomach when he realised. In the Army there is no greater sin than losing your weapon. Gifting a loaded HK53 to the IRA was also likely to be frowned upon.

The likelihood was that the gun had slipped off the helicopter's boom early in the journey, so we were despatched to scour the neighbourhood close to the barracks. Our search didn't quite extend to knocking on doors in the middle of the night and asking, 'Have you by any chance seen our rifle – and if so, please could we have it back?' But we covered the ground as best we could. It felt like a pretty forlorn cause from the beginning, to be honest. And, indeed, no stray weaponry was to be seen. However, later that afternoon, the missing rifle was handed in by a scrupulously responsible old lady. Apparently it had dropped through her garage roof. Her car wasn't in the garage at the time, fortunately, so only the roof was damaged. I believe the Army fixed it up for her. The rifle, however, was a complete write-off. But at least it was now accounted for.

I had become the proud owner of a car myself by this time. I'd had the chance to put some money aside, and I had seen something I really fancied for sale outside a run-down little garage in Omagh. It was a BMW 3 Series – the 320 Motorsport version, in black with some serious alloys and lowered suspension. A sucker for that kind of thing, I lost my heart to it the moment I spotted it. It also seemed to be going for an extremely reasonable price, so one afternoon I left the barracks and went and bought it.

Returning behind the wheel to Lisanelly, I was held at the gate for slightly longer than I would have expected. Eventually, one of the soldiers came out of the office, looking concerned.

'Oh, hello, boss – it's you,' he said. 'Can I just ask you, boss, why you're driving a tagged vehicle?'

They had run my prized new possession through the computer and it had flashed up as a security concern, the property of a known IRA terrorist, recently arrested and jailed. Which was a touch embarrassing.

Great car, though. I ended up taking it back to the UK, where I discovered it was known to the English police as well. While trying to re-register it, I had to report with it to a police station. Unsurprisingly, the chassis number didn't quite match their records. A while after that, one Sunday night, heading back to Middle Wallop from my parents' house, I hit a patch of water on the M27, spun three times and slid down the central barrier backwards before coming to rest in the fast lane. Not much traffic around that night, thankfully. I picked up the pieces that had come off the vehicle and limped back to the Officers' Mess with the bumper poking out of the boot. That car was trouble.

In February, halfway through my attachment, I had two weeks' R&R and nothing to do with it. My fellow platoon commander and great friend Charlie Sulocki suggested I fly to Bermuda with him. He had a cousin out there. We grabbed some last-minute flights and had an extremely nice, sunny break away from the pressures of wintry Northern Ireland. Our flight home was via New York and just as our connecting flight from Bermuda landed at JFK, so did a massive East Coast snowstorm. It shut the airport down for three days. We were trapped.

'Do you know anyone in New York?' Charlie asked me. I certainly didn't. But Charlie got on the phone to his dad, who found a friend who was willing to put us up – a wealthy car dealer who seemed to own a number of Nissan dealerships around the city. Even more crucially, we found a taxi driver who, with cars stranded in the snow everywhere you looked, somehow managed to drive us downtown to Charlie's dad's friend's extremely well-appointed apartment. For the next three days, we trawled New York in the snow until the airport reopened. We caught some grief from the company commander for being late back, but what could we do?

It was Charlie who joined me for an armed shopping trip to Belfast one weekend. We hadn't intended it to be armed: we'd intended it to be a simple visit to a city that neither of us had been to. But when we went to the guardroom to sign ourselves out, someone there said, 'OK, but what weapons are you taking?' Four years earlier, two British Army corporals had found themselves caught up in a funeral procession at Belfast's Milltown Cemetery and had been dragged from their car, beaten and shot. Since then, no one was recommending that Army officers – even Army officers dressed, as we were, in our civvies – went anywhere far from base without taking precautions. So Charlie signed out a 9-millimetre pistol and concealed it in a holster under his jacket.

It felt odd, to say the least, to be doing the tour of the city and browsing around Marks & Spencer, etc., with a loaded gun for company. Still, the only complication we met with was when Charlie decided he would like to try on some clothes. Rather than risk revealing to other potentially startled shoppers what he was carrying (and also dangerously exposing ourselves), we decided it would be better if I wore the gun while Charlie shopped. Our chosen venue for the handover was the loos in a small café where we stopped for coffee. The plan was that Charlie would enter the loos, I would give him a couple of minutes in one of the cubicles to get out of the holster, and then I would go in, take it from him and put it on under my own jacket. Easy.

So Charlie went off to the loos and, after a reasonable pause, I followed him in there. Were other people in the café watching me over their coffee cups as I crossed the room, or was I just being paranoid? Ignoring those feelings of mild anxiety, I pushed through the door to the loos – and thereupon discovered that in fact it wasn't a loos; it was a loo, singular. One door, one toilet – and now two men, with the door firmly closed behind them and locked. Anyway, we switched the weapon, and then left, trying to look as nonchalant as possible and not meet the eyes of the other customers who, if they hadn't been watching us before, were certainly watching us now.

What with the patrolling and the armed shopping expeditions, the eight months of my attachment quickly passed and, before I knew it, I was getting ready to leave and head back to Middle Wallop and the relative calm of the Army Air Corps. To mark the conclusion of my stint with the Royal Green Jackets, I was given a company dining out. I sat at the table, thinking back over the experience I had gained, and the lessons in soldiering I had learned, and reflecting how none of this might have come to pass if my superiors had ever twigged what had gone on with those two Land Rovers in the boggy hills above Fermanagh.

After the dinner, the sergeant major stood up and made a speech in which he thanked me for my time with the unit and bade me farewell and good luck on the road up ahead.

And then, almost as an afterthought, he added, 'And by the way – don't think we don't know what happened on that night patrol.'

You can't get anything past these people. Don't even try.

CHAPTER SEVEN

*Learning to hover, splashing down in the dunker,
and getting into a bit of a hole*

A rmy Pilots' Course 350 began in June 1993. There was
ground school: the basics of meteorology, rules of the air,
navigation, principles of flight. We were also expected to learn
the technical aspects of our aircraft in far more detail than we
ever had previously. Before progressing on to helicopters, we were
put through our paces during thirty hours of basic fixed-wing
training. We were back in the venerable Chipmunk, and as I
climbed into the cockpit for my first sortie I was greeted by the
familiar smell of engine oil, fuel and cordite from the cartridges
used to start the engine.

The fixed-wing phase was progressing well. I was encouraged by
the post-flight debriefings that I was making the grade. Each sortie was given a colour code in addition to a few scribbled comments.
Green meant all was going well – blue, above average. Brown
equalled must try harder. And red was an outright fail. Notch up
enough browns or a red and you would be on 'review', sent for an
interview without coffee with the chief flying instructor and given
a couple of extra hours' flying as a last chance to prove yourself.

When my instructor, Chas Boyack, an ex-RAF Red Arrows
pilot, showed me my colour chart, I was pleasantly surprised to see
a healthy sprinkling of greens and blues. However, the first solo

long navigation sortie was looming – probably the most daunting challenge for any basic fixed-wing student. I should point out that these were the days before smartphones and no training aircraft were fitted with GPS. Navigation was achieved with strictly a map, a compass and a clock. Fly on a heading at 500 feet, adjusted for wind, and hope to God that you hit the right place at the right time.

We had a radio, of course, and air traffic control could provide a location-finding service if really needed. If you were genuinely lost, you were supposed to own up and make a 'PAN' call on the emergency frequency, one step away from a 'MAYDAY'. But air traffic also provided a 'Training Fix', for pilots or instructors who wished to practise the lost procedure even while fully aware of their position. I heard many requests for a 'Training Fix' over the radio during my time at Middle Wallop and I suspect that behind at least half of them was a trembling solo student, wondering where on earth they were and if they had enough fuel to get home. But why would any self-respecting student own up to being lost when they could obtain the information they needed from air traffic while passing it off as a training exercise?

My good friend Toby Everitt was on the course ahead of me and had already completed his solo long nav. I decided to ask him if he had any tips and how he got on. 'Oh, just fine,' he casually replied. 'I never really went anywhere.' In response to my puzzled expression, Toby explained that he wasn't overly confident about getting around the route, so he'd elected instead to go to Lopcombe Corner, an obvious road junction just out of sight of the airfield, and do lazy figures-of-eight for an hour and a half. Toby made it back to the airfield on schedule with no one the wiser and headed back to the officers' mess just in time for afternoon tea.

Toby later transferred to the Royal Navy, becoming a senior pilot on the single-seat Harrier strike aircraft, so I guess he mastered the whole map and compass thing at some point.

Fixed-wing flying was fantastic fun, especially since our thirty hours of instruction included aerobatics and the exhilaration which came from mastering loops, rolls and stall turns – even better with an ex-Red Arrow pilot to teach me. However, I had joined the Army Air Corps to fly helicopters. For as long as I could remember, I had dreamed of that feeling of hovering and low-level nap-of-the-earth flying that only helicopters can achieve. Now, at twenty-one, it was finally time.

The type of helicopter in which we were taught was a Gazelle. People nickname the Gazelle the 'screaming chicken leg', on account of its shape and its high-pitched, whining engine. It's a fair description although, personally, I prefer to think of it as the helicopter world's equivalent of the MG Midget sports car: basic, raw, sporty, a pilot's machine calling for pure piloting skills.

You sit low with your legs stretched out in front of you, in almost a racing driver's position, and strapped in with a five-point harness. The Gazelle is a reconnaissance aircraft, which means it has a large Perspex bubble and a skinny instrument panel with small dials, switches and controls to allow maximum forwards visibility. You feel quite exposed with all this Perspex around you and when you transition away for the first time you are much more aware of your height above ground than in the Chipmunks we had been flying. But there is a beauty and elegance in the way that it flies. Nothing is automated and once you master it you feel absolutely 'connected' to it. The Gazelle becomes an extension of you, relying on that proprioceptive feel in order to be well flown, your body moving in tune with it. In larger, more complex aircraft you often lose that raw connection as it gets dulled by autopilots and stability control systems working away in the background. Not with the Gazelle.

My rotary wing instructor was Alan Wiles, an ex-AAC major who had previously served in the Royal Marines as a sergeant pilot, seeing action in Aden, Radfan, the Falklands and Northern Ireland, earning himself an Air Force Cross along the way. Alan is the only man I know who can compete with Yuri Malenchenko, my Soyuz commander, for his ability to emanate a sense of calm. When I eventually became a helicopter instructor, I learned how there is a fine balance to be struck when giving students the responsibility to control an aircraft. Most students are slightly underconfident, and often the instructor has to push them a little, carefully gauging what they think they are capable of and nonchalantly empowering them to give it a go. Alan was an absolute expert at this. I remember on just the second sortie, him telling me to go and start the helicopter, get the rotor blades turning and he would come and meet me on dispersal in a few minutes.

That feeling of sitting in the pilot's seat, alone, going through the pre-start checks, starting the gas-turbine engine and hearing it whine into a frenzy, then advancing the throttle lever and watching the rotor blades slowly turn until they become a blur and you finally feel that you have control of that spinning disc above you – it was pure magic.

Pure magic if you get it right, of course. A few years earlier, one poor sod, for reasons known only to himself, had pulled up on the collective lever and managed to inadvertently get himself into a hover. Since hovering hadn't yet been covered, that flight didn't last too long. The instructor walked out to a scene of carnage on dispersal; one written-off Gazelle and a miraculously unscathed student looking rather sheepish.

There's a real art to hovering a helicopter. The control inputs required are so imperceptible and frequent that you can't actually think about what you are doing – you just have to feel it. It doesn't help that every small movement of one control affects all the others, so you are constantly fighting against your own inputs. In the early stages this often leads to a 'pilot induced oscillation' – basically where the student can't keep on top of the helicopter and puts in ever increasing control inputs which escalate to the point at which the instructor has to take over before disaster strikes. Highly amusing to watch, not so funny when you're the student sweating at the controls.

The trick is to start off one control at a time. First, keep the helicopter pointing in the right direction with the pedals. Next, just control the height with the collective lever. Finally, the really hard part, keep the same position over the ground using the cyclic stick. Now try all three together. It usually takes about ten hours of flying before students can master the basics of helicopters and are let loose for their first solo. Which is why Alan took me completely by surprise when he jumped out of the aircraft at the end of one sortie and told me to take the helicopter round for a circuit myself and he would meet me back in the crew room. I was still a couple of hours of instruction away from the much anticipated 'first solo'. Still, if Alan had faith in me then I just needed to believe in myself.

The loss of 80 kilograms or so of instructor in the left seat made quite a difference in such a light helicopter. It hung lower to the right and took off sooner than expected. However, I wasn't really paying much attention to any of this – I was by myself, I was solo. When I landed back on dispersal my cheeks were aching from grinning the entire way round the circuit. I had gained the honour of first solo on my course, which my crewmates were equally delighted about. Not that they cared about my success. They simply knew what it meant for them. That night, in line with long-standing tradition, I would be ringing the bell in the bar and buying all the drinks.

*   *   *

Naturally, as well as learning how to fly a helicopter, you had to learn a few things about what to do if you ever crashed one. That's why the dozen of us on the Pilots' Course were soon climbing into a minibus and heading off to Yeovilton in Somerset to experience for the very first time the delights of the dunker.

You might not assume so from the outside, but the fact is you've got a very strong chance of surviving a helicopter crash into water. Even with an engine failure, it's possible to enter autorotation and make a successful landing on the surface. A Chinook has the buoyancy of a boat and could probably sit happily on water for a short time while you got everyone out. The Apache, by contrast, is heavy and narrow and sinks like a brick. In the ocean, you would be 50 metres under before you got the door open, which is why the preferred technique is to explosively jettison the canopies on the way down. If you're in an aircraft that's going to go over, then the best idea is to get it over with as quickly as possible: put it into the water, kill the engine, kill the fuel, roll the aircraft, get the blades stopped and then swim out.

You'll need some training, though, which is where the dunker comes in. Based at the Royal Naval Air Service's Underwater Escape Training Unit, the dunker is basically a purpose-built swimming pool in a big white-walled warehouse. It boasts a cylinder that can be adapted to resemble a Sea King cabin, or a Lynx or a Puma, as required, and a crane to lift it up and lower it into the water. Students climb into the cylinder and strap themselves into their appointed seats. Normally, as aircrew, we would just train in the front seats, but when you're going through the Pilots' Course, there's a dozen of you, so you rotate through the different seat positions, depending on the configuration. Once inside, you adopt the position: basically, if you know you're going to crash in a helicopter and you're a passenger, one hand goes to your harness release and the other goes towards the exit. So, if the exit is on the right, you grab your harness with your left hand and hold tight to the seat on your right. That way, even if you're upside down and tussled around and it's dark and all kinds of chaos is going on, that right hand is going to take you towards where the exit is. So, with everyone ready in the cylinder, the crane begins to drop you towards the water and (very important, this) you take a deep breath.

The first dunk will be a relatively simple one, to ease you in – an upright descent into the water. Unbuckle and go. For the next run they will lower you into the water again but this time, once you're in, they will turn the cylinder upside down. Again, unbuckle and go. And then, once you've come through that, they might do the same thing again, but this time with the lights out. Pretty soon they'll be twisting you around in the air before you reach the water, to get a bit of disorientation going on. Each time, two Navy divers accompany the cylinder underwater. Once in place, you get the tap on the head from one of the divers. That's your cue to release your harness and head for the exit. Eventually, as aircrew, you'll be going down in the front seats with the doors on, learning how to smack them out and jettison them as soon as the cabin has filled with water and then getting out.

It's very good training, but some people hate it. The prospect of being suspended under water in a metal box can understandably provoke some strong emotions. We had a sergeant on our Pilots' Course who was one of the older students, with experience behind him as an air trooper, and who generally gave every appearance of being more switched on than the rest of us. When he was strapped into the dunker behind me and we were waiting to be lowered in, he abruptly developed the most acute case of verbal diarrhoea that I had ever encountered. I thought he was putting on a performance as some kind of joke, but when I turned round to look at him I could see that he was genuinely terrified and babbling uncontrollably. Still, better verbal diarrhoea in the dunker than the other kind, definitely.

I did the Pilots' Course before they introduced the Short Term Air Supply System – a canister a bit like a SodaStream bottle, which you can breathe from and which, in the unlikely event that you are breathing normally, will grant you about two minutes of air. I learned how to work with those later, but all of my Pilots' Course dunking was done with held breath. It felt like a lot longer when you were down there, but most of the time everybody would be clear within about thirty seconds. Even with all the necessary exertion, that wasn't too much of a strain on the lungs. However, there was one run where I was in the number six seat, the last to exit – the shortest straw, for obvious reasons. But on this occasion the straw proved to be even shorter than it might have been because next to me, at number five, was Corporal Theo Street. Theo was a Royal Engineers diver. This was a guy with thousands of hours of diving on his record who could comfortably hold his breath underwater for whole minutes on end, and certainly for far longer than I could.

So we hit the water and the evacuation starts. I'm sitting there with one hand on my harness, the other towards the exit, and I can feel the others going out ahead of me. But not Theo. Theo doesn't move at all. The cylinder has become still now. It's got to be his turn. Also, quite soon, the breath I have taken will run out.

Concerned, I twist my head to look across at Theo. Theo is staring at one of the Navy divers, who is also in the cylinder with us and holding his breath too, and both of them have smiles on their faces. It appears they are having a competition, as professional divers, to see which of them can hold his breath for longer. What? They want to play this game now? While I'm still in here, sitting at number six? I start punching Theo on the shoulder with increasing urgency and eventually give him a massive shove. At that point he relents and casually floats out ahead of me and I get to the surface behind him with my lungs stinging. Boys and their fun, eh?

When I wasn't inadvertently caught up in Royal Engineers v Royal Navy honour battles, I felt pretty comfortable in the dunker. And just as well because I ended up spending quite a lot of time in there. As Army aircrew, who don't tend to be flying over water particularly frequently, we would do a refresher at the UETU about once every two years. Navy helicopter pilots, obviously, are in there more frequently – maybe every six months. But as I moved on to become a helicopter instructor and then a helicopter test pilot, I found myself making the trip to Yeovilton many times. On one job, working on Apache and testing the ease, or otherwise, of evacuation through its canopies, it practically became the office: four dunks in the morning, lunch break, four dunks in the afternoon, with pauses in between for evaluation. I got pretty inured to dunking.

But the more the better, perhaps. In August 1995, shortly before my tour of duty in Bosnia, an Army Air Corps Lynx doing routine reconnaissance training accidentally crashed and went upside down in the Adriatic, 5 miles south of Ploče in Croatia. Its commander escaped but swam back down to try and help his trapped crew. He didn't make it back up and, tragically, he and three other soldiers died. Only one person survived. Swimming to the surface with only cuts and bruises, he was rescued by a Croat fishing boat. That soldier happened to be an airtrooper whose job at one time was to drive the minibus taking students on the AAC Pilots' Course to Yeovilton for their dunker courses. Because he loved the dunker, every time he went there he asked if he could join in. He had apparently done day-sessions in the dunker multiple times. It paid off, clearly, and yet again highlighted the value of training.

* * *

During the Pilots' Course our weekends were our own. After a week of helicopter flying, with its exhausting pressures and adrenaline rushes and its need for constant, needle-sharp concentration, it stood to reason that by the time Saturday came around we just wanted to be able to sit quietly somewhere with a cup of tea and a good novel.

Actually, that's not quite true. We would drive around the country to go potholing.

David Amlôt was the chief protagonist here. A fellow student on the course, David was an extremely good rock climber and kayaker. At his suggestion, he, Toby Everitt (he of the pioneering solo flight around Lopcombe Corner) and I would jump in the car when classes finished on a Friday afternoon, drive over to the Gloucestershire and Herefordshire border and spend the next two days kayaking in the Wye Valley, and climbing and abseiling on the Wye Valley cliffs and Symonds Yat Rock.

David was the best climber among us, so he would go up ahead and top-rope us, and Toby and I would climb up to join him. And then we would all abseil down. Descending from a pinnacle one day, we were rather startled to see a couple of heads pop out of the sheer rock face. These two people emerged like gophers, apparently from nowhere. We chatted to them and they explained that below the surface on which we were climbing lay a potholer's paradise, a whole subterranean world of caves, tunnels and chambers. We had a quick look in with a torch and decided that this was for us: next weekend, we were expanding our activity range to include potholing.

I shake my head in bemusement now, and even wince slightly, to reflect on exactly how unprepared we were for this adventure as we set out that Friday night for the Mendips, quite late and determined to get straight into it as soon as we arrived. Helmets? No. Wetsuits? No? Proper head-torches? No, just mini Maglites. Supply of spare batteries? Are you kidding? That said, we weren't completely gung-ho about it. David had dug out from somewhere a rudimentary written guide to the cave network. And before we set off, we rang a pal of ours, Charlie Howard-Higgins, and told him what we were up to. If you're planning on disappearing down a hole in the ground, it always pays to let somebody know. The deal with Charlie was that we would ring him when we got back above ground, which we reckoned would be around midnight. If he didn't hear from us, Charlie was to give it a couple of hours and then raise the alarm. I suppose it was reassuring to know that, in the event that we got into difficulties, someone about 100 miles away would be on the case. We just had to hope he didn't fall asleep.

Such, clearly, is the sense of invincibility casually enjoyed by a bunch of twenty-one-year-olds – not entirely unaware of the risks and the dangers as such, but very sure that those risks and dangers are unlikely to apply to them personally.

It was raining quite hard when David, Toby and I reached our chosen hole and we were quite relieved to duck into it and get out of the wet. Just as we had assumed, being in the dark and following a path through this hidden limestone network, shining our torches up at the roofs of its damp and glittery ceilings, was tremendous fun. Eventually, when we had been exploring for some time, David paused to read aloud from the guide that we were following.

'When you reach the pebble crawl,' David read, 'if there is a noticeable flow of water through it, you should abandon the cave. A flow of water at that point indicates a high risk of flooding.'

Fair warning. Good to know. Just as soon as we came up against this pebble crawl, we would make sure to monitor it very carefully for …

I think it dawned on all three of us at about the same time. Hadn't we already passed through a section answering very closely to the description 'pebble crawl'? Indeed, hadn't we done so about half an hour ago?

Wasn't that the short, narrow tubular section, with the pebbles in it, where we'd had to get down on our hands and knees and crawl for about 15 feet or so before we could stand up again and continue?

That place where the water, as I crawled, had been mid-way up my forearm and with a definite current to it …?

It seemed like this would be a good moment to turn around and head out.

With a new urgency about our pace, we threaded our way back through the pitch-black tunnels and caves the way we had come. All of us were now feeling some panic starting to rise. I was remembering as we walked how hard it had been raining when we had arrived – and wondering whether that ought to have sounded a louder alarm with us than it had.

Eventually we were at the entrance to the pebble crawl. We anxiously shone our Maglites down into the tunnel. The water had risen massively, and was moving strongly. Thankfully, though, the tunnel wasn't yet full. When I got down onto the pebbles, there were just enough inches of clearance for me to crawl through with my chin tipped up and my mouth just out of the water.

We all crawled through the water and pulled ourselves out at the other end, and then we headed on, sodden, back to the surface. Another twenty minutes and who knows how long we might have been stuck down there. That little episode shook us up. It made us realise that if we were going to do this kind of thing, then we needed to be more serious about it. Straight afterwards, we got ourselves properly kitted out, with helmets, wetsuits and head-torches. We got hold of the proper guide books and read up a bit. We learned how to do presses and duck dives. We showed the underground environment a bit more respect and proceeded a little more wisely.

The potholing (along with the climbing and kayaking) continued for the year of my Pilots' Course. We got pretty adept at it – taking ropes and harnesses and carabiners into caves with us, so that we could do technical rope work down there and really get to places. I loved it and, although I didn't know it at the time, it was laying some useful ground for me. What I learned, mucking about in the limestone below the Mendips and Herefordshire with David and Toby, about self-control and claustrophobia and about operating in confined spaces, would stand me in good stead years later, in certain other tight environments.

One of the things we always did, when we first got into a cave and were 20 or 30 metres inside, was to switch off all our lights and just sit there for a couple of minutes in the pitch black. Obviously that helps your eyes acclimatise to the darkness, but I also found I could use that couple of minutes to absorb where I was and go into a different zone. It was as if you were pressing a reset button and saying, 'OK, this is the new norm for the next two hours – get used to it.' And then the torches would go on and we would head off, feeling pretty serene about it all. Panic can come very easily in caves – as I found out during that pebble crawl incident. There might be moments when you think you are lost; the extreme closeness of the surroundings enters your mind, and anxiety begins to rise. But if you've got yourself into that calm zone beforehand, you're in a frame of mind that damps down the agitation before it even starts.

I think about when the moment came in December 2015 and I climbed into the Soyuz capsule ahead of launch. That little metal can on the top of the rocket might as well be a tiny crawl space. You're strapped tight into a pocket of room in a seat that's been moulded to fit your body – and you're going to be there for about twelve hours. And just to monkey with your mind a little further, you're positioned on top of a 300-tonne bomb that's about to blast you into orbit. If you were to crawl in there thinking about how cramped it was and how trapped you were going to feel, panic would be close to the surface from the start.

So instead you embrace it: you decide that it feels nice and comfy. You convince yourself that it's a desirable place to be: that you love the cosiness of it and that, in fact, there is nowhere else you would rather be. For me, those moments of settling into the capsule were very much like the two minutes of lights-out after entering a cave: a reset point at which I told myself, 'Right. This is where my body is right now, and I am perfectly comfortable here.'

It's not for nothing that the European Space Agency's astronaut training sends you to live for seven nights in caves in the mountains of Sardinia. OK, on that trip I would be in the company of professional cave-explorers. But even then I was grateful for the lessons that I had learned informally twenty years earlier with my adrenaline-mad mates.

Back on the Pilots' Course, things were proceeding more smoothly. I moved on from basic rotary to advanced rotary. Basic rotary was essentially learning to fly the aircraft: circuits, sloping ground landings, flying in confined areas – becoming a helicopter pilot. Advanced rotary was about going beyond that and becoming, specifically, an Army pilot: map reading, reconnaissance, battlefield tactics. You started doing tactical formation flying, working in two ship formations and sometimes three ship formations, learning mutual support, practising 180 turns and crossovers. Then there was all the academic work that under-pinned it – the structures of Army units, how you work with artillery, armoured reconnaissance platoons and so forth. And once you passed advanced rotary, you got your wings.

That was the proudest moment of my life up until then: my wings ceremony – the qualification I had dreamed of and worked towards. All my family came for the presentation. I had started seeing a friend of my sister's called Helen whom I had met while I was at Sandhurst and spending weekends in the student union at Keele University, though we had only got together after I came back from Northern Ireland and started the Pilots' Course. Helen came, too, for this day-long event at Middle Wallop with a parade and a formal lunch and the presentation.

Later we would drink until late at night in the officers' mess bar, but for now the senior presenting officer pressed my wings – the Army Flying Badge – onto the strip of Velcro that I had pre-attached to my uniform (you sew the badge on properly later). I stood back with satisfaction and watched our instructors put on a flying show in our honour – a pair of Gazelles doing a synchro routine, a Lynx performing backflips, loops and rolls.

None of those are moves in a helicopter that you have come even close to covering on the way to your wings. That level of proficiency, I well knew, was way off in the future at this point, making that flying display a celebration, and also a quiet nudge – a little reminder that I had come a long way as a pilot but that there was still a lot more out there to learn if I kept going.

Growing up in Whitley Close, in the village of Westbourne near Chichester – perched on Grandad's Triumph with my sister Fiona.

Even aged two I clearly wanted to 'Go', I just hadn't quite worked out where to . . .

My first visit to Buckingham Palace with the Whitley Close gang on a rare trip to London.

Life had dealt me two potential disadvantages for coping in the jungle which is the playground at school. First, I was small. And second, I was ginger. With those twin blessings, I had to develop a thick skin for banter. But I made good friends and enjoyed school.

Family camping trips gave me a taste for the outdoor life. Here I am with Mum, Dad and my sister Fiona in the Ardèche.

Eventually we graduated to a caravan – with me in a tent, naturally.

The beauty of the Yorkshire Dales beckons – and some challenging solo walks for a young teenager. I have to give my mum and dad credit; they weren't potholers, kayakers, rock climbers or orienteers, but they were happy to encourage me in all of that reckless stuff when I showed an interest in it.

I couldn't wait to get into the river and start some real exploring. Little did I know then that one day I'd be sitting in something doing Mach 25!

'So you get to dress up as a soldier AND you get to go in a kayak? That felt like an irrefusable offer to me.' I'm sure when I joined the school's Cadet Force, at the age of thirteen (front row, far right), my parents would have been wondering whether this was just the next thing that I was going to try on for size and soon discard. But I fell for it hook, line and sinker and eventually became senior cadet.

My lovingly assembled radio-controlled aircraft starter kit, shortly prior to its disastrous maiden flight. Thankfully I had better luck flying the real thing.

My 'C', 'D' and 'E' grades in Physics, Maths and Chemistry did not win me any awards at Chichester High School, but I did receive my colours as a proud member of the school's hockey team (standing, far left).

Following a sponsored event for charity, where a few of us cadets spent a cold and miserable night on an inflatable dinghy in the Chichester canal, the RAF Search and Rescue team winched us out in a Sea King helicopter. The event left a sizeable impression on me and must have too on Paul 'Haz' Hunter (middle), my partner-in-crime during those youthful experiments with explosives. Paul went on to be a RAF Search and Rescue Winchman. Being plucked up and spun away in the sky amply confirmed my early impression that helicopters were where it was at.

Who knows where the inspiration that will carry someone onward for the rest of their life might come from? For me it was the Operation Raleigh Alaska Expedition in the spring of 1991.

And to pay for it – I decided to try out as Chichester's answer to Tom Cruise in *Cocktail*, working as a barman at The Nag's Head with the endearing Frenchman Olivier Barbedette.

Here I am now, in my forties, having travelled to space. Yet I still find myself from time to time thinking back to that trip I made when I was nineteen and reflecting on how much it shaped my attitudes and altered my aspirations.

Operation Raleigh took in forestry conservation, river surveying, construction work in a lagoon, camping, kayaking, mountain climbing and some glacier exploration thrown in. I met so many wonderful people and we shared so many extraordinary times together. In my diary, towards the end of the trip, I would write 'the best time of my life'.

Humpback whales, sea lions and otters would be popping up next to you, eagles would be soaring overhead and all around you was this classically beautiful backdrop of Alaskan mountains. And then, at the end of the day, you would pull the kayaks onto a beach, drag some dry wood together, get a bonfire going and cook up the night's meal. It was bliss.

After the excitement of Alaska, I was ready for my chosen career path – the British Army. At Sandhurst I was taught the discipline, organization, leadership, teamwork and even tidiness that would serve me well when I became an astronaut. For the first three weeks of the course we spent every evening in the platoon room shining or 'bulling' our boots. Some people seemed to find bulling therapeutic, but for me trying to achieve the requisite shine was an endless struggle.

Celebrating completion of Sandhurst's final exercise with One Platoon, Alamein Company. I'm in there somewhere . . .

At the Sovereign's Parade with my grandparents, marking the end of a year's intensive training.

My last day as an officer cadet – tomorrow I'd be Second Lieutenant Peake, Army Air Corps.

PART 2: SKY

CHAPTER EIGHT

Upside down rooms, vanishing tyres,
and rapid descents

In July 1994, I went to the quartermaster's stores at Middle Wallop, collected four wooden boxes and returned to pack up my room in the officers' mess. I was shipping out to my next posting – Gütersloh in northern Germany, where I would join up with 1 Regiment Army Air Corps and the esteemed 652 'Chosen Few' Squadron. I would be based in Gütersloh for four years, but in that time I would accept a succession of temporary postings which saw me return to Northern Ireland and spend time on duty in Kenya, Cyprus, Bosnia and Canada. I would continue doing what I loved, which was flying Gazelles; I would become an aircraft commander and then a flight commander; and I would meet the woman I would fall in love with and eventually marry. What with one thing and another, I have lots of reasons to be thankful for Gütersloh.

I went there knowing nothing about the area and very little about Germany, which I had visited only once before, on a camping holiday with my parents. To the best of my recollection that trip did not involve us in spending a great deal of time in North Rhine-Westphalia. It's safe to say that Germany has far stronger tourist magnets than the north German plain. Indeed, the old joke about Norfolk probably applies: the north German plain is not flat and boring – it's slightly undulating and boring. Gütersloh itself was a small industrial city with some quaint timbered buildings in the centre; some big German business concerns were quartered there (Bertelsmann, Miele), but it had little in the way of excitement to offer a twenty-two-year-old.

On the surface, our military base didn't seem much of a hotspot, either. It was a run-down former Luftwaffe airfield which had subsequently been home to the Royal Air Force. But they in turn were gone, and 1 Regiment Army Air Corps had just moved in, along with three regiments of the Royal Logistic Corps, the Army's supplies, movement and maintenance arm. Hangars which would once have been home to Harrier jets were now being used for truck storage and general warehousing, leaving just one hangar at the far end of the airfield which was where our helicopters were, slightly pressed for space, with a tarmac dispersal and a patch of grass to land on.

Yet I have to say, despite this unpromising setting, the atmosphere I found myself plunging into at Gütersloh was utterly fantastic. The lieutenant colonel in charge of 1 Regiment AAC at this time was Arthur Gibson, a lovely guy with an endearing touch of the old-school flying club about him. He and Will Mellows, my squadron commander, believed firmly in making fun a high priority, were strong advocates of the virtues of adventure training, and ensured that a real work hard/play hard ethos was in operation at all times.

So that was one key element of the Gütersloh chemistry. The other was the demographic of our fellow tenants. In those days, when many parts of the Army were closed off to women altogether, the Royal Logistic Corps tended to have a higher population of female officers than any other regiment. Consequently, our blended community on the Gütersloh base consisted of lots of single officers, in a ratio of almost 50/50, women to men, which was practically unheard of in the forces. This excellent social mix was housed in about six accommodation blocks, facing each other across a greensward, in an arrangement which seemed to lend itself effortlessly to drinking, partying and a cracking social life all round.

The environment also loaned itself effortlessly to practical jokes – and perhaps always had. The officers' mess had originally opened in 1937 as a club for German fighter pilots and a room in its tower was known as the Goering Room. Hermann Goering, commander in chief of the Luftwaffe, visited the place before and during the Second World War, and allegedly liked to hold court up there, where he would apparently regale airmen with tales of his prowess as a solo fighter pilot. Often sensing that his stories of outstanding heroism were provoking scepticism in his audience, Goering would frequently find himself saying, 'If what I'm saying isn't true, may the roof fall on my head' (or words in German to that effect). Indeed, he became so famous for that rhetorical flourish that some highly resourceful airmen decided to arrange it for him.

They secretly chopped into the ceiling above Goering's favourite table and set up a system of levers, ropes and pulleys so that the next time Goering sat there inviting the roof to fall in on him, a portion of it actually did open up and swing down in the direction of his head. Another version of this story says the anecdotalist who got tricked was not Goering but an obscure elderly commandant. Either way, long before the AAC and the RLC turned up, Gütersloh had clearly been a place where people were prepared to go the extra mile for a prank. That didn't change over the four years that I spent there.

I should confess that I have always had a bit of a thing for practical jokes. It dates back to childhood when, with my dad as an eager conspirator, I would lay a line of coins along the top of the door to my sister's bedroom so that as she entered she would bring down a shower of change on herself. And I think we came quite close to losing my mum the night she watched a scary movie in the sitting room, rose from the armchair afterwards and turned to meet a shrouded figure behind her – her dressmaking mannequin, dolled up for the occasion.

Happily for me, it turned out that practical joking is rife in the military, and that Army life offers both the scope and the materials for some really quite advanced endeavours in this area. The subject was certainly studied very hard at Gütersloh. For example, I had seen thunderflashing practised at Middle Wallop, but I saw it taken to whole new levels in Germany. Thunderflashes are grenade simulators – bangers, essentially, wrapped in cardboard, used during exercises as mock hostile fire. These create a large but ultimately harmless explosion, but lobbed into the room of someone who has made the mistake of retiring to bed early, they provide a pretty incredible noise to wake up to.

Then there was what we might call Thunderflash Level Two, which involved putting the thunderflash inside a pillow and thereby generating both an astonishingly loud noise in the sleeper's bedroom and at the same time a blizzard of feathers. Very satisfying.

I pause here to record the infamous mistake made by a Cavalry officer of my acquaintance, who foolishly attempted to pioneer Thunderflash Level Three – two thunderflashes placed in a large bag of flour, liberated from the mess kitchen. He had imagined that this would amusingly coat the chosen room in a dusting of white powder. But he had neglected to remember that flour, in fine powder form, is highly flammable, and what he had in fact assembled was a bomb. The first thunderflash had the desired effect of evenly dispersing the flour; the second thunderflash, detonating a fraction of a second later, had the undesired effect of blowing the windows out. In fact, such was the force of the explosion, it literally raised the roof, cracking the lintel and causing actual structural damage to the building. It certainly woke up the occupant, who was subsequently driven to hospital for a quick check-up. As a consequence of this episode (which, if truth be told, came on top of a couple of others), the Cavalry officer in question was asked to seek another career. Be careful how and when you thunderflash, is the lesson.

Then there was the 'upside-down room' – the practice of entering a person's room in their absence and turning absolutely everything on its head, including the wardrobe. This was a Gütersloh speciality. And while you were in there ... bedrooms in Army quarters tend to have sinks. (There will be a shared bathroom block somewhere down the corridor.) So another favoured practice was to creep in and remove the U-bend from the pipes underneath. That way, in the morning, the unwitting occupant might fill his sink to shave, then pull the plug and empty the contents onto his feet and across the floor of his room. Needless to say, the results were even more rewarding if the occupant was someone who got up in the night and used the sink to pee in. (Unsavoury, I know, but a common practice – or so I'm told ...)

If you dish it out, you've got to be ready to receive it. I emerged from the mess one morning to find my car on bricks. I had traded my lovely but cursed black BMW 3 Series for an altogether less glamorous silver Nissan Almera, and now all four of its wheels were missing. It took me quite a while to locate them and bring them back together again. One was in a tree. Another was waiting for me at work when I cycled the mile and a half down to the hangar that morning. A third eventually turned up in the officers' mess where it was being used as a fruit bowl. The fourth had been deposited with the adjutant in Regimental Headquarters where I had to go and put in a formal request for it.

Even so, I may have got off lightly. Charlie Howard-Higgins had some bits of old salmon and kipper fed through the ventilation ducts of his car so that as he drove around in the cold north German winter with the heater on, he was continually assailed by the smell of rotting fish. It proved impossible to shift.

Was this, and a hundred similar incidents across my four years at Gütersloh, the responsible behaviour of mature adults? I guess not, really. But it was certainly amusing and I would also argue that it was necessary. You were in the Army, after all. You knew that your job was going to put you in the way of danger at some point, bring you up against some harsh realities. In that context, it wasn't just inevitable that people would experience a more intense than average drive to create some light relief here and there; it was essential.

Only two months into my tour, my squadron had to confront harsh reality in its most brutal form. In September we were despatched to central Germany for a two-week regimental exercise. We lost two good members of 652 Squadron on that exercise – Sergeant Les Beresford and Corporal Andy Beck. The first I knew that anything was amiss that day was when I landed after flying a sortie and saw the whole of the squadron formed up on dispersal. I hurried to join them and in due course the squadron commander came out and informed us what had happened. Les and Andy had dropped off some passengers at Mendig airfield and taken off again. They had climbed away to about 600 feet when their Lynx shed a rotor blade, causing it to fall out of the sky into a field.

I can still recall the quiet that fell across our group as we stood there and absorbed this. I had only been with the squadron for two months but I had worked with those two guys day in, day out, socialised with them in that blithe Gütersloh atmosphere, drunk with them, got to know them. It was extremely hard to reckon with.

The exercise was immediately called off. Our squadron was spared the task of securing the crash site; the other two squadrons stepped in to do that work. We returned to Gütersloh. Pending an investigation into the accident, the Lynx fleet was grounded, not just in the British Army but around the world. Gazelles resumed flying a couple of days later, but our regiment didn't fly its Lynx helicopters again for the best part of three months. I remember there being a lot of pressure for the fleet to resume flying but Lieutenant Colonel Gibson really put his neck on the line and pushed back hard against that, insisting on the most thorough investigation. I admired him for that. The investigation would reveal that the plastic coating surrounding the metal tie bar, which holds the rotor blade to the rotor shaft, had degraded, causing the tie bar to fail.

Many of the squadron were able to attend the funerals. A young officer, freshly arrived in the regiment, I now stood saying goodbye to two colleagues and witnessing the incomparable suffering of parents burying their son.

* * *

Not long after the tragedy at Mendig, when very little flying was going on, the chance to get away and do something different for a while came up and I seized it. Our regiment was forming a Nordic ski team, so I volunteered. So did two of my officer friends, Toby Everitt and Michael 'Jimmy' James. We were complete novices when it came to Nordic skiing – the cross-country endurance branch of the sport, as opposed to the more glamorous downhill version – but, always up for a challenge, we thought we would give it a go.

Also with us, and part of our team of eight, was Gav Rowley, an RAF flight lieutenant who was on exchange to our regiment and who was the senior officer among us. Some years earlier, Gav had been unfortunate enough to be involved in a terrible accident in which he had been struck by a Volvo truck. His facial injuries had required reconstructive surgery and Gav had spent months in hospital recovering. Needless to say, then, his nickname was Ovlov, which is the imprint a Volvo nameplate might leave on an unfortunate pedestrian. Military banter for you. Anyway, Gav was chosen to head up the team and we had a sergeant with us who had some experience of Nordic skiing and acted as our ski instructor and team coach.

The idea was to do a month of intensive training in Norway, get the team up to speed, and then head to France in January and defend the honour of our regiment at the Army and Inter-Service Winter Games. We were going to prepare for the biathlon, which is cross-country skiing with rifle marksmanship thrown in. So we loaded a minibus and a canvas-backed Land Rover with skis, guns, ammunition and (knowing in advance about the price of drink in Norway and what it could be bartered for) as much cheap whisky as we could find, and drove up from Germany to Denmark before taking the ferry across to Norway.

Now, this was the winter of 1994, before either Norway or Denmark had signed up to the Schengen Agreement on unrestricted borders. So at this point we were crossing international lines with two vanloads of guns, ammo and alcohol. What could possibly go wrong? And who could possibly object? Fortunately nobody bothered to inspect the minibus and, after some anxious moments in the various queues at passport control, we journeyed on, unarrested, to our accommodation – a pretty rudimentary chalet out in the wilds, a few miles west of Lillehammer.

That month of training was tough work, and absolutely brilliant conditioning. Gav was intense about it, wanting us to do as well as we possibly could – and quite right because otherwise why bother? Our sergeant had us up at 6.00 a.m., out jogging in temperatures which tended, at that point of the day, to be hovering around the minus 15° C mark. It was bitterly cold, with no moisture in the air, just that bracing, frozen, sub-Arctic crispness, and we would be out there in a pair of Ronhill jogging bottoms, trainers, a puffer jacket, a hat and gloves, doing dawn runs to loosen up. Then it would be back to the chalet for a quick breakfast and out again for technique training all morning. We'd take a quick lunch break and in the afternoon we would do endurance training followed by circuit training, after which we would return to the chalet, have supper and crash out, exhausted. Next day, same thing, through to Friday. So we were really working hard, putting in endless kilometres on skis, and we got very, very fit in a very short space of time.

On Fridays, though, we would give ourselves a night off and we'd jump in the freezing cold, canvas-backed Land Rover and head to Lillehammer for the bars and clubs. It wasn't always a straightforward journey down to town, on twisty icy roads in the dark, and indeed, one time, after a slight misjudgement on a corner, the Land Rover ended up rolled over onto its side in a snowbank. No real damage, fortunately, either to us or the vehicle, except that thereafter the windscreen wipers stubbornly refused to work. Being young and resourceful, we solved the problem using a loop of baling twine which we wrapped around the stem of each wiper and all the way around the windscreen. This cunning rig-up meant the person in the passenger seat could now grab a hold of the twine and, by moving it left and right, operate the wipers on behalf of the driver. Of course, the system required the side windows to be open so everyone inside froze the whole way. But at least you could see where you were going. And it was never that warm in that Land Rover anyway.

Down in the nightspots of Lillehammer, we seemed to have our pick among the local Norwegian girls because, oddly, for reasons we never really worked out, the local Norwegian guys didn't seem very interested in drinking, dancing, or having a good time. We, on the other hand, after a week of training, weren't interested in anything else. A lot of the time we were quite well away before we arrived, having learned the value of loading up cheaply at the chalet before venturing out. One time, as we were climbing back into the Land Rover after another night of jollity, one of the corporals with us announced, full of bravado, 'Don't worry about me, boys. I'm all right for the night.'

We looked at him in his jeans and trainers, his white T-shirt and thin leather jacket. It was a long way back to the chalet in sub-zero temperatures.

'Are you absolutely sure?'

'Yeah, no worries,' he said, with a wink. 'I'll see you tomorrow morning.'

At about 4.30 a.m., I was awoken by a knock at the door. Outside stood that corporal, now fully converted into a block of ice after a long walk through snow into the mountains. It turned out that by 'You can walk me home' the corporal's new friend had meant nothing more nor less than 'You can walk me home' – painful enough to hear anywhere, but especially painful after midnight in wintry Norway when you don't have any transport, a mobile phone or even a proper coat.

We finished our month of training, packed the minibus again and then – physically in great shape, if technically a little raw – went down to Les Contamines-Montjoie in the French Alps for the month of competitions. In the first phase we would compete against other corps and regiments of the British Army. If we qualified, we would move on to the Inter-Service competitions.

Getting a good night's kip ahead of these contests was clearly crucial if we were going to perform at the top of our game. Alas, sleep was proving more problematic than we had hoped. For accommodation at Les Contamines, we had a tiny chalet not much bigger than a caravan, with a small kitchenette, a living area in which there was barely room to shake a cocktail, and bunk beds. Just to make it even more crowded, we were also hosting some other guests – a family of rats. It made it a little easier that a kind of time-share system had spontaneously evolved whereby we had the use of the place unobstructed during the day, and the rats had the run of the facilities by night when we were in bed sleeping ... or trying to sleep. To tell you the truth, the constant scratching from our nocturnal housemates didn't bother me very much. But when Toby, who seemed particularly unimpressed by these additional inhabitants, woke one night to find himself face-to-whiskers with an extremely large rodent on his pillow, he decided to take matters into his own hands.

The following night, Toby was ready. And thus it was that, at some ungodly hour of the morning, I awoke to the unmistakable crack of a rifle shot. As I came to, I saw Toby sitting on his bunk, gun in hand with a smile on his face. In the corner lay the bloody remnants of one of our recently departed tenants. The rest of the rat family seemed to get the message straight away; they packed up and moved out and were never heard from again. We all slept pretty well after that, including Toby.

Whenever you get a group of competitive military teams together you can expect a lively social life and Les Contamines didn't disappoint in that area. However, under Gav's management, we were also continuing to take the skiing seriously, and we easily qualified for the next phase – the Inter-Service competition in which we would now also be pitted against the best that the Royal Navy, Royal Air Force and Royal Marines had to offer.

This next phase necessitated a move to the famous resort of Val D'Isère, just over 140 kilometres away. There, our new digs proved to be a great improvement – bigger, brighter and, most importantly, free from furry animals. While we unpacked our gear and settled in, Gav wisely decided to do a stocktake of our kit. As he ticked off the items, it seemed all was in order except just one thing – a box containing two thousand rounds of .22 ammunition.

It's still not clear quite how that item got overlooked as we packed up the caravan-size flat in Les Contamines. It seems we must have responsibly stowed the ammo deep in a locker at some point – and then, less responsibly, left without remembering it. In the Army's eyes, losing ammunition is nearly as bad as losing a weapon. Accordingly, despite the lateness of the hour, Gav and I immediately jumped in the Land Rover and set off back to Les Contamines. The weather was foul – sleet and blizzards all the way – and, of course, this was the Land Rover with the bust windscreen wipers. So now, while Gav edged his way gingerly through the snow, I sat for three hours in the freezing cold, with the windows open, tugging a damn string left and right so that we could see where we were going.

Meanwhile, back in Val d'Isère, Toby was frantically trying to get hold of the chalet operators, out of hours, to try and explain why we urgently needed to be readmitted briefly to the chalet – and doing this to the best of his ability without at any point using the word 'ammunition'. As Gav and I discovered, the chalet was now occupied by a family of four. They were very understanding, that nice family, about two soldiers knocking on their door in the middle of the night and asking to search their holiday home, and to this day they probably have no idea what was in that box and why we needed it so badly.

Alas, the glory of gold eluded us at the games in Val D'Isère. I guess we were always more about the stamina than the technique and we got outclassed in the end. However, we were winners of the Novice Team prize so we did return to Gütersloh with more than just pride. All those early starts and Gav's intensity had paid off at least to that extent.

*   *   *

I mentioned before how Arthur Gibson, the lieutenant colonel in command of 1 Regiment AAC in Gütersloh, was a strong advocate for the virtues of adventure training. That was why, apart from going on check flights with the squadron helicopter instructor Chris Keane in order to familiarise myself with the local area and local rules, so much of my first summer in Gütersloh had been spent kayaking and hiking. That was also at least partly why I had been able to spend the winter months learning Nordic skiing. And it was why, the following April, in 1995, I ended up repeatedly throwing myself out of an aeroplane in order to qualify as a free-fall parachutist.

Lieutenant Colonel Gibson was very keen that all young officers who came to his regiment should learn to parachute – a directive not entirely unconnected with the fact that Arthur ran the AAC parachute display team. With Arthur's overt encouragement, I reported to Bad Lippspringe Airfield and signed on for the accelerated free-fall course.

In some respects, these were not optimum conditions in which to learn to parachute. April in north-west Germany isn't particularly balmy and we were jumping out of the aircraft at 14,000 feet, where the temperature is seriously sub-zero. We were also falling through cloud and frozen rain, so as you descended you could look forward to flecks of ice hitting your face at 125 miles an hour. It was like being pebble-dashed. But I loved it – loved the madness of it and the huge adrenaline rush it gave you. Diving through the sky was an incredible way to see the Earth below you – indeed, the best view of the Earth I'd ever had, up to that point.

I even got a thrill out of the feelings of terror that come with the sport by default. My instructor was Dick Kalinski, one of the senior NCO pilots in my squadron who had seen action in the Falklands conflict. Dick had done over 2,000 jumps, so I asked, 'At what point do you not get scared jumping out of an aircraft?'

He quickly put me straight. 'If it ever didn't scare me, I'd stop doing it.'

Dick took me for my first tandem jump and straight after that I was jumping out at 14,000 feet with an instructor on either side of me and building my independence from there. On what must have been about my third or fourth jump, I was going up in the plane with some more experienced guys who were jumping out at 12,000 feet and doing four-way, which is where you practise synchronised manoeuvres on the way down. Those guys left us and we carried on up to 14,000 feet. They always used to try and take the new students as high as possible, because it gave you some vital extra seconds of free-fall and more time to practise your manoeuvres.

Very soon after jumping and starting to go through my skills and drills, I remember looking sideways and catching sight of what seemed to be an opened canopy. But it couldn't possibly have been, surely. We were too high up and nobody would want to drift down gently under a parachute from that kind of height, in this kind of temperature. I didn't think any more about it, having far too much to concentrate on.

At the bottom, though, I learned that my eyes hadn't deceived me. One of the guys in the four-way who jumped before us had got his foot tangled in his buddy's parachute which had accidentally deployed the reserve parachute at 12,000 feet. It took him about thirty-five minutes to get down. He was almost hypothermic by the end and had no feeling in his fingers. Apparently, after about fifteen minutes, realising the extent of the ordeal that lay ahead of him, he had toyed with the idea of cutting away the chute, doing some more falling and then deploying his main parachute. But he thought, 'Hang on, no – I've got an operating parachute here. The sensible thing to do is to suffer.' Good decision, certainly. But suffer he definitely had.

The last jump we novices had to complete was a low one, from 5,000 feet – what they call a hop-'n'-pop. That's where you jump out and you're so low that you only have time to count to about five before you deploy the parachute. Now, for some reason I wasn't terribly good at stabilising myself in the air at the start of a fall. This hadn't been much of a problem at 14,000 feet, where I'd had plenty of time to sort myself out and get in the right position eventually. But this time, I came out of the aircraft and somehow assumed entirely the wrong position, upside down. I was on my back like a flipped-over tortoise, staring up helplessly at the aircraft. Meanwhile the seconds were ticking down.

There was nothing for it. I deployed my chute and watched with clenched teeth as the cord shot up between my legs. Fortunately the escaping canopy did not, as I had a split second to fear, wrench my gonads clean off on its way through. Instead it whipped me round faster than I had ever been moved, from being on my back to being the right way up in a flash.

At the bottom, Dick said, 'What on earth went on there?' Bent double with whiplash, I mumbled something about failing to get stabilised. Of course, I had to go straight back up and repeat the jump. Fortunately, that time I managed to get myself round the right way.

So, that was it: I was now the proud owner of a Category 8 licence to jump. The whole adventure had been a blast – with a nice social element to it, too, because you got to do quite a lot of drinking coffee and chatting while you were packing up your parachute and waiting for your next jump. That, plus the thrill of the jump and the fall … I could see why people became devoted to it. At the same time, it wasn't a sport that made me think, 'This is it – I'm going to go and get a thousand jumps under my belt and become a seriously qualified free-fall parachutist.'

However, what I did do subsequently was help the AAC parachute team with its Friday practices. In the absence of a plane, we would take the doors off a Gazelle and remove one of the front seats. Then I would take them up to 10,000 feet over the airfield on a Friday afternoon and they would perch on the edge and jump out. Did I long to join them? Not too much. I was always happier flying aircraft than jumping out of them. Later though, as a test pilot, when wearing parachutes became routine on high-risk trials, I would be extremely grateful for having had the chance to cut my teeth jumping in Germany, ever aware that today might just be the day I needed to bail out.

* * *

In the spring of 1995, life in the officers' mess at Gütersloh was as lively as ever, but things at work with 1 Regiment AAC were much slower. The Lynx fleet were only just getting off the ground following the accident with the tie bar, and the sorties were sporadic. At the same time, my desire to be up there, flying, improving and growing as a pilot, was as high as ever. I was starting to feel a bit frustrated, which is why I jumped when the chance arose to do a short flying tour in Northern Ireland, starting in May. It would be my second time there, of course, after my attachment with the Royal Green Jackets following Sandhurst, but this time I would be going there with my own corps. And I would be airborne.

At that time, one of the tasks for the Gazelle fleet was to provide overwatch for Belfast city. This involved working closely with the RUC and intelligence services, often flying them in the back of the aircraft whilst communicating with personnel on the ground, directing operations and generally keeping an eye out for trouble. Known as City Flight, it was interesting work, requiring an intimate knowledge of Belfast's streets and often calling on the ability to react at short notice as trouble started. More importantly from my point of view, it guaranteed long hours of flying which I was yearning for.

When I arrived in May 1995, the change in tempo from Germany was remarkable. Within a couple of days, I had been given a quick aerial tour around the province, issued with my HK53 assault rifle (these are shorter than the Army's standard issue SA80 and would slot neatly into our armoured helicopter seats) and found myself on duty, hovering over Belfast at 3,000 feet. At that height we were out of range of small-arms fire or rocket-propelled grenades and the threat from surface-to-air missiles was deemed to be fairly low. From that relatively comfortable vantage point, I got to know Belfast better than my home town. There were parts of the city where you had to be particularly vigilant – the Ardoyne, Falls, Shankill, Whiterock to name a few. We would cover the soldiers and police on the ground using sights and sensors that were attached to a stores boom on the helicopter and which, along with some additional armour protection, made it heavier than any other Gazelle I had flown – another new challenge to adapt to.

There were also a couple of new flying manoeuvres that had to be mastered. I had good friends alongside me for this. Two of my closest pals from 1 AAC had already headed to Northern Ireland. There was Jimmy James, who had been part of the Nordic ski team and was now commanding City Flight; and Philip McCabe, a charming, gregarious Irishman, who can talk faster than any man I know. Our city base was Palace Barracks, to the north-east of Belfast, and landing there necessitated something called a rapid descent. There are variations on the theme, but basically you needed to get from 3,000 feet to ground level in the shortest possible time whilst presenting a difficult moving target should anyone wish to have a pop at you. You would inevitably end up close to the ground travelling in excess of 130 mph.

One day when I was flying with Philip, he decided to show me his version of a rapid descent. Pausing only to utter the always worrying words, 'Watch this,' he proceeded to turn the aircraft upside down, then point the nose directly at the ground and pull it into a tight 2g turn where we made several revolutions. I cast my astonished gaze across to Philip at the controls. He looked more like he was stirring a bowl of soup than flying a helicopter. We pulled out before hitting the ground though, which is always a bonus, and the whole thing was mightily impressive, even if there was a look about Philip that hinted that parts of the manoeuvre may have been less under his control than others.

As time went on and my confidence as a new pilot in handling the aircraft increased, there was some fun to be had with the rapid descent move. A competition soon arose between the City Flight pilots as to who could record the fastest speed over the roof of the NAAFI at Palace Barracks and still be able to stop in time to land on the rather small and confined parade square. The competition must have run for a good couple of weeks before the adjutant popped over to see us and ask why, recently, he had been seeing helicopters come screaming past his office window standing on their tails. Whether or not he believed our explanation (I think we said something about trialling a new defensive flying manoeuvre) I have no idea, but we decided to calm things down a little after that.

Well, only a little. The summer of '95 was stinking hot and there wasn't much relief to be found from the relentless sun while sitting under a Perspex bubble in a high hover. For a while we decided to take all the doors off the helicopter to encourage some airflow through it. It's not everybody's cup of tea to hover at 3,000 feet all day, looking straight down at the ground. You feel quite vulnerable in a small helicopter and feelings of vertigo are never far away – and even more so with no doors. Which is why, one day, whilst hovering over the city, I decided to play a joke on Jimmy.

He was leaning over the side of the helicopter, looking through a pair of stabilised binoculars. They were on a strap around his neck and he was wearing his five-point harness, of course, although it was on the inertia-reel setting which, much like a car seat belt, allows you to move forwards and backwards and will only lock with rapid motion. Looking across at Jimmy's concentrating form, it suddenly occurred to me that it might be amusing to give him a real eyeful of the city far below. Without warning, I put the helicopter on its side, the side Jimmy was looking out of, tilting it through a full 90 degrees to give him the maximum benefit of the view. The reaction was better than I could have hoped for. The inertia reel let him go a little further out of the door than I had expected before locking. Dropping the binos, thankfully caught by their strap, Jimmy managed to grab the side of the helicopter and haul himself back into his seat whilst at the same time screaming language at me of the most colourful kind. I nearly wet myself laughing.

Of course, there were serious days, too. Grabbing an early breakfast one morning, I was surprised to read in one of the papers that Lee Clegg, a British Army soldier who had been jailed for life in 1993 for his involvement in the shooting of a teenage joyrider in West Belfast, was being released on early parole. I was just going on duty with Staff Sergeant Steve Porter after taking a day's break at the regiment's main base at RAF Aldergrove. We hadn't received any prior warning of the decision to release Clegg, which obviously had the potential to be inflammatory. As Steve and I climbed into the Gazelle, we both knew that we were in for a long and busy shift.

Approaching Belfast that morning flying over the Black Mountain, all we could see was the smoke already rising from the city. Cars were being hijacked and set ablaze; there were angry parades along the Falls Road with Molotov cocktails being hurled at the gathering convoys of riot police. I watched as a bus was hijacked, the driver and occupants ejected before it was set ablaze, blocking a major road junction. We spent nearly seven hours over the city that day, watching, reporting, liaising with those on the ground. It turned out to be the most violent day in Northern Ireland that year.

In my five months with City Flight I had clocked up some 200 hours flying and now had over 500 hours in total. I had gone to Northern Ireland feeling like a rookie pilot, with a cautious approach to flying, knowing that the aircraft I was in control of could far exceed my flying capabilities. I returned to Germany a different pilot – far more confident and capable. I had a feeling taking root that maybe, if I stuck at this, I could really have something to offer the world of aviation.

CHAPTER NINE

Rough nights on Dartmoor, tough times at sea, and the interview from hell

In the whole of my mission on the International Space Station, orbiting in weightlessness at 27,600 kmh for six months, I knew the sensation of space sickness only once. It was on my first full day on the ISS, and I had got up in the morning and was cracking straight on with unpacking bags of cargo. That work involved a lot of diving in and out of the recently arrived Cygnus supply spacecraft and spinning myself around, and after a short while my body, which was still in the very earliest stages of adjusting to microgravity, decided to have a word with me.

It was a peculiar feeling, though, and unlike any other sensation of sickness I had ever had: there was no nausea, no wave of queasiness, no desire to sit down for a moment – just the strong and very specific realisation that it would be a good idea for me to empty my stomach right now. I grabbed a sick bag, threw up into it, and then went straight back to work on the bags. I was never sick in space again.

Give me space sickness over seasickness any day, then. Is there such a thing as seasickness without debilitating surges of nausea? I'm not sure there is, and I think I would have found out. Everything I know about seasickness I learned in October 1995 on the 315 Combat Search and Rescue Officers' Course – seventeen days of classroom lectures, drills and practical training on all aspects of aircrew combat survival, in which four hours in a single-person dinghy on the choppy sea was just one of the testing experiences on offer – possibly one of the lighter moments, actually. It was certainly lighter than the 'Conduct After Capture and Resistance to Interrogation' exercise, but we'll come on to that shortly.

Incidentally, if the sea wasn't choppy enough for the instructors' liking while you were out there in your little emergency inflatable, they would happily churn it up with a few circuits in a Rigid Raider. They were very thoughtful like that.

The CSRO course stays in my mind as an extraordinary set of learning experiences in my education as a young army officer, and also as a forceful demonstration of the value of making training real. After all, you can be lectured about seasickness and its debilitating effects while sitting in a classroom, and maybe come away with a pretty strong sense of it. But it's when you've been in a dinghy for four hours and you're vomiting over the side that you really think, 'Yeah, OK – now I understand.'

This the instructors well knew. And the same went for what was, essentially, the over-arching theme of the CSRO course: the will to survive. On the course, we examined all sorts of case studies of people who had been in these incredibly tough survival situations, some military, some civilian, with the odds stacked against them, and yet who had emerged alive. And the one constant thing that seemed to differentiate them from people who hadn't made it, was their will to survive – which is a hard thing to quantify, and an even harder thing to teach.

Yet if the training approaches reality closely enough, there are things that can be learned even about the will to survive – and certainly ways in which that will can be encouraged. This was apparent in the four-hour dinghy session. There will be some people who, when hit by seasickness in that dinghy scenario, will simply shut down and kind of cocoon themselves, preoccupied by how bad they are feeling. Those are the people who will probably stop thinking about where they are, stop bailing out the water, stop wondering whether they are drifting, where the sun is, what time of day it is ... In other words, they will stop maintaining their situational awareness, which is the thing that is going to help you stay alive. The point of going through that hardship out in the dinghy was to learn a strategy for staying focused on the important things, and not on how you were feeling. Various endurance courses since then, including on my way to becoming an astronaut, have reinforced the same lesson: whether it's sickness, injury, fatigue, cold or hunger, it's basically something that is trying to interfere with your efficiency as you attempt to function. So you have to recognise it, first of all, then apply a means of overcoming it.

More often than not, the solution is to set yourself a routine. In that dinghy scenario, we were asked to imagine that we had parachuted into the sea. In preliminary sessions in the swimming pool, you practised releasing yourself from your parachute in the water, inflating your life jacket, getting your dinghy off (it's flat-packed and attached to your back or your bum), inflating it and scrambling inside, and then pulling the canopy over your shoulders for protection. That was a precursor to going down to the coast and doing the same thing in the Atlantic, this time having rolled into the water off a speedboat at 25 knots – enough to batter you around and disorientate you as you enter the waves.

A very clever piece of design, those flat-packed dinghies. They've been constantly refined and modified, but they are essentially unchanged since they were developed in the sixties – historical artefacts, almost. They're a bit like an old Soyuz capsule in that regard. If it ain't broke ... That little arrow-shaped float is probably good for at least forty-eight hours of survival. You've got a couple of foil packets of water in your kit and some energy tablets – white cubes which look like firelighters and taste quite like them, too. Then there are your signal flares in a small plastic wallet, and your metal firing pen, and your bigger day/night flare which gives off smoke. You've got an entire survival environment in one small, tight pack. It's been brilliantly thought out.

Anyway, during the exercise, once I was Velcroed inside, peering out through the little clear pane in the canopy as I lurched about on the waves, I found that it was possible to get beyond the nausea by having a routine to go to. Every fifteen minutes, I was going to check my dinghy. Is it inflated? Is it leaking? Is my sea anchor out? What's the wind doing? What's the weather up to? It just gave me a tool – something to anchor my concentration on, rather than dwelling on the queasiness that was coursing through my system and feeling sorry for myself.

The same thing works for fatigue in endurance exercises. You can battle against tiredness by finding other things to think about, either going through routines, or setting goals. Deep into a navigation exercise across, say, the Brecon Beacons, your ultimate goal is, of course, to get to the end. But when parts of your body are screaming at you to stop, that big goal isn't necessarily going to get you through the next five minutes. So you try and set closer goals, and the more tired you become, the closer you set them: making it to the next 5k checkpoint, say, making it to the top of that hill, or that treeline down there, and then let's see what's next. Set the goal, achieve it, set another.

Clearly a whole night of forced marching – or any other large and demanding project, for that matter – can seem endless, conceived of whole; but broken into chunks it becomes more manageable. What you are doing, while you are thinking about those smaller chunks, is distracting yourself from what your body most wants to do, which is just to stop and lie down and have a twenty-minute nap. And instead of that, you're keeping going. Those moments of extreme discomfort are about where you go inside yourself and about cultivating, if possible, a mental strength that keeps you absolutely set on doing things. Some people will have a more natural disposition towards that kind of thinking than others. But it can be trained for. I found that out on the CSRO course.

I took the course on the recommendation of Chris Keane, my terrific qualified helicopter instructor in 652 Squadron, who had basically been mentoring me since I arrived in Gütersloh and with whom I had shared some thoughts about eventually qualifying as a QHI myself when the time was right. Part of what a QHI does is assist pilots with their regular re-certification, including dinghy and survival drills, so a QHI needs to be a CSRO. If I was serious about taking my helicopter flying to the next level, it would be a good box to have ticked.

So, with my joining instructions in hand, I returned from Gütersloh to the UK and reported to the Royal Air Force School of Combat Search and Rescue at St Mawgan near Newquay in Cornwall. I was one of eighteen on the course – a great bunch, as it turned out. As ever, on a Tri-Service event, there was a lot of banter flying around between Air Force, Navy and Army, but a lot of great camaraderie as well, which would only grow as the experiences of the week unfolded.

Initially we were doing a mixture of classroom sessions and outdoor practical exercises. In the morning you might have a lecture on navigation or survival psychology. In the afternoon you would be out at sea, learning how to get winched out of the water by a rescue helicopter. Top tip: let the hook hit the water before you grab it. An awful lot of static electricity builds up when that winch is on the way down and the last thing you need, on the verge of rescue, is to shock yourself rigid.

On another day you would be out in the countryside doing limited knowledge orienteering. You would be given a small amount of time to look at a large-scale aviation map of the Cornish moors, and then the map would be taken away and you would be dumped somewhere in a group of three and have to work your way back, using what you remembered and your own big-picture navigation skills. That part of the course appealed to me enormously.

There were sea drills at night, and classes on Arctic survival, desert survival and jungle survival, with their different niche skills. And we were drilled in the four elements of post-crash survival, which are: protection, location, water and food, in that order. So, number one is protection, because the first thing that's going to kill you is the elements. Number two is location, because you need to be found and you're going to need to identify your location to other people, whether that's by radio or smoke signal or whatever other means is to hand. (Your greatest chance of being found, incidentally, is to be found quickly. The longer you are lost, the greater your chances of remaining that way, not least if you are drifting away from your crash location. So, another top tip: get yourself noticed as soon as you can.) And then three and four are water and food – water before food, of course, because we can survive weeks without food, but we can only survive days without water.

And then, ultimately, we were deemed prepped and ready for the climactic phase of the course, an outdoor exercise leading up to the practical Resistance to Interrogation and Conduct After Capture elements. And that's where things became properly demanding.

We were now taken up on Dartmoor and shown rudimentary survival skills and taught how to live off the land. We practised building shelters, studied methods of lighting fires and discovered how to catch, kill, skin and cook rabbits. The latter was not something I had ever wanted to take up as a hobby, but it turned out that if I had to do it, I could.

And soon after that, we were on the run. The scenario was that we were aircrew who had just crashed in enemy territory. We were uninjured and we had forty-eight hours to get ourselves around a set of rendezvous points at the last of which we would be recovered by a search-and-rescue team. We had our life vests with us, meaning we each had two water pouches and some of those delicious firelighter tablets that I mentioned earlier, but no other food. We also had a pretty rudimentary map, a compass and a knife.

So, an orienteering mission, basically – one of my favourite things. The snag was, there would be OPFOR – opposing forces – out on Dartmoor, hunting us down. We would have a forty-five-minute head start, and then OPFOR would come looking for us. With dogs. The part of OPFOR on this occasion was to be played by some members of Her Majesty's Parachute Regiment. I think there was a Para company on exercise in the area at the time, and the CSRO course organisers said, 'Look, we've got a bunch of aircrew officers on an Escape and Evasion exercise – would you like to run after them with some dogs and, if you find them, give them a good kicking?' And I'm guessing, being Paras, they jumped at the chance.

Our task was to try and evade capture by OPFOR for the whole of the forty-eight hours, using all the cunning, guile and newly acquired knowledge at our disposal. We were divided into pairs and I felt myself fortunate to be teamed up with a really top-notch RAF Search and Rescue guy with whom I had already been getting along well. It was mid-morning when we were set loose and the pair of us knew that the best tactic for evading detection was to hole up for the day and do your moving at night-time, under the cover of darkness. Unfortunately, as we looked around us, Dartmoor, being moorland, didn't seem to be offering a great deal in the way of cover. It did, however, offer a massive area of prickly gorse. For the next half-hour, scratching ourselves to shreds in the process, my comrade and I crawled into the heart of this gorse, where we knew no sane person would ever come to look for us, and more than likely no Para, either. There we hunkered down and waited for nightfall.

Without a torch, and with a minimal map, moving towards our big-picture checkpoints in the dark was complicated, to say the least. Just to add a frisson, you could occasionally hear the dogs barking off in the distance. At one point the dogs started up, about 1 kilometre away, and the accompanying shouting seemed to indicate that some of our colleagues were getting captured. That drew a lot of the OPFOR in the direction of the noise and the pair of us were able to use it as a distraction and head the other way. We seemed to be doing OK.

The deal was, if you got captured early on you would be given some 'prisoner handling' for an hour or so and then released back into the wild. But we were keeping out of trouble. We were ducking in and out of the gorse, moving when we could, grabbing sleep here and there, and ticking off those RV points, and the Paras and their dogs weren't getting anywhere near us. The longer time went on and we remained undetected, the more determined we became to see it out. That kept us going through the tiredness and the hunger. As the end of the forty-eight-hour period neared, we closed in on that final RV, knackered, starving and very scratched, but also extremely elated and proud of ourselves.

At which point we got ambushed. All the fugitives were making their way into that one concentrated area around the exercise-ending RV, and the Paras had obviously decided to hang out there and let the prey come to them. In a heart-sinking moment, we heard runners behind us and turned and saw two Paras chasing us down. There was nowhere to go.

Knackered, starving, scratched and now captured – that was a truly deflating moment. But all was not necessarily lost. Our training had taught us that your best chance of escape is commonly in the moments directly after capture. There is sound thinking behind that. It's possible, for instance, at that early stage, that you've not yet been fully restrained, stripped and searched; more-over, if you do manage to get out, you probably still know roughly where you are. Twenty minutes later, on the other hand, you could have been trucked somewhere, deeper into enemy territory, you could have had your possessions taken away from you, your clothes removed ... your chances of a successful escape could be greatly diminished and getting slimmer with every minute.

So, with all this in mind, no sooner had my RAF colleague and I been detained in a kind of centralised holding area, than we cast each other a furtive glance and then broke hard for the exit, like Steve McQueen in *The Great Escape*, except without the motorbike.

It didn't come off. Both of us were intercepted and the only reward we received for our effort was a good kicking from the Paras as they frogmarched us back to the holding area.

The exercise now moved into the Resistance to Interrogation phase, which was to unfold over the next thirty-six hours. Now, before I get into some of the lurid details of this section of the course, I should carefully stress that it was 1995 when all this happened, and no doubt different regulations on health and safety have come into play in the quarter of a century since then. Nevertheless, I also want to underline that, even in 1995, this exercise took place under rigorously controlled conditions. There were cameras monitoring everything; there was a codeword we could use if we had had enough and wanted to be removed from the training; there was a designated staff member present who you knew was not part of the exercise and to whom you could speak at any point and have a conversation about whether you wanted to continue in the scenario or exit. There was a duty of care, in other words, and a set of closely followed procedures. It was thought through and extremely well run and professionally delivered. We weren't just getting knocked around for fun here.

But were we getting knocked around? Yes. Back in that initial containment area, after our failed escape attempt, my comrade and I now had our hands tied behind our backs and hoods thrown over our heads. Our captors brought the dogs up close to the hoods, barking. I remember their hot breath on my face through the sacking. And then I could feel a stream of warm liquid being directed over me. What was that liquid? I know what I was meant to think it was ... I just had to do my best not to think too hard about it.

Shortly after that, still hooded and still with our hands behind our backs, we and our fellow prisoners were loaded into a truck and taken down to the coast. I assume we were in the vicinity of Plymouth Harbour because the next thing I knew we were being forced into a confined space in the bottom of some kind of boat. We were forced to sit cross-legged and we had to hunch over because the ceiling above our heads wasn't high enough to allow us to sit up straight. The floor and walls felt slimy and stank of fish – I'm guessing it was the hold of a trawler, where the fishermen would throw their catch. Cooped up like this, we were then taken out to sea for three hours on the lurching waves until everybody was either feeling violently sick or actually vomiting.

Eventually we were brought back ashore and now, as I sat there in my hood, I could hear scrabbling and big splashes – the sound of people being pushed into the water. I probably don't need to say that it's a very unpleasant thing to contemplate being pushed into the sea off the side of a boat when you are handcuffed and hooded. And it's even more unpleasant when it actually happens. As it turned out, the water was shallow enough that I wasn't going to drown. But I didn't know that as I fell through the air. Again, it was a controlled and ultimately safe environment; it's just that you couldn't be entirely sure that it was controlled and safe, so your experience was pure shock.

Dragged out of the water in which we hadn't drowned, we were then taken to the Resistance to Interrogation training facility. Here we were separated from one another, I was taken into a room, my hood was removed, and the more formal eye-to-eye part of the process began. I was stripped naked in front of a bunch of people, men and women, and told to bend over – all with a view to my intimidation, embarrassment and humiliation. Then I was thrown a thin green coverall to put on.

Meanwhile my clothes were being searched. One of the interrogators, checking over my flying boots, ripped out the insole from one of them and then stuffed his hand deep inside it before wincing and hastily withdrawing it. Obviously one of the little metal staples that had been holding the insole in had gone under the interrogator's nail. Kind of ironic, when you think about it: the torturer tortured. I couldn't help smiling. Big mistake. We had been thoroughly drilled on the key tactics to adopt under questioning, and I was already learning the wisdom of one of them: don't antagonise your interrogators.

'Oh, you think that's f— funny, do you?' said the interrogator, up close to my face.

I received my second shoeing of the exercise.

Other essential tactics under questioning: willingly divulge only your name, rank, number and date of birth. To every other inquiry, the answer is, 'I cannot answer that question.' Today, more information can be divulged using a system of 'controlled release', but back then it was just 'the big four' as it was known. If food is offered, take it. If warmth is offered, take it, too. But give no information, to the best of your ability, for at least twenty-four hours. You're buying time. The more time goes by, the longer there is for the military plans you know about to be changed, and the less value any information in your possession will have.

My further punishment for the boot-related insolence was to be barefoot for most of the rest of the exercise. I was in and out of the interrogation room numerous times over the course of the twenty-four hours. Training protocols dictated that prisoners can only be interrogated for a total of eight hours in any twenty-four-hour period, but that didn't mean the hours of 'prisoner handling' in between could be spent peacefully relaxing and gathering your strength. In a cell with a concrete floor, I was ordered to assume stress positions – made to adopt a press-up position or, with loud 'white noise' blasting in my ears, instructed to sit with my hands on my head and bawled at if my arms began to drop. At one point in the middle of the night I was taken out and doused in freezing cold water, then returned to the concrete cell.

My sockless feet felt like ice blocks at this point; my green coverall was drenched. I was as uncomfortable as I had ever been. It was at this point that I was marched back, shivering, into the interrogation room. It seemed very warm in there. And now it appeared that I was talking to someone who, in stark contrast to the screamers and shouters that had been mercilessly questioning me before, seemed calm, reasonable – apologetic, even.

'Come in, sit down,' he was saying. 'I'm sorry that you've been treated like this. We really don't like to do it this way.'

He handed me a clean, dry coverall.

'You seem to have nothing on your feet,' said the instructor, wincing sympathetically. 'How has that happened?'

A pair of socks and some new boots appeared. A mug of hot tea was suddenly in front of me. And some biscuits. I eagerly put on my new clothes and reached gladly for the tea. ('If food is offered, take it. If warmth is offered, take it.')

At that point the door opened and, propelled by a couple of guards, in came my RAF comrade – stark naked, shivering, practically hypothermic and with his name-tag tied around his penis.

'Look,' my interrogator said to him, gesturing at me. 'Your colleague here has told us everything we need to know and see what he's got – fresh clothes, tea. Do what he did, and you'll be treated the same way.'

With that, my comrade was led straight out again. And no sooner was he gone than the tea and the biscuits were whipped away, I was instructed to take off those clean, dry clothes and the warm socks and boots, and I was handed back my cold, wet coverall to put on instead. And then the questioning resumed.

Cruel? Certainly. Realistic? I don't doubt. Indeed, I'm sure you could face this and far worse as downed aircrew in enemy territory. And the degree of realism was the point. We come back again to the importance of making training real, and therefore making it count. I've always felt this about flying in simulators – that there are amazing benefits to be accrued from spending time in simulated environments, but also obvious limits to what you ultimately take away. When you are confronted with an emergency in a flying simulator, there's inevitably going to be some small voice somewhere in the back of your mind saying, 'If this all goes wrong, in twenty minutes' time I'll be sitting with a cup of coffee, having a debrief with the instructor. I certainly won't be dead in a field.' The best training is the kind that finds ways to silence that small voice.

Yet how do you silence that voice in a training scenario without putting students at too great a risk? That's a difficult balance to achieve. Later, during astronaut training, we would experience situations in caves and underwater for extended periods that successfully managed it. To my mind, that Royal Air Force CSRO course managed it, too. Yes, it was hard and the treatment was rough. But there was a sound and important principle behind it: the pursuit of realism. At the end of the day, a kick in the ribs is real. Indeed, it doesn't get much more real.

I thought it was a brilliant course – an amazing opportunity, in fact. Let's face it, you weren't going to be given the chance to pick up this particular set of life skills every day. Not unless you were extremely unlucky, that is. I would even go so far as to say that I enjoyed it – or, at any rate, that I enjoyed having done it. I came out of it qualified as a Resistance to Interrogation instructor, meaning that I was now in a position to teach aircrew techniques for handling enemy capture scenarios, so it was an important step towards my eventually qualifying as a QHI. I also came out of it feeling proud that I had made it through. Over and above the interrogation game-play, I felt I had been asked some extremely searching questions. When you don't know what's happening to you, how do you cope, stay calm, continue to think clearly? The course had conducted a fairly thorough search for my breaking point and hadn't found it. And it meant that when eventually, as a newly recruited ESA astronaut, I came to face the belligerent Jeremy Paxman in a gruelling *Newsnight* interview in 2013, I was ready for him. (We'll get to that prickly encounter in due course.)

In the debrief, I was told that I had made one error. Soon after that moment when my RAF colleague had arrived naked and interrupted my tea and biscuits, they left the pair of us alone together, cuffed but unhooded, in a small wooden room with no windows. My comrade had asked me, 'What did you say to them?' And I told him that I had actually said nothing and the whole thing was a set-up. That was my one big mistake. We shouldn't have conversed with each other. You're meant to assume the enemy will be listening in at any time, and ready to gain something from anything you say, even if it's just ascertaining that the two of you are friends, which is something they could use against you.

Poor guy, though. My RAF friend had spent a very large part of that interrogation scenario completely naked. His sin? He had refused to bend over in front of the assembled mixed company when asked. Absolutely wasn't having it. We all have our non-negotiables.

It was around 10 o'clock at night when the exercise ended and I was finally dismissed. I went back to my room, had a shower and got my things together. I then got in the car and drove straight to a twenty-four-hour McDonald's where, utterly starving, I spent at least £15 on burgers, fries and milkshake. It may have been one of the most satisfying meals I have ever eaten.

CHAPTER TEN

The call of the wild, the 1000-hour chat,
and the door that became a wall

It's shortly after dawn in Laikipia County, Kenya. The sun is already beginning to beat down on the African bush, the smell of warm grass and dust is in the air, and I am in a helicopter, herding elephants. Which is not, perhaps, why you sign up for military service with the Army Air Corps but, boy, I'll take it.

This is Exercise Grand Prix, and I am engaged on my first duty of the day which is to take up a Gazelle from our base at Nanyuki, at the gateway to Mount Kenya National Park, and to clear the firing range at Mpala Farm. Later that morning the Argyll and Sutherland Highlanders will be coming through here on a live firing exercise, and nobody wants any elephants getting caught up in it. Nor any giraffes or zebra, for that matter, who also feature in my early-morning sweep.

It's incredible to witness these animals in the wild for the first time. Your zoo-reared Western mind is instinctively looking for the fence, the edge of the enclosure. And then you realise there isn't one. Up here in the Gazelle, we are keeping, I think, a respectful distance. We're not trying to start a stampede, after all – just drop a polite hint that it might be best if these animals spent the next few hours a couple of hundred yards to the left. But one huge bull elephant doesn't take it well. I can see him down

163

there, up on his hind legs, flapping his ears, angrily bellowing his disapproval and looking like he would happily swat me out of the sky with his trunk if he could. I find myself reflecting that it's easy enough for me to be so brave right now in my single-engine helicopter, but if this engine fails, that elephant will see me off in no time at all.

Firing range cleared, I fly back to the Nanyuki golf course for breakfast.

Kenya was perhaps the best six-week detachment of my Army life. It was the spring of 1996 and I had been promoted to captain and made flight commander, in charge of 652 Squadron's Gazelle flight. I was only twenty-four, which is pretty young for that role, and certainly young to be trusted with sole command of two helicopters and two dozen soldiers on exercise in Kenya. I may have owed at least part of that show of faith in my abilities indirectly to Ian Curry, who had just turned up in the regiment as flight surgeon – 'specialist aviation medicine', to use the proper term, a niche qualification rolling together military pilot and doctor in one extremely impressive bundle. Having recently completed his Pilots' Course, Ian was posted to 652 Squadron's Gazelle flight to complete a tour as a pilot and consolidate his flying experience. He and I instantly hit it off and ended up doing some great flying together – and he would eventually be the best man at my wedding. But he was also a major, which actually put him at the same rank as our squadron commander, and the additional security of having a senior officer with military experience around probably helped settle any residual nerves about giving me the lead in Africa. Whatever, I was going to run with it.

We packed up two Gazelles into the belly of a Hercules and rigged up our hammocks across the back of the plane so that we could sleep, swaying gently, all the way down to Africa (via Cyprus and then Riyadh for refuels). Kenya was, and remains, a big training area for British Infantry, with a permanent base at Nanyuki, running two or three of these large-scale exercises per year. On this occasion, we were there as aviation support. Our primary role was casualty evacuation, or CASEVAC as we call it – providing rapid recovery and rescue and getting injured soldiers to medical care quickly. We were also there for transportation – humping and dumping with underslung loads. Or we might be called upon to fly the commanding officer around, if he needed to go and visit his troops, deployed around the countryside. In between times we would get to do some tactical flying ourselves.

And what flying. No hazards, no wires, no pylons, everything at treetop level with skids through the bushes. And then the beautiful scenery, flying round Mount Kenya itself, going over the Aberdare Mountains to Lake Naivasha and supporting the troops over there ... All flying should be like this.

That said, flying at night in Kenya was a challenge – like nothing I had ever experienced before. In the UK, and elsewhere in Europe, even on the darkest night there would always be a bit of ambient light somewhere. In Kenya, it was black as ink – nothing to see. If there was some moonlight to work with, you could use your night vision goggles and get by. But, in the absence of the moon, if you weren't flying on NVGs – if you were 'mortal flying', as we called it – the darkness was just phenomenal. Almost as dark as space itself. There was no way to tell where the black of the sky stopped and the black of the land began. Which, needless to say, somewhat complicates the process of landing a helicopter.

We would practise landing using a 'NATO-T.' This is where ground troops position lights in the shape of a T to give you a visual fix in the sea of darkness. There are strict regulations on the size and position of the T and the distance between its five lights so that pilots, when they get familiar with the process, can form a sense of how close they are to the ground and whether the approach is too steep or too shallow. In the absence of battery-powered lights, the solution on this exercise was to use Benghazi burners – old cans filled with sand and fuel, then ignited. The first time I tried this in Kenya, one of our new airtroopers had set out the NATO-T in advance. Confronted with the problem of laying down the T in a confined area amid bushes and shrubs, the airtrooper had decided to make it quite a bit smaller in order to squeeze it in.

Not such a good idea. I couldn't seem to land. During each approach, I kept looking down at this Benghazi T that, judging by its size, still appeared to be way below me, and then I'd be hearing the helicopter's radar altimeter alarm giving me a low altitude warning, causing me to pull up again. I must have been on my third or fourth attempt at putting the aircraft down before I realised that I was dealing with an unhelpfully gnome-like T. Moreover, the airtrooper had been a bit over-zealous with the fuel and, on account of the shrubbery down one side, I ended up landing so close to the T that flames were licking up the side of the helicopter near the tail rotor – also not a great idea. I shut down as quickly as I could, while the airtrooper put a wet blanket over his work. Other landings, fortunately, went more smoothly.

The Infantry battalion was quartered in the base at Nanyuki, but we set up our helicopter landing site just across the road from there, on the edge of the golf course. I say 'golf course', but this was Kenya, so think sand and short yellow grass rather than groomed fairways and Augusta National. However, not far up the road was the Mount Kenya Safari Club, a five-star hotel, all villas and landscaped gardens. One of the first things the departing unit had pointed out to us during handover was that the Safari Club had a helicopter landing pad. So, too, did the Aberdare Country Club, another high-class spa resort which was conveniently on the way to Lake Naivasha. And so did the Samburu Simba Lodge, en route to Archer's Post, which was another centre for the exercise. And it turned out that all these enchanted places were entirely happy to have a helicopter drop in for a visit – happy, too, for the helicopter's crew to take a quick dip in their swimming pools and enjoy a refreshing fruit cocktail if they were so disposed.

And we certainly were disposed. In fact, we took to wearing swimming trunks under our flight suits so that we could be ready to take advantage of the facilities whenever the opportunity arose. We would be doing service runs to Archers Post, nicknamed Archers Roast, where it was 5 to 10 degrees hotter than on our relatively balmy plateau at Nanyuki, and where the camped-out Infantry were lighting fires at night to keep the lions away and knocking their boots out every morning to check for scorpions; and flying back we would be hopping out for a swim and a fruit juice on the verandahs of some of Kenya's most coveted destinations.

There was work to be done, too, of course. One day Ian, my friend the flight surgeon, took a call at base from the troops over at Lake Naivasha where there was a suspected case of appendicitis. We needed to be sure that the emergency was genuine because Lake Naivasha was 200 kilometres away; there was a lot of cloud around that day and flying would be tricky. But Ian was persuaded by what he was hearing that we needed to risk it. So he and I got airborne and flew over the Aberdare Mountains. The cloud was thick and low, sitting tight to the peaks, and we really had to push it to get through, squeezing over ridgelines where the cloud was less thick and more misty. It was strange to be grubbing around at 50 feet above ground level and yet over 10,000 feet above sea level. Eventually we were on the other side of the range and heading down to the lake.

When we landed we found our patient doubled up in agony. Ian examined him and established, within about thirty seconds, that the problem was not appendicitis: it was constipation. The soldier was simply bunged up by good old Army catering. (Army rations are actually designed to stop you needing the loo too often. Sometimes they work too well.) Ian prescribed some laxatives to get the poor bloke moving again and, somewhat frustrated, we climbed back into the helicopter.

By now the clouds had fully descended onto the mountains. We weren't supposed to fly in cloud in Kenya because there was no facility to do a radar approach. I got on the radio to Corporal Pete Balcombe back at Nanyuki who confirmed that the weather on the other side of the mountains was patchy, but with plenty of blue sky through the gaps. So, knowing we could get home, we decided to go above the clouds instead and clear the mountains that way.

However, as quickly as our Gazelle could ascend in the afternoon heat, the cloud was rising too, bubbling up all around us and forcing us higher and higher to stay on top of it. We were eventually at 13,500 feet, which is about 3,500 feet higher than the recommended height to fly without oxygen. So that was becoming a bit of a worry – as was the fact that, ahead of us, the cloud was now thickening over our landing site, too. I radioed down to Corporal Balcombe and asked, 'What's the cloud base?' He said, 'You've got at least 1,000 feet of cloud base.' So, knowing that we should be well clear of the mountains, we had to trust the aircraft instruments and drop the helicopter down through several thousand feet of cloud and just hope that there would be clearance underneath. Eventually, after an anxious spell in the mist, to Ian's and my immense relief we broke through the cloud and into visibility with about 1,000 feet of clearance. That was definitely one of those times when I experienced what we might call 'the pucker factor'. And all for a case of constipation.

Another time, an Infantryman had broken his leg, and Ian and I flew up to collect him. Unfortunately, in order to fit his stretcher in the Gazelle we had to take out all the seats but one, meaning I had to leave Ian behind and fly back solo. It was night-time, and I would be using night vision goggles, and you would never normally fly on NVGs on your own, and certainly not when the only other person in the aircraft is completely incapacitated. I had to take us over the mountains, look after the navigation (still no GPS in our aircraft at that time) and do all the radio work going into unfamiliar airspace over Nairobi airport. That was a lot of pressure, all in all, for someone who was still in the early years of their flying career. It was another one of those challenging sorties which you get to the other side of and which, as a pilot, give you those little notches of confidence to take forward with you.

Something else I learned about Kenya: if you've got a helicopter, everybody loves you. It's the ideal transport for the terrain, in a place where vehicles aren't always easy to come by, and we often found ourselves being asked to do favours for local people – passing messages, delivering things and even locating missing vehicles and livestock. We obliged when we could. One time we were approached by a team doing wildlife conservation who were going out to dart some rhinos for tagging purposes. They explained that this was going to take a long time to do in a Jeep: could we help them out? I made a call back to Germany to ask for authorisation and got it granted. The next thing, we were taking the doors off the back of one of the Gazelles, going up with the researchers and their rifles, and then low-flying over the scrub while they leaned out the back and fired darts into rhinos. Quite a scene. I certainly hadn't seen much of that sort of action at Middle Wallop. Or even in Gütersloh, actually.

Meanwhile, home life on our golf course encampment went from strength to strength. Having our helicopters on the other side of the wire from the main Army camp gave us freedom and a little more leeway to create our own environment. Some of the engineers actually slept over there to keep an eye on the aircraft. It made for a hot party spot on a Friday night – especially when we started funding those parties with a bit of wheeling and dealing on the side.

Nothing dodgy here, I swear. But our aviation fuel was being delivered in big 50-gallon drums. You would shove a Zenith pump into the drum and electrically pump the contents straight into the aircraft. The thing was, the Zenith pump would never quite get all of the fuel out of the drum. You would tip the drum around and waggle the pump about, like trying to suck the last drops of a drink out with a straw, but no matter how hard you tried, there would always be a little bit left at the bottom. That fuel would then stand around and, overnight, the air in the drum would cool and condense causing the fuel to become water-contaminated – no use for aviation. After about twenty refuels we had all these dribs and drabs of contaminated fuel sitting around the place.

Now, fortunately, we had an Army logistics guy supplying the fuel who was in touch with his inner Del Boy. This officer had been in Kenya before we arrived and had built up a little network of local contacts who, he knew for sure, were in the market for a bit of unwanted fuel. What's more, as he pointed out, the farmers loved those 50-gallon metal drums – would cut them in half and use them for water troughs and all sorts. So they, too, were a commodity, if we fancied flogging them.

I checked with the battalion commander. Would he mind us being a bit creative with some Army surplus? His response was, as long as no money left Kenya in anybody's pocket, he didn't really mind how we got rid of our unusable fuel and our metal drums. So I went back to Del Boy and told him that everything was cushty.

Bingo: we now had a ready supply of cash which we were obliged to spend. And the obvious thing to spend it on, of course, was booze for our Friday-night parties. Those parties became legendary. The invitation was open, beer flowed freely, a fire burned, and songs were sung. They were so legendary that John Hurt, the actor, turned up one night. I think he had a place near the Mount Kenya Safari Club, but he strolled in to see what the fuss was about and stayed for the evening.

Hanging out with film stars at parties? This wasn't really in the Army script, any more than herding elephants by helicopter was. But that was Kenya – a magical period for me, doing some amazing flying, in some amazing scenery, and at the same time having the responsibility of being in command of my own flight, managing the aircraft, the maintenance, the flight hours, the team, and for the first time really feeling like I was running my own show.

Incidentally, even after spending out on the parties, there still seemed to be some money left over from our fuel and drum business. So we invested it in our living quarters. We had ordinary 12x12 military tents that looked like all the others in the Army camp. But when you pulled back the flaps, we had wooden slatted floors with rugs, chests of drawers, proper wooden beds with pillows ... Meanwhile all the other Infantry boys were bedding down on their metal-framed Army cots. We even employed a couple of locals – the two Steves – who came in to help with the cooking, cleaning and washing. Well, you've got to be comfortable, haven't you?

But not too comfortable. Being too comfortable can get you into trouble. I learned that in Kenya, too.

As a pilot, experience is your friend. And yet it can sometimes be the opposite if you allow it to be. There's a mental adaptation which comes with experience. The first time you fly at 100 feet, it feels incredibly low. But then you do it a few times, and that feeling goes. Then 50 feet feels low – until 50 feet, too, no longer feels that way. And then it's 20 feet, and then it's 10 feet … But you need to stop yourself becoming numb to it. You need to know when you're flying too low.

I had gone to pick up three of my guys who had stayed out overnight with an artillery unit. They got in the back of the helicopter, and there were two of us in the front, five being the maximum a Gazelle can carry. I took off and I thought, just for the fun of it, I would give the artillery unit that we were leaving a farewell flyby – a little cheerio display. I pulled up into a steep wingover, came back and did a very low-level pass, next to the artillery's guns. Then I rose away and off we went.

Even at the time, I was thinking to myself, 'That was stupid. Five people in the aircraft, heavy, hot … It really doesn't take much for things to go wrong. Why push it? And for what? A showy departure?' It was the worst thing I've ever done in a helicopter and the thing I am most ashamed about. I got cocky.

Later that night, one of my experienced pilots, Sergeant Nige Day, who had been a passenger for my ill-judged flypast, took me to one side and gave me a talking-to. I think he'd had a couple of beers. It was more like being spoken to quietly by a close friend than an exchange between a sergeant and the officer in charge. I'm not saying Nige was being so pally that there was no discipline or respect for rank when he spoke to me: there was, absolutely. But it was brilliant, the frank way in which he did it. And what it brought home, in particular, was that someone I was supposedly in command of had had cause to look at me and think, 'No, you just went wrong. And I can't go with you on that.' And I could do nothing but feel ashamed and apologise and tell him that I completely agreed with him.

I think of it as my 1,000-hour chat. Every pilot needs to have one. If someone doesn't give it to you, you should give it to yourself. You've reached 1,000 hours of flying, and suddenly it feels like you know everything and can do everything. What could flying ever show you that you hadn't already encountered in that time?

And it's the point of maximum risk – the point at which you risk blowing everything. That's when you need the slap in the face, the wake-up call, the reminder. And hopefully you go on from that to become a better pilot. It has stayed with me, that incident in Kenya. And ever since, when tempted to push it just for the hell of it, I always think, 'Hang on. Don't let this be a Nige Day moment.'

After Kenya, it was back to normality at Gütersloh – or what passed for normality there. Africa might have been behind us but it was soon in our minds again when my colleague and friend Captain Jim Richards was posted to Kenya for a similar six-week exercise. Remembering how much we had enjoyed making our tents comfortable in Nanyuki, we decided to give Jim the same opportunity by sending his home comforts to him from Germany. Without telling him first, of course. Using any spare space available on the cavernous Hercules supply plane that left every week, we despatched Jim's possessions to him, bit by bit, marked for his attention. His TV, his bookshelf, his wardrobe ... pretty much the entire contents of his room went out to Africa in the end.

Meanwhile we were preparing a small surprise for Jim on his return. The long, wide and rather samey corridors that were a feature of Gütersloh's accommodation blocks were lined with wardrobes. People shoved them out there to free up some space in the less than generous living quarters. It then became a standard trick to conceal the doors to people's rooms by pushing wardrobes in front of them. Long experimentation had shown that someone returning from the bar after a lengthy evening could be quite easily disoriented by such a move. But for Jim, we upscaled this trick significantly. When he got back from Kenya, he found a wardrobe where his door ought to have been – but he was then baffled to push that wardrobe aside and see only wall behind it.

He kept pushing: no door. Still only wall. That was because we had paid a local builder to brick up Jim's doorway and then plaster and make good, right down to the skirting board. The guy, who was called Mario, had done a lovely job, actually. Each of the eight of us involved in this scheme had to find about £50 to pay Mario for his time and labour, but it was worth every penny. Inside the wardrobe we had thoughtfully left Jim a sledgehammer so that he could knock his way back in. He was soon to be seen outside, locating his room's window, and then pacing out the distance from the corner of the building so that he could come back into the corridor and start knocking the bricks down in the right place.

You'll perhaps be wondering what the mess manager made of officers undertaking these structural adjustments to the property. In fact, we had gone to see him beforehand to run the notion past him. He cleared it straight away. He thought it was a great idea.

Only in Gütersloh.

CHAPTER ELEVEN

Friday night pot-shots, a lift for the Brigadier,
and daylight robbery on the CASEVAC ladder

In June 1996 my regiment deployed to Bosnia to play a part in Operation Resolute. It was a relatively calm time in which to be heading there. By that point, the worst of the Bosnian War was over. The Siege of Sarajevo had ended in February, most of the fighting had stopped, and it was the phase where the UN coalition forces, including the British Army, were damping down flare-ups where they occurred, trying to assist in the clear-up process, isolating and removing mines, and generally attempting to create a situation in which the locals could rebuild their lives after years of conflict and misery.

But that didn't mean I didn't get shot at a couple of times. One night we were flying down the stunning twisty gorge that leads to the city of Banja Luka. Think Cheddar Gorge on steroids. We were getting down low on approach and therefore definitely in the small-arms threat environment, and some tracer fire started arcing through the sky in our general direction. It was nothing sustained, nor particularly threatening, and it didn't really have a hope of damaging us. We flew on and landed unscathed. Other crews reported experiencing the same thing from time to time, and it would happen to us again, too. There were a lot of AK47s around, and ammunition seemed to be in plentiful supply. It was most likely just people loosened up by a good session of Friday-night drinking, who had heard a helicopter buzzing overhead and decided to take a pop for the hell of it. Clearly there were some people on the ground who didn't want peace to reign right then in Bosnia, who felt they still had a job to finish. But most of the population we encountered welcomed us and, except at the wrong end of a beered-up Friday night, you were unlikely to feel like a target. I felt a greater threat in Northern Ireland than I did during my six months in Bosnia.

Our regimental headquarters were in Divulje, near Split Harbour, on the coast of Croatia. We were billeted there in a tall, thin, red-brick building, a former Yugoslav People's Army barracks where we shared rooms. I can report that there are much less beautiful spots in which to be camped than an Adriatic fishing port. Decent NAAFI, too. Split was where our aircraft would go for longer-term maintenance and air testing, but we would only tend to be there for two or three days at a time, and then we would go up country to the Bosnian locations that we were supporting – Banja Luka and the town of Gornji Vakuf. There we would do tasking for any of the military units in Bosnia at the time who needed helicopter support. For the Gazelles, that meant a lot of ferrying people to and fro, either senior officers or logistics personnel. Sometimes obs posts, up on the mountain tops, needed resupplying or required crews to be changed out. And sometimes we would be doing observation, covert night ops with Special Forces, or reconnaissance, keeping an eye on the movements of certain factions in accordance with intelligence. It was a rich mix of work.

By now Will Mellows had been replaced as 652 Squadron commander by Major Chris Collett. I think it would be fair to say that Major Collett was far more interested in his Lynx flight than his Gazelle flight. For him, I often felt, if it didn't involve Lynx, it wasn't really happening. That could have been frustrating for me.

On the other hand, it was also liberating. I thought, 'Well, I've got my flight. We're a great bunch. They know what they're doing, they're very competent, and we get on brilliantly together. We can just crack on with it.' Since we were the only Gazelles flying in Bosnia at the time, I often ended up reporting to Lieutenant Colonel Gary Coward anyway. He was the commanding officer, the one ultimately managing the regiment's assets, so if I needed permission for anything, he was as good an authority as any. Lieutenant Colonel Coward was one of those commanders who you could learn a great deal from – a dynamic and intelligent officer who went on to become a Lieutenant General Sir Gary Coward.

Evidence of the war's severity was everywhere. Gornji Vakuf was riddled with bullet holes. Abandoned observation posts and defences abounded. Mines seemed to be everywhere – far and away the greatest threat to our lives. We had to be very careful about where we put our helicopters down, using only recognised landing sites, and never landing off the side of the road or even on the road itself, unless you knew that it was being heavily used by vehicles. But the flying was stunning. It was more mountainous than any environment I had flown in up to then – way more challenging than the Aberdare Range in Kenya and a proper test for the mountain flying techniques I had been shown in Wales during the Pilots' Course. The weather, too, would add some spice. I was going up and down the Neretva River Valley which runs from Croatia into Bosnia and which is a narrow gorge that is often capped over with clouds. It could feel like you were pushing your helicopter through a tube, and doing it at night, on NVGs, merely added another level of difficulty. We worked long days, occasionally doing over seven hours of flying. Seven is normally the limit, and you would have to get an extension from the CO to be able to do more, which was all good for the log book, of course.

The rule was, no flying in the clouds. We call that Instrument Meteorological Conditions, or IMC, where you're taking your cues from your aircraft instruments alone. It wasn't that we weren't qualified for IMC, just that there wasn't sufficient radar availability over Bosnia to make that safe, so everything had to be VMC, Visual Meteorological Conditions. That meant a lot of grubbing around below the clouds in the scud, getting bumped around in the valleys and gorges. As we transitioned from autumn into wintertime, you got used to a bumpy ride.

That said, one night I was flying back from Sarajevo on NVGs with Corporal Price when there was a massive bang. In the dark, that's especially alarming. Was the engine or gearbox about to give up? I checked the temperatures, checked the pressures – everything was reading normal. The rotor speed? Normal again. We were still flying, maintaining altitude, the aircraft wasn't falling apart. Corporal Price and I were looking at each other, extremely disconcerted. Neither of us had any idea at all what it could have been. But we weren't far from base so flew on, and landed as soon as we could.

At that point we discovered that my side of the canopy was thickly smeared with blood, guts and feathers. A bird strike at night? That's unusual, and extremely unlucky. At least the thing hadn't come through the Perspex, which might have given us a problem. I still have no idea what bird that was – apart from one that regretted it, obviously.

You came to know your locations pretty thoroughly, and that's always an advantage, especially in bad weather. One morning I was flying into a base at Sanski Most to pick up a passenger and I'd had bright, clear weather all the way until I was practically there. The base itself, though, was in a river valley, and utterly shrouded in fog at this point. I radioed ahead to Corporal Balcombe on the ground – the same corporal who had helped me out of the clouds in Kenya and who often seemed, serendipitously, to be on the end of my dodgy radio calls.

'I'll be in for a refuel in five minutes' time.'

Corporal Balcombe seemed a bit surprised.

'Are you sure about that, boss? You do realise I can't see to the opposite side of the parade ground here ...'

'Absolutely sure. Get the fuelling point ready.'

The thing was, I had flown this last bit of the route several times in the daylight. I could see the road below me, on the edge of the fog, and I could visualise the route it took from there. So I simply had to follow the road, keeping an eye out for the bottom of the pylon (its top was lost in the mist) and making sure I flew under the electrical wires before rejoining the road and following between a couple of buildings to a tall chimney. That in turn led me onto the military base where I landed and casually walked in to look for breakfast. Corporal Balcombe was pretty slack-jawed, as was my waiting passenger, who had given up on all hope of his lift getting through in these conditions. The fog had been around since dawn, after all, and it was now about 11.00 a.m. Both of them looked at me as if I had just completed some kind of massive, heroic, fog-bound odyssey, rather than a long journey in the sunshine followed by a couple of minutes across extremely familiar territory. I may have disabused them, I can't quite remember.

Familiarity with the geography is everything, though. We heard about pilots on detachment before us, who had started in Bosnia in the winter and had flown into this valley several times with no concerns at all. It wasn't until later on in their tour, when the weather got better, that they noticed there was a set of wires strung high across the valley. Those wires had been covered in cloud for their first weeks, and they had been flying underneath them without even knowing they were there. That story still gives me chills.

So, in a different way, does the memory of the day I got lost over Bosnia with a brigadier on board. I lay blame for that on my slightly chaotic pal Philip McCabe – the man with the stunning rapid-descent skills, as displayed in Northern Ireland. Philip was now serving as an operations liaison officer in Šipovo, supporting another military unit there. The plan was that I would fly in to Šipovo, where I had to pick up another aircraft and a passenger and fly them back to Split, prior to taking a couple of days' leave; and Philip would take over my aircraft for a short tour in order to keep his flying hours up.

So I landed in Šipovo and handed over my aircraft to Philip, ready for him to start his duty. But he was looking a little worried. 'I haven't had time to get my nav bag together,' he said.

Now, let me at this point explain how important a nav bag is to a pilot in an operational theatre. We're not talking about a wash-bag here, with your deodorant and shampoo in it. When you are part of a multinational military force, working within a NATO structure, in an international environment, things obviously become a lot more complicated than if you are simply out there pottering around the north German plain. For instance, there are airborne early-warning aircraft with whom you have to clock in every time you fly: you're going to need passwords for that. You're going to need a whole set of codes with you to show that you are NATO forces. You'll need to have another set of passwords ready in case you are shot down and Search and Rescue crews are required to come out and rescue you. Basically there are packs and packs of information, all the right maps, all the right numbers, all the right everything, and all to be stowed in your nav bag. And Philip had none of it.

I couldn't believe it – except that it was Philip, so I sort of could. But what's a friend to do? I thought about it for a minute: ahead of me lay a straightforward flight to Split, in perfect flying weather, in an aircraft that finally had GPS installed. And once I reached Split, I was going on leave. It wasn't ideal, but …

'OK, Philip. You can borrow my nav bag. Give it back to me in Split at the end of your duty.'

So Philip set off with the pure gold which was my nav bag, and I tried to put it out of my mind and got airborne in the other aircraft with my passenger – a Royal Engineers brigadier, as it turned out, whom I had flown before. Straight after take-off, I consulted the trusty GPS, nudged the helicopter 2 degrees right as instructed by the display, and settled in for the flight.

I was going along smoothly, solo-flying on a lovely day, having a good old chat with the brigadier in the back, just occasionally glancing at the GPS and keeping us on course. Which always seemed to involve going 2 degrees right, oddly. But that was what the GPS was telling me, and if you couldn't trust the GPS, what could you trust?

After about twenty minutes, and several more glances at the display, I was growing properly concerned about the amount of '2 degrees right' that I was doing. Did this make sense? The sun was now coming from a completely different direction than I felt it should be and, according to the compass, we were heading west. To get from Šipovo to Split, shouldn't we have been going more or less south? I glanced again at the GPS – still saying '2 degrees right' – and felt my stomach slightly sink. The GPS must have frozen on take-off. I looked down at the land below: nothing that I recognised. So now I was lost, somewhere over Bosnia, with a bust satnav, a brigadier in the back and (curse you, Philip McCabe) no nav bag.

I must have gone rather quiet for a couple of minutes because the brigadier now said, 'Is everything all right at the front?'

'Absolutely fine, sir,' I replied. 'No problems at all. I'm just trying to work out when we'll arrive. I think we've got about another 30-to-40 minutes to go and we'll have you back, no problems at all, sir.'

Meanwhile, my mind was whirring: I knew I'd strayed right, so if I just kept flying, presumably I would eventually hit the Adriatic. And if I did that and turned left, then somewhere along the coast, surely, I would find Split. Big-picture flying – there's a lot to be said for it.

Just to blag it out even further, I called the AWACS aircraft and gave them an early call, as obliged, that I was 'shortly to be transitioning from Bosnia to Croatia and out of theatre'. I mean, I wasn't quite sure when that was going to be happening, if the truth be told. But at least I would have made the right radio call when it did happen. Now I just needed the fuel to hold out.

The ploy worked. Sure enough, I eventually hit the Adriatic – a beautiful sight in the sunshine at the best of times, but even more so on this occasion – and I turned left and flew along for another ten minutes, at which point Split Harbour popped up and in we went. I'd like to think the brigadier was none the wiser, but the wry smile as he said farewell hinted that he may have been all too aware that I had just got away with a close call.

Flying VIPs always brings its own pressures. At some point during this tour, HRH Princess Anne was due to visit and a Lynx helicopter had been designated as her possible mode of travel. I found our engineering officer on dispersal one morning poring over the aircraft and doing quality control checks. I asked him what he was up to.

'Oh, just making sure this aircraft is safe for an upcoming VIP visit,' he said.

'Aren't all our aircraft safe?' I asked.

'Of course they are,' he replied. 'It's just that some are safer than others.'

Ask a silly question …

Emergencies weren't commonplace but they did occur from time to time. If you fly helicopters for long enough it's inevitable that you will pick up your fair share of malfunctions. The problem with helicopters is that, unlike their fixed-wing counterparts, they don't naturally want to fly. An aeroplane, with minimal control input, will glide through the air in perfect balance. A helicopter, by contrast, remains aloft by beating the air into submission using brute force – in fact, by employing a combination of forces all acting in opposition to one another. When things go wrong in helicopters, it is not unusual for them to stop flying, immediately and disastrously.

I'm sure this is something that Lindsay Morris, our assistant adjutant, was contemplating as I flew her from Sarajevo to Split one day. Lindsay was sitting in the back of the Gazelle that I was flying solo. We had just reached the Adriatic coast and were enjoying the stunning scenery as we headed north towards Split. That was when the helicopter's hydraulic system decided to fail, sending a big jolt through the aircraft. This, in itself, is not the direst of emergencies. The Gazelle can be flown without hydraulics by using brute strength to move the controls, which just takes some getting used to. However, hydraulic failure is considered serious enough to warrant an amber 'master warning' caption. At the same time, the aircraft's low-altitude warning had been triggered and for some unknown reason I couldn't cancel the annoyingly loud alarm.

So, a massive jolt, warning lights, a piercing alarm ... Poor Lindsay was watching this all unfold from the back. I'm not sure she was reassured, exactly, when I suggested that she remove her boots as a precaution. (We were still over water and a ditching, although highly unlikely, wasn't out of the question.) Hydraulic fluid can also be extremely flammable when sprayed as a fine mist towards a hot engine. Knowing that the hydraulics had failed pretty spectacularly and that fluid was likely to have sprayed everywhere, I tried to sound as casual as possible while asking Lindsay if she could keep an eye out for any signs of fire back there. Except I couldn't casually ask her – I had to shout over the noise of the bloody alarm which I still couldn't cancel. All in all, it spiced up an otherwise routine morning sortie. Lindsay was brilliant – she took it all in her stride, although I think the cup of tea that was waiting for her at Split was more welcome than usual.

Bosnia was a great tour in terms of building experience, both in flying and in leadership with my team. It also set me to work for the first time in an international environment. At Gornji Vakuf, we were co-located with Dutch pilots from 299 Squadron in the Royal Netherlands Air Force, who were superb guys to be sharing a camp with – not least for their creativity as builders. Within minutes of arrival, they had started work on the construction of what they called 'Coyote Avenue', an in-camp recreation centre that would eventually include a capacious bar, a TV room and, incredibly, a CORIMEC swimming pool. The pool was freezing cold, but that didn't stop us using it, not least at times of celebration: birthdays, 1,000 flying hours, any excuse, really. The Dutch also had the best coffee – a brew that put the British Army Nescafé to absolute shame. We found ourselves stopping by there a lot.

Occasionally, the Americans would drop in, too. We got the call late one summer evening, requesting a refuel, and shortly afterwards some Kiowa Warriors, some Apaches and some Blackhawks descended. It was pretty late by the time they were ready to move out again, so we suggested they stop over. And then we all headed to the Dutch bar – where, of course, the 'two cans only' rule that applies to all personnel in theatre was strictly adhered to by one and all. At least, I think it was. Afterwards, on a warm night, people were strewn around the helicopter landing site in their sleeping bags, crashed wherever they could, or sleeping in their aircraft.

Those six months of my life seemed to pass very quickly and soon we were packing up and heading out. One of our aircraft was going to need somebody to fly it back from Bosnia to Wattisham in Suffolk to be swapped out. Christmas was fast approaching and nearly all of the guys, who had been away since July, wanted nothing more than to get home to their families as fast as possible. But a young officer called James Champion, who had recently joined the squadron, and myself were both single at the time and not all that fussed about missing Christmas. So we volunteered to get the helicopter home.

The arrangement was that we would simply be following a Lynx crew, who had just delivered an aircraft from Wattisham to Split and were making the return journey. It was all last minute, and James and I had minimal time to return to Split and pack before we hooked up with the Lynx crew and then set off together. We'd be flying over the Adriatic Sea, so we had donned our full immersion suits and life preservers, just in case our single engine failed and we got dunked, and off we went in the wake of this Lynx. Pretty soon the weather, which hadn't been all that great to start with, began to get worse, and continued doing so. We were starting to grub around and the horizon was beginning to get all milky, which you never like to see. It was becoming impossible to tell the difference between grey cloud and grey sea.

At this point the Lynx crew decide that they're going to climb into the freezing cloud and go on instruments and then do an instrument recovery at Vicenza in Italy. Which is all very nice for them, of course, in their Lynx, but here in our Gazelle we've got no icing clearance and are not authorised for instrument flying, having not flown in cloud during the whole tour of Bosnia. So we're stuck down there in the shitty weather, with our maps out, in foreign airspace, having to scoot so low that we're only about 50 feet above the water. I don't think anyone has ever been as relieved to spot Venice as I was that day. We scudded around the city and made it north of Venice airport where miraculously the weather cleared a little. That enabled us, with a sigh of relief, to get into Vicenza, where the Lynx crew, of course, were already sitting in the warm with their feet up.

There was just enough time to tell them what a bunch of tossers they were for abandoning us in the middle of the Adriatic when a fog descended – a fog which would persist for five days. It was a proper north Italian winter special, and we couldn't move. On the bright side, Vicenza is a US airbase, so at least we were able to hit the American PX store each day – buy jeans and get the Christmas shopping done at dollar prices. By the time we had finished, our purchases practically filled the back of the Gazelle. The rest of the time we spent sightseeing in Vicenza.

The fog gave up eventually, and our journey resumed with a leg to Nice, then Lyons in France. Like every sortie on this trip, it was stressful in its own special way. This time, as we approached Lyons, the fuel gauge was right down at the bare minimum for landing. By the time we put down, we had spent an edgy few minutes with the engine literally sucking on fumes, which is always a horrible feeling. From Lyons I tried to take us out to the west coast of France so that we could head up to England from there but, on account of the weather, which was not being helpful, and on account of the need to stay under the clouds at all times, we ended up having to back-track and then do big-picture flying in order to pass around Paris. Finally we found ourselves on the north coast, at Le Touquet, a small airport near Calais, where we landed for a refuel.

This trip was supposed to have taken three days; we were now on day nine. And just to compound the agony, I had run out of fuel chits. How the hell were we going to get home now?

I rang the pay officer in Gütersloh and sought his advice.

He said, 'Do you have a credit card with you?'

Well, yes, I did. After my recent trips to the PX on the American base in Vicenza, it was currently weeping quietly in my wallet.

'Will I get reimbursed?' I asked, weakly.

Apparently I would. Which is how, in December 1996, I ended up paying with my own money to fill up an Army Gazelle and a Lynx at an airport in France. I trembled when I handed over that credit card. But in fact, it wasn't as much as I'd thought it was going to be. The bill came to about £250. OK, that's close to £500 in 2020 terms, allowing for inflation. But even so, for two helicopters? Bargain. (And the pay officer wasn't bluffing: I did get reimbursed.)

We made it across the Channel to Wattisham Airfield in Suffolk without further misadventure. And then James and I got the aircraft changed out and flew back to Gütersloh. We arrived that evening just in time for the regimental Christmas party which, naturally, we attended and which, equally naturally, proved unrestrainedly raucous. I dimly remember walking away from the officers' mess at about 4.00 in the morning. Unfortunately, the following day at 9.00 a.m., I was required to be on parade in full service dress for the end-of-tour medals ceremony.

A horrendous ten-day journey across Europe, followed by a long night of Christmas drinking ... I remember, at the parade, standing out in front of my flight, swaying backwards and forwards slightly, my face a shade of pale green, just concentrating my gaze as hard as I could on this yellow line on the ground in front of me, and thinking, 'Don't faint, don't faint ...' When the ceremony was through, I limped back to my room and crashed out for the rest of the day.

*   *   *

In the summer of 1997 I flew to Calgary in Canada, and then on to the town of Medicine Hat to take part in a military exercise at BATUS – the British Army Training Unit Suffield. This is where the British Army run exercises all through the year. Because the available area is larger than Salisbury Plain, it's possible to work on a large scale and exercise a whole battle group. I wasn't with 652 Squadron this time; I was accompanying another flight under the command of my friend Captain Jim Richards, the man who entered this story earlier when his doorway got bricked up in Gütersloh. By this time, even though I was only twenty-five, I had more than 1,000 hours under my belt and was one of the most flown flight commanders in 1 AAC, so I went out to help and be an experienced hand on that exercise. As in Kenya, on Exercise Grand Prix, the helicopters were there to do CASEVAC, but this time we were also exercising our own military role, doing reconnaissance and working with the tanks and artillery and all the other military hardware on the ground. Having done lots of the training for that role in isolation, over in Germany and elsewhere, it was a good chance for me to put it all together with the rest of the Army in the field. It was also, as it turned out, a chance to work in extreme heat. We had a max temperature in the Gazelle cockpit one day of 45° C. I had no idea Canada could get that hot.

With casualty evacuations, the key was always evaluating the situation that you were being asked to go in and relieve. We'd had that experience in Kenya where Ian Curry and I had responded to an urgent call-out in bad weather to an appendicitis case, only to discover when we got there that the guy had constipation. Not to deny the pain of constipation in that instance, but certainly the risk of the flight had outweighed the need on the ground. The last thing you want to do, flying in response to an accident, is make the situation worse by becoming an accident yourself. There have been multiple cases where helicopters have been despatched on search-and-rescue missions or casualty evacuation, only to fall foul of those same conditions that caused the problem in the first place.

Fielding the calls for CASEVAC, it was important to remove yourself from the drama to some extent. Very often those calls would be coming in from very stressed people on the ground, insisting that the need was urgent. Yet before you simply jumped into the Gazelle and set off, you had to weigh up the conditions, the weather, terrain, whether it was night-time. How much risk were you accepting for yourself and your colleagues? You've got your co-pilot alongside you, and a doctor in the back, so that's three people going to the rescue of one. You didn't ever want to turn it into a situation where three people die and the next morning somebody goes to hospital by truck to get their arm set.

In Canada, though, for the first time I appreciated it from the other end of the spectrum – how there could be troops on the ground calling in who, far from stressed, were misleadingly cool and collected about everything. They would be telling you very calmly about a casualty, a blue-on-blue and only very gradually would it emerge that this casualty had already lost two pints of blood and was still bleeding profusely, meaning you really needed to be in the air straight away. Sometimes those situations could be as hard to evaluate as the ones where you were hearing sheer panic coming down the line at you.

Just to make it a little more interesting, some of the pilots in Canada started the CASEVAC ladder. Points were scored by the evacuating crew depending on the severity of the injury, whether it was night or day, whether it was bad weather or good weather, whether the flight was on night vision goggles, and so on. The harsher the conditions your evacuation met, the more points you got. Which I suppose might sound a touch macabre, given that there was some genuine suffering going on beyond these antics, and people putting themselves in the way of actual danger. But, again, it just comes down to the military way of dealing with adversity.

What I can certainly tell you is that I quickly roared to the top of the CASEVAC ladder, and looked set to take the honours. The evacuation that put me there involved a Trooper who had got his leg trapped in a Challenger tank. The traversing turret had caught his leg and broken his femur quite high up. We flew in and landed right next to the tank where the casualty was now lying on the ground. The troops had managed to get him down off the tank, which must have been excruciatingly painful for the soldier, yet he was being phenomenally composed about it. The female Army doctor with us assessed it straight away as an emergency hospital case, but didn't want to set the leg or risk moving it too much because of the break's proximity to the femoral artery. The job then was to find some way to fit this soldier into the Gazelle without unduly changing the position of his leg. It turned into some kind of weird 3D jigsaw puzzle, with all of us trying to work out how best to accommodate myself, the doctor, the casualty and this broken limb. Eventually we managed it by taking a front seat out and arranging the patient with one leg down in the baggage compartment and the other going out towards the side window. (It didn't have to hang out of the window, thankfully; he was entirely inside the cockpit by the time we departed.) Then I flew him to the hospital at Medicine Hat and we stretchered him in for treatment. Even without sedation, the patient had been completely calm all the way, but as soon as they got him into the hospital they decided to insert a catheter. That was the point at which he screamed. It made me realise how little I ever wanted to be given a catheter.

That adventure carried me right to the summit of the CASEVAC ladder, and the prize was as good as mine, I thought. But what do you know? Right at the last minute, somebody got shot in the butt, at night, in bad weather, and another crew went out and picked them up, thereby leapfrogging me to the top of the table. Sport can be very cruel.

CHAPTER TWELVE

Eyes meet across a crowded Mess, milk gets spilled,
and lemmings take to the air

In October 1997 I returned to Germany after the battle group exercise in Canada. I was now nearing the end of my four-year tour and I already had my posting order back to the UK: shortly after Christmas, I would be moving to RAF Shawbury in Shropshire to begin the Qualified Helicopter Instructors course. The future was shaping up. The Gütersloh years had been immense – memorable parties, lifelong friendships, busy operational tours and more flying experience than I could have dared to hope for. But it was time to move on and, for me, that meant working towards the goal I had formed soon after arriving in Germany – becoming an instructor. As I drank in the remaining weeks of my tour, my focus was already shifting to the next stage of the journey.

One afternoon, around 4.30, I wandered in my flight suit over to the officers' mess for afternoon tea, a Gütersloh ritual: toast and jam, tea in little pots, the newspapers – all very British. At one of the tables, a young female lieutenant in combats lowered the paper she was reading and cast a glance across the room at me. That was the first time my wife and I set eyes on each other.

Rebecca had just begun her first posting with the Royal Logistic Corps. She had been commissioned at Sandhurst in April, had gone on to do the Troop Commander course, and had arrived in Germany in September while I was away. She had chosen Gütersloh because she wanted to be somewhere that people hung around at the weekends and didn't, as she had found at Sandhurst, slip off home leaving the place feeling a bit flat. Slipping off home for a brief visit hadn't been an option for Rebecca at Sandhurst: home was ten hours away in Perthshire.

Our first proper meeting was soon after that tea-time when the pair of us found ourselves part of a small group who were charged with arranging an 'upside-down room' for an unsuspecting new arrival. On this occasion, in addition to the usual inverting of the furniture, the upside-downing included removing all the CDs from the poor victim's extensive collection and returning them to the wrong cases. That was Rebecca's job.

I certainly hadn't been looking for a serious relationship with anyone, but I was very quickly smitten. Yet over the next few weeks I tried to ignore my feelings. Rebecca had just arrived in Germany, I was just about to leave – there was no future in it. And anyway, I doubted my feelings were reciprocated.

There was another issue, too: I had a girlfriend at the time. I happened to be dating our regiment's assistant adjutant, a military policewoman. I think both of us knew the relationship wouldn't outlast my upcoming posting order. But still: it was another reason not to open up to Rebecca.

So my attraction to Rebecca was strong but I continued to try and ignore it. And I also managed to fail to notice the strength of her attraction to me. Because it turned out I was wrong and those feelings *were* reciprocated. Whether it was my old insecurities returning to haunt me again, or the dreaded legacy of my boys-only education, the final assessment of my behaviour during this phase is unavoidable: I was rubbish. Practically the whole mess would have agreed. All my mates were telling me to hurry up and make some kind of move with Rebecca, given that I so obviously wanted to and given that she clearly wasn't going to rebuff me. All her friends were telling her the same, but I had a girlfriend so the first move couldn't really be hers to make.

Eventually, despairing of my hopelessness, friends decided that the matter needed to be taken out of my useless hands. Matchmaking would have to occur. It was like a plot from Jane Austen, if Jane Austen had spent any time in the Army and had taken up a posting at Gütersloh. At the officers' Christmas dinner, Andrea Luck, in charge of the seating plan, placed Rebecca and I opposite each other and my girlfriend at the far end of another table.

What do you know? I finally plucked up the courage to make my feelings known to Rebecca. Unfortunately, and to my lasting shame, I had not first plucked up the courage to update the other person who needed to know. The overlap was, I swear, merely a few hours, but it was enough to provoke some understandable anger on the part of my ex-girlfriend, anger which expressed itself in true Gütersloh style. I returned to my room to find that a couple of litres of milk had been thrown around inside it, ruining my TV, stereo and speakers. Also, many of my CDs were cut into shards, and a guitar on which I had taken to plucking away idly in my downtime was now going to be even harder to get a decent tune out of. Well, fair enough. This had not been my finest hour. I'm only grateful that the new car I had recently acquired – a rather lovely BMW Z3 two-seater – was safely locked in a garage.

So finally Rebecca and I began doing the obvious thing and seeing each other, sharing those special, heart-warming Gütersloh moments together – as, for instance, when Rebecca had too much to drink and retired early to her room, only for some of her friends to perform a fairly rare version of the upside-down room:

an upside-down room with the occupant actually in it. When I went to check on her, I found her lying on the floor under her duvet, having crawled out from under the sofa which had been overturned on her bed. I did what I like to think any chivalrous man would have done in the circumstances: I went straight to the ringleader's room and undid the U-bend on his sink – Rebecca's avenging knight, with a monkey wrench. Such moments of gallantry can bond a couple. As did heading to Scotland to spend that New Year together with Rebecca's friends – my first proper Hogmanay, complete with whiskey, reels and industrial-strength hangovers.

And then it was 1998 and I was leaving Gütersloh for the UK, and Rebecca, not that long after, was heading to Bosnia for a six-month tour of duty, entirely setting the pattern for our courtship. Over the course of the next two years, at the mercy of Army schedules, we would rarely be in the same country as one another, let alone in the same room. We would snatch weekends when we could, with me, late on Friday nights, driving from England to Germany, fuelled by cans of Red Bull, and then haring back again for Monday morning. Or Rebecca would shoot over to the UK, seizing bits and pieces of R&R or post-tour leave, as and when they arose. In the whole of 1998 we must have seen each other for the equivalent of less than a month in total.

It was the kind of arrangement which would have taxed the strongest of relationships, let alone a brand new one. And yet, with us, it was its own kind of proof. Somehow, the more circumstances contrived to force us apart, the more we knew we should be together, and the more certain we became that we would be.

Meanwhile my poor handling of the start of our affair would return to haunt Rebecca not once but twice. During her Bosnia tour, the officer I had been seeing when we got together was posted out to join the unit. Everyone was billeted in the usual small rooms in CORIMEC prefab buildings, and there was a spare bed in Rebecca's. You can probably guess who was allocated to it. Rebecca had to have a quiet word in advance with the quarter-master and say, 'I'm not sure that's going to work ...'

Furthermore, at the end of the tour, Rebecca attended a barbe-cue given during the handover phase to the incoming unit, who had just arrived from England. She and another officer fell to talking about Gütersloh.

'I hear it's pretty wild there,' said the new arrival.

'Well, it's certainly fun,' said Rebecca.

'Yeah, crazy stories come out of there,' the arrival went on. 'Apparently there was this pilot who got his room trashed with milk because he was dating an RMP who lost it when he got together with someone else ...'

There was a lot of silent wincing around the place as Rebecca's face reddened. It seems we had become, not just an item, but the stuff of Gütersloh legend.

* * *

I will always be proud of what I achieved on the Qualified Heli-copter Instructor course at RAF Shawbury between March and May 1998. Bear in mind that I hadn't lived in the UK for four years, so I was woefully out of practice and there was a whole pro-cess of re-acclimatisation to be gone through, putting my colleagues at a distinct advantage. Nevertheless, I knuckled down and put in the hard yards and my performance ended up speaking for itself. At the weekly Thursday curry nights in Shrewsbury, I went from being able to manage a mild chicken tikka masala at the beginning of the course to graduating with honours with a hot jalfrezi by the time it finished, just two months later. It only shows what you can achieve if you apply yourself.

My partners on those steam-releasing (in more senses than one) Shrewsbury trips were a mix of Royal Navy, Army, Air Force and Royal Marine pilots – eight of us in total, all of whom had arrived at RAF Shawbury in Shropshire, the home of the Defence Helicopter Flying School (now known as No. 1 Flying Training School), in order to learn how to instruct. Our first task, though, was to convert onto a new helicopter – the Eurocopter AS350, known simply as the Squirrel. The Squirrel had replaced the Gazelle as the military's basic training helicopter and, although still single-engine, it was larger and more modern and better reflected the front-line helicopters that students would most likely end up flying. Although I would go on to fly over thirty different types of aircraft, this was only the second type of helicopter I had flown. I had been a staunchly loyal Gazelle pilot up to now, and that aircraft would always hold a special place in my heart, but I was excited to be trying something different.

It rapidly became apparent that being a good flying instructor requires three distinct skills. First, you need to be able to fly accurately yourself: you won't get away with the 'do as I say, not as I do' approach when you're up in the air with a student in a helicopter. Next, you need to have a good patter. Your patter is the ability to talk about what you are doing, when you are doing it. It may sound easy, but actually it's a real accomplishment, requiring a lot of practice and the negotiation of many pitfalls: don't say things ahead of time, or after they have occurred; don't try to impart too much information, but don't miss out anything vital, either; and remember to say how you are doing something and why you are doing it. You need to be able to speak coolly and calmly while a million other thoughts are running through your brain. Finally, you have to be able to recognise when the student is doing something wrong, what it is and how to put it right.

To get going on this, we would first fly a sortie where we would receive the 'perfect' lesson as a demonstration from our instructor. Next, we would fly with another student instructor and each take turns teaching the exercise. Finally, we would fly with our instructor again, but this time we were expected to deliver the perfect lesson while they played the part of the student. Sometimes they would be a good student, but more often they would not, in order to throw you difficult situations to deal with.

In addition to the flying, we also had to prepare and deliver ground school lessons. Instructors were expected to be able to teach the whole flying syllabus – meteorology, principles of flight, navigation, air law, the works. Getting all of that into your head was highly demanding and required some immersive studying.

And then, towards the end of training, came the notorious 'lemmings day', which possibly speaks for itself. The lesson we were learning to teach here was engine-off landings – in other words, how to get yourself safely back on the ground when your helicopter's one and only engine has decided to pack up. A lot of people probably don't think that's even possible, but I'm here to tell you that it is. I'm also here to tell you that it can be taught. As long as you commit to throwing yourself at the ground quickly enough, you can keep the rotor blades turning by using the force from the air as you descend. This is known as autorotation – the same principle by which a sycamore seed spins as it falls from the tree.

During autorotation, you can still control the helicopter but you can't escape the fact that you are coming down pretty fast. The key to a successful engine-off landing is getting it all right just before hitting the ground. You have to pull the nose up, which both slows the helicopter and puts more energy into the rotors.

Then you have to stop the descent with a short, sharp 'check' on the collective lever. This uses the energy in the rotors to give you some lift – but you have to use it wisely. Check too soon and you'll find yourself at 50 feet with the rotors stopping and a mighty drop beneath you. Too late and you'll hit the ground anyway. And at the same time you'll need to push forward on the cyclic to stop the tail rotor striking the ground, and then cushion the landing with any collective lever you have left. All in all, there's quite a bit to think about.

As a front-line pilot you would only practise this manoeuvre a couple of times each year and only ever under the guidance of an instructor. It's all about weighing the risk. After all, engine failures don't occur very often and yet there have been several aircraft written off from practice engine-off landings that didn't quite go to plan. Instructors, of course, have to put themselves through the engine-off hoops far more regularly.

As such, the engine-off landing is a bit of a holy grail for a new instructor. After all, if you can be held responsible for teaching those, you're coming along quite well. And accordingly, it's a big moment when 'lemmings day' arrives. This is the point at which two instructor students are let loose for the first time to practise teaching engine-off landings to each other. The blind leading the blind, I suppose you could call it, although the idea is that, by this stage, you more or less know what you're doing. However, when the time came for my group to start pairing up, I'm not sure who was wincing hardest – the instructors peering out nervously from the safety of their offices, the engineers looking on from the hangar, or the students in the cockpit.

Remarkably, and much to everyone's relief, the eight of us emerged from 'lemmings day' unscathed, along with the aircraft, and soon after we were going our separate ways as newly qualified helicopter instructors. I was posted back to Middle Wallop, as a flight commander teaching students on advanced rotary – the phase where the focus is more on tactical military flying. It was rewarding work, in a place I knew well. What's more, my morning commute was just a five-minute walk from the officers' mess to the hangar. I felt very lucky.

However, that commute soon got longer. As a further distraction from Rebecca's absence overseas, I decided it was as good a time as any to get my motorbike licence. Once that was achieved, my journey to work every day went from being a five-minute walk to a ten-minute extended lap of the airfield on Hampshire's winding roads aboard a Honda CBR600. Not quite Tom Cruise and Miramar Naval Air Station perhaps, but I'll admit to having hummed the theme to *Top Gun* once or twice on the way in.

While I was a rookie instructor, I was only allowed to teach the most basic engine-off landings – entering from straight and level at 70 knots from 500 feet. However, after a few months in the business, assuming you pass a check ride with the chief flying instructor, you can progress to teaching advanced engine-off landings. This is where you might enter from high speed and very low level, or from a high hover, or far away from a landing site and thereby employing the 'maximum range' method.

It was on one of my first outings after passing my check ride that I found myself at 3,000 feet, just south of Andover, returning with a student to the airfield at the end of a sortie. The airfield was still some way off and I decided it was a good time to introduce my student to the 'maximum range' engine-off landing.

By way of a preliminary, I asked the corporal I was flying with if he thought we would make it to the airfield if the engine failed.

'Not a chance, sir,' came the reply that I was hoping to hear.

'You'd be surprised,' I said confidently. 'Let's give it a go.'

As it happened, we would both be surprised. No sooner had I shut down the engine than the first twinges of doubt about this project entered my mind. The airfield, away in the distance, was moving up the windscreen as we descended, a sure sign that we were falling short of the intended landing point. No need to panic yet, though. We still had plenty of height as I adjusted to 95 knots, our 'maximum range' speed.

Still not good enough. Next, I started raising the collective lever to slow the rotors just a little and get a small amount of lift in return. You have to be careful at this point not to slow the rotors too far. If you go below a certain point, you'll never recover the rotor speed and the result will be catastrophic. That's what the alarm is there for – the alarm that had just been triggered and which I was now hastily switching off.

'Don't worry,' I said to the corporal, who was looking at me rather doubtfully. 'The alarm is quite conservative – we can still slow the rotors a bit further.'

Normally, when you demonstrate an exercise as an instructor, there's an awful lot you have to say and not enough time in which to say it. It turns out that's not the case with the maximum range engine-off. Time hung heavily. As silence descended in the cockpit – silence from the engine, silence from the rotors and silence, most definitely, from the student – I found myself casting around for topics of conversation to take both our minds off the possibility that this landing was anything other than a shoo-in.

'So, how's the course going?'

'Food in the mess OK?'

'Any plans for the weekend?'

Somehow the conversation wasn't quite igniting. However, with about 500 metres to go, I sensed that I was in with a chance of making it, helped only by the fact that, in 1998, our airfield perimeter at Middle Wallop was a low hedge and not the 8-foot-tall wire fence that it is today. As we passed narrowly over the hedge, I felt my buttocks clench involuntarily, as if physically trying to lift the helicopter a couple of extra inches by sheer gluteal force alone. And then suddenly we were down, skidding over the perimeter track and onto the grass – still well short of the designated landing area but, holy smokes, we had made it.

All in all, I thought it was a pretty convincing demonstration of the art, although I did point out to my student that you probably wouldn't want to cut it much finer than that. Indeed, any finer and we would have spent the afternoon picking bits of hedge out of the tail rotor.

Word gets around, of course. The following morning, I did the pre-flight brief with another student on the same course. He asked me if we would, by any chance, be doing engine-offs as part of the sortie. He sounded rather nervous about the prospect. I wonder why.

Through all of this, Rebecca and I were urgently finding time to be together when we could. One weekend when we were both free, I was determined to join her in Germany but found myself staring at the obstacle of a regimental dinner at Middle Wallop on the Friday night. Those dinner nights were extremely frequent but I figured that these days, as an instructor, I could probably get away with missing one. So I approached the commanding officer, explained that my girlfriend was in Germany, that she had just done six months in Bosnia and that we hadn't been seeing each other much, and I asked him if he would mind me skipping the Friday dinner on this occasion so that I could get away a little quicker.

The commanding officer said he very much would mind.

I hadn't realised that dinner was quite so important. Oh well. I didn't drink all evening and, as soon the Queen had been toasted and we were allowed to leave the table, I bolted, changed out of my mess gear, jumped in my car and drove straight to Germany, arriving there at around 5.00 a.m.

Neither of us was in a hurry to get married at this stage. If our long-distance relationship could stand the test of operational tours and months apart and mad midnight dashes to Germany, then a few more months of dating wasn't going to change the way we felt about each other. Then, as always, the British Army got involved and mixed things up a little. I was told that I had been selected to replace my old potholing friend David Amlôt as the British Army Exchange officer serving with the 1-227th Aviation Battalion in Fort Hood, Texas. It was a highly sought after and prestigious three-year posting – flying Apache helicopters with the US Army's most technologically advanced division, the 1st Cav. I felt a real surge of excitement at the thought. With a week to go before Christmas, my posting officer casually mentioned that this tour could be accompanied (meaning married), or unaccompanied – it was my choice. I asked for some time to consider and was told to come back with an answer after Christmas leave.

I did some serious thinking. In my mind, I was probably six months away from proposing to Rebecca. I knew that she was the woman I wanted to spend the rest of my life with, regardless of any posting. But I also knew that this was a wonderful opportunity for us both to travel and make an adventurous start to our life together. If I turned the posting down, proposed six months later and casually mentioned that we could have been enjoying three years in the US together, Rebecca might well have thought me mad. So I decided that I should first propose to Rebecca. I knew it would seem a little premature – but maybe not *that* premature. If she accepted, I could then move on to question two, and ask her how she felt about upping sticks and joining me in the middle of Texas for three years. Oh, and by the way, could she possibly give me an answer on that within a week?

Curse the British Army.

I'm not sure 'romantic' is a word that any of my friends would use about me. But even I recognised that there were some moments in life that called for an injection of romance, and that a marriage proposal was one of them. Moreover, given that the timing of it had been somewhat forced upon me, I wanted my proposal to be as special as possible so that my true feelings about the moment would be unmistakable. Here, for once, fate had played into my hands. For the Christmas holidays, Rebecca and I had booked a skiing holiday in St Anton in the Austrian Alps with some mutual Army friends. Mountains, snow, New Year's Eve fireworks … it was going to be the perfect, heart-melting setting in which to ask her to be my wife.

And, indeed, I'm sure it would have been unbeatable if flu hadn't struck our chalet. And not just any old flu: one of those properly punishing ones that seems to affect every square inch of your body. Both Rebecca and I succumbed and on New Year's Eve we were huddled under blankets, shivering with fever and listening miserably to the noise of the partying seeping up from the streets below. It was a long way from the idyllic scene that I had imagined and, with both of us reaching for the paracetamol and the Lemsip, it certainly didn't seem like the right time to propose. I went home with the big question still un-popped.

But now the clock was really ticking. We returned from St Anton to the UK with the holidays about to end and decision day for that US posting looming. Romance? I was going to have to improvise. And so it was that I proposed to Rebecca, not on a balcony overlooking the Austrian Alps while fireworks plumed in the skies above us, but in my room in the officer's mess at Middle Wallop, after a takeaway curry, a bottle of red wine and a movie. Moreover, it took me a while to finish steeling myself for the vital moment and by the time I had done so, either as a consequence of the wine or the curry or the movie, or maybe a mixture of all three, Rebecca had fallen asleep on the sofa. This meant that in order to ask her to marry me, I had to gently nudge her awake first. I cursed the British Army again, I cursed the flu ... But actually I didn't give a damn – Rebecca said yes.

Then, of course, I had to move on to question two. 'That's absolutely wonderful,' I said. 'Now, this is completely separate to the question I just asked you about whether or not we should get married – there's this opportunity to spend the first years of our married life out in America on this posting ...'

I made clear that I loved what I was currently doing – that I didn't need to go anywhere else. But if it was something she could see us doing together, we could do it. It was a huge decision for her. She jumped in with both feet. 'Let's do it. Let's go to America.'

The next day, both slightly shell-shocked, we headed into Southampton on a shopping trip for two laptop computers, and avoided virtually any mention of the previous night's proposal.

But then, of course, we assumed we would now have a bit of time to sort out the details at our own pace. So much for that. A couple of days later, Rebecca headed back to Germany, only to hear that her battalion was to be posted at short notice to support the developing situation in Kosovo.

Some days you are the pigeon, and some days you are the statue. I was beginning to get fed up with being the statue.

* * *

On Patrol with the 1st Battalion, the Royal Green Jackets during the Troubles in Northern Ireland. Within weeks of leaving Sandhurst I was sent to gain some leadership experience. At the age of twenty, I had thirty riflemen under my command. They were streetwise and hardened in ways that I definitely wasn't; fantastic soldiers who absolutely knew their stuff, but were also ready to push the limits with me, a fresh-faced young officer.

A tea break nearly cost me my career in the Army, when under my watch two armoured military Land Rovers got stuck in a bog in County Fermanagh.

I did most of my urban patrolling in Strabane, which for a while knew the distinction of being the most bombed town in Northern Ireland.

After returning from Northern Ireland, I began my Pilots' Course at Middle Wallop. First, we were put through our paces during thirty hours of basic fixed-wing training in the venerable de Havilland Chipmunk.

I had joined the Army Air Corps to fly helicopters. Now, at twenty-one, it was finally time. We were taught on the Gazelle, nicknamed the 'screaming chicken leg' on account of its shape and its high-pitched, whining engine. Despite its nickname she was a beauty to fly.

Potholing with David Amlôt during my Pilots' Course would often be fraught with peril, but the experience has come in handy more than once during my time as an astronaut.

In June 1994 I was presented my 'wings' – the Army Flying Badge – by Admiral Sir Jock Slater. The proudest moment of my life up until then.

In July 1994, I was posted to Gütersloh in northern Germany for four years, but in that time I would accept a succession of temporary postings which saw me return to Northern Ireland and spend time on duty in Kenya, Cyprus, Bosnia and Canada. This photo of me enjoying a cuppa with fellow City Flight pilots was taken in Palace Barracks, near Belfast.

'Look Mum – no hands!' Flying the Gazelle over Belfast with Jimmy James, my good friend who I nearly lost out the door one day.

As an officer in 1 Regiment Army Air Corps, you were expected to learn how to freefall parachute – here at the airfield with fellow army officers and good friends Lindsay Morris and Kirsty Atkinson.

Winter 1995 and fitter than I have ever been in my life, competing at the Inter-Service Nordic Ski Championships – we picked up best novice team for our efforts.

Gütersloh was a haven for practical jokes – we arranged for Jim Richard's door to be bricked up whilst he was away in Kenya.

Kenya was perhaps the best six-week detachment of my Army life. It was the spring of 1996 and I had been promoted to captain and made flight commander. One of my first duties of the day was often to shoo elephants, giraffes and zebra from the firing ranges!

Left: With Ian Curry, who later became my best man.

In June 1996 I deployed with my Gazelle flight to Bosnia for six months, as part of NATO's Implementation Force (IFOR).

Below right: With James Champion, shortly prior to our epic ten-day flight home in the Gazelle from Bosnia to Germany via the UK.

In January 1999, I proposed to Rebecca, not quite as planned – on a balcony overlooking the Austrian Alps while fireworks plumed in the skies above us – but in my room in the officers' mess at Middle Wallop, after a takeaway curry, a bottle of red wine and a movie. Thankfully she said yes!

Later that year, Rebecca deployed to Kosovo and I was posted to Fort Rucker, Alabama to learn how to fly the fearsome Apache attack helicopter – and become a gun pilot.

Saying farewell to the 1-227th Aviation Battalion in Fort Hood, Texas. Our time with the 1st Cav was marked by good times and great friends. It was hard saying goodbye after three exciting years of travel and adventure. Little did we know we'd be back in Texas eleven years later, this time working at NASA's Johnson Space Centre.

There is life before children and life after children – every parent knows this. I'm often asked: Did spaceflight change my perspective? Absolutely it did. But did it change my perspective as much as becoming a father? Not even close.

In 2005 I became a Test Pilot at Rotary Wing Test Squadron at MoD Boscombe Down, Wiltshire. Here I conducted some of my most challenging and dangerous flying, including high-altitude trials in Arizona and icing trials in Nova Scotia testing the Apache, and a mission to Afghanistan with my buddy Andy Ozanne (below right) testing the Mi-17 helicopter to assist Special Forces.

Less than forty-eight hours after being told I had made it through their demanding astronaut selection, I was at ESA's HQ in Paris for the big press announcement. With fellow 'Shenanigans' (from left to right): Luca Parmitano, Alex Geist, Andy Mogensen, Samantha Cristoforetti, me and Thomas Pesquet.

Survival training to be an astronaut – in case of an unplanned landing.

Getting to grips with the Soyuz spacecraft – and a manual all in Russian.

With Samantha, about to observe open heart surgery as part of our medical training.

Spacewalk training – physically and mentally challenging. One last look at the checklist.

Rebecca headed off to Macedonia, en route for Kosovo, in late February 1999. For her, it was an incredibly exciting time – swept up in a rapidly evolving situation that was headlining global news and with her unit at the spearhead of NATO forces. This may have seemed like a small-scale conflict between the Federal Republic of Yugoslavia and the Kosovo Liberation Army, but in reality it was the stage for a much bigger showdown between NATO and Russia, culminating in a tense stand-off between the two sides after Russia beat NATO in the race to Pristina Airport. A potential global conflict was avoided, thanks in no small measure to the diplomatic skill of General Mike Jackson who, rather than taking the airport by force, as authorised by US General Wesley Clark, decided that a stroll in the company of the Russian two-star General Viktor Zavarzin and a hip flask of whisky, was the more prudent approach. I was at a dinner night in Washington DC with General Jackson a few years later when he spoke at some length about this event. As it happens, Rebecca was one of the first NATO units to arrive in the vicinity of Pristina Airport – witnessing an uneasy scene where Serb soldiers were sitting around with Kalashnikovs waiting to be extracted, Russians were organising defensive positions around the airport and NATO troops were waiting for an order.

Back in sleepy Middle Wallop, all I could do was watch the situation developing, hearing about the artillery strikes hitting NATO positions in Macedonia and hoping that Rebecca was safe. It was an unusual position for me to be in ... the one left at home, waiting for news and hoping everything would be OK.

That said, I had important matters to deal with, too – matters just as serious as the future security of the Balkans. I had an engagement party to host. Rebecca's and my families had not yet met, so we had put a date in the diary for a gathering in April at Rebecca's parents' house in Comrie at which we could celebrate the engagement and introduce the two families to each other. Of course, events in Kosovo partly scuppered that plan. Come the day, Rebecca could be present only in the form of her engagement ring (which she had yet to see, let alone wear) which sat on the coffee table while I made awkward attempts to introduce everyone.

All that aside, the afternoon was quite a success – not on account of my silky-smooth social skills but largely because both family groups hit it off straight away. I think it may have helped that in the Five Nations rugby championship (in its last year before Italy joined to make it the Six Nations), England lost by a solitary point to Wales that day, thereby handing Scotland the championship on points difference. This was the only time I have been able to reconcile myself to an England rugby team being beaten. I firmly believe my family would have met with a far cooler reception north of the border if it had finished otherwise.

These were the days long before Skype or Zoom calls. Rebecca and I didn't have mobile phones at this point, and we didn't even have email. I couldn't call her, but she could ring me using her military call card which gave her ten minutes each week. She would go to a derelict factory's office block where there were three semi-private phone cubicles for the officers and soldiers to use. The plan was that she would call her parents' house that afternoon during the party, and at least she would get to hear everyone shout their greetings to her.

The phone didn't ring, though. The cake and the sandwiches disappeared, the guests came and went – still no call from Rebecca. I was beginning to worry that something big had kicked off. Either that or she had gone off the idea of marrying me. Neither of those was a particularly welcome thought. Three long and anxious hours after everyone had left, Rebecca finally managed to get through, having been dealing with ammunition, food, water and fuel supply issues all day. I was still there with her parents. Obviously it was an emotional call. Quite unfortunate to miss your own engagement party. Even more unfortunate to miss it twice.

I finally managed to give Rebecca her engagement ring a couple of months later, when she returned to the UK briefly for a week of R&R. After all our snatched weekends, a full week felt like a real luxury. Even so, I often wonder about the value of R&R during a long tour. On the one hand, it is a chance to see the people you love and get a break from the stresses of military operations. On the other hand, it can take a lot longer than a week to make the mental adjustment back to normality and it's sometimes hard to come down only to have to pick yourself back up again. And if it's a war zone you're heading back out to, you're letting yourself in for the mother and father of all 'Sunday night feelings' when it's time to pack up and head out again. As unsentimental as it may sound, sometimes maybe it's better just to plough through.

Assuming it had been practically possible, would I have welcomed the chance to pop home from the Space Station for a weekend? I knew, when I was up there, that I was in a good place; completely mentally focused, constantly alert to danger, yet feeling healthy, rested and able to function effectively with some spare in the tank. Coming back would have thrown everything up in the air again. I wouldn't have wanted the disruption and I suspect my family wouldn't have wanted it, either. I missed them desperately while I was up there and there were plenty of times when I ached to see them. Yet in many ways it was probably better that nipping back wasn't an option.

Anyhow, one day Rebecca's boss in Kosovo told her she would be able to go out on R&R for a week starting the next morning, and even though there was no notice, it felt like too good an opportunity to turn down. So, at the end of the working day, Rebecca rushed to stuff a few things in a bag and get ready. At the time she was in charge of a platoon of sixty on a military camp with no running water. Bathing involved sitting in a large plastic bucket with her knees up while a helpful female corporal tossed buckets of water over her head; and shaving involved dangling a leg out of the tub and wielding a razor blade by the light of that corporal's head-torch. Bath-time complete, she ran for a Chinook leaving Kosovo for Macedonia, and flew from there back to the UK.

It happened to be the weekend of 'Music in the Air', an annual summer event at Middle Wallop. So, a day that had started for Rebecca in a rain-soaked campsite in a war zone ended with her sitting with me on a picnic blanket on a warm Wiltshire evening, sipping champagne and watching hot air balloons and a display of vintage aircraft to the sound of the London Pro Arte Orchestra. Surreal doesn't begin to describe it, really.

Once Rebecca had made the adjustment, though, it was a rare and precious week together and it culminated – quite fittingly, I thought – in me promoting my future wife from lieutenant to captain. Ironing her uniform for her on the night before she went back, I secretly buttoned on a set of my captain's pips. The next day, back in Kosovo, Rebecca unknowingly wandered around the place as a captain for quite a long time before someone asked her, 'Ma'am, did you get promoted while you were away?' You can get punished for wearing the wrong rank, so I'm not sure how well the joke went down. But it had to be done.

There was still time on Rebecca's Kosovo tour for me to send her a care package containing that new-fangled item, a mobile phone, and, with it, a European SIM card. Finally, for the first time, we were able to text each other. I also put some tangerines in the parcel because I thought it would be nice for her to have some fresh fruit. With the package sent off, I sat at home in Middle Wallop, warmly imagining how pleased Rebecca would be about that little extra touch.

Over in Kosovo, Rebecca was staring at those tangerines and thinking, 'I'm marrying a man who doesn't know I don't like oranges.'

CHAPTER THIRTEEN

Deep in the heart of Texas, all the way to Mother
Rucker in a dodgy motor, and if you ain't Cav, then...

On Rebecca's return from Kosovo, we managed a couple of weeks together before I had to depart for the USA to begin my three-year exchange posting to the 1st Cav. It was September 1999 and the plan was that Rebecca would join me soon after we were married the following April. I flew into Washington DC, met the military attaché, attended a drinks reception at the British ambassador's house and completed my 'inprocessing' with the British Embassy. And then I headed down to Fort Hood, Texas, and checked in to the sumptuous luxury of the three-star Hampton Inn in Killeen, which would be my home for the next fortnight while I completed my handover from David Amlôt. Meanwhile, Rebecca was posted from Germany back to the Army Training Centre, Pirbright, serving as a platoon commander and training new recruits.

I already knew David really well from our potholing, kayaking and climbing adventures during the Pilots' Course. We had also served together in Northern Ireland and I had been at his wedding to Norma, a notoriously colourful occasion which I will have cause to come back to a little later. He and Norma were now living in suburban Harker Heights, just over from Killeen, in a house that I would inherit when they left and in which Rebecca would eventually join me. It was a simple place in a fairly non-descript area, but it was nice enough and spacious by comparison with most British Army married quarters. I arrived on a Friday night and first thing the next morning, David picked me up and took me in for my first look at Fort Hood.

As a British Army captain arriving from Middle Wallop, it was hard not to be taken aback by the enormity of the place. Fort Hood is one of the largest military bases in the world, sprawling out across 200,000 flat acres and with about 50,000 people living and working on it at any time. You could drive for miles along Battalion Avenue or Tank Destroyer Boulevard, passing endless artillery units, tank battalions and armoured fighting vehicles. It represented one of the greatest concentrations of US military hardware and yet, that first Saturday, in shorts and T-shirts, David and I drove onto Robert Gray Army Airfield unchallenged, parked, casually wandered over to an Apache helicopter and climbed up to sit in the cockpit so that I could get a feel for it.

That light-touch approach to security would change in a split second not very long after this, while I was still in the country – on September 11th 2001. However, that day we were more than free to look around, and then every day for the next fortnight, on a more formal basis, I would go in to work with David and he would show me the ropes and help me get to know the place. I began to meet my new US battalion – the battalion commander, the company commanders and so on. Eventually, David left me to it and returned with Norma to the UK, although not before he had sold me his car – a giant, whale-sized Ford Taurus.

It looked sound enough. I said to David, 'I'm going to need this car to get me up and down the I-10 to Alabama. That's a twelve-hour drive. Is it up to it?'

'Absolutely,' he said. 'Completely reliable.'

His word was good enough for me.

On the Monday morning, my first official day at work, I put on the British Army flying suit that I would continue to wear throughout this posting, but now with a new 1st US Cavalry patch sewn to its right shoulder. It was the combat version of the division badge – a black silhouette of a horse's head above a thick black diagonal line on a green background. A piss-taking Marine instructor would later say to me, 'It's the horse you'll never ride and the line you'll never cross.' To which I gave the expected Cav retort: 'If you ain't Cav, then you ain't shit!' It really didn't sound quite right with an English accent. But I felt very proud to wear that patch nonetheless. I was aware, too, of the pride in the badge of those who wore it all the time – a pride in the 1st Cav's aviation history, its links with Hueys and Cobras in Vietnam and its indelible association with *Apocalypse Now*. I was proud, too, to be given the traditional broad-brimmed black 1st Cav hat, although the truth is I look terrible in hats. I stuck to my Army Air Corps beret.

And thus attired, I used my new, completely reliable Taurus to drive over to the airfield and hook up with my battalion. As I left the car and walked towards them, some of these soldiers seemed to be sniggering, which I thought was a little rude in the circumstances. Then I worked out that they weren't laughing at me – they were laughing at the car.

'You didn't buy that off Captain Amlôt, did you?' one of them asked.

I confessed that I had.

'It was always broken. He was forever tinkering with it. He was famous for it. I'll give it a week before it claps out on you.'

This wasn't the best news, and naturally I was a little wary of the car from that moment on. But actually it outlasted that first week and was still running a few days after that when I went to pick up Rebecca from the airport at Killeen. She was on leave and had flown in via Houston so that we could spend some time together and she could look around and get a taste of what lay ahead for her in April when she came out permanently.

I was a little anxious about those first impressions. I wanted them to be great and nothing but reassuring, of course – exciting, even. But I knew that Killeen Airport was not an especially attractive place to pitch up in. I was grateful at least that Rebecca's plane was touching down late at night. It meant that our drive back to the house could take place under the cover of darkness. As it happened, the dark only served to show up even more brightly the strip joints and tattoo parlours, and the drive-up alcohol stores with their barred windows and sprinkles of broken bottles outside them that lined the route back towards Harker Heights.

And then there were the locusts. Those didn't help, either. They were swarming pretty thickly that night in the heat of a Texan summer. We pulled in to a grocery store to pick up some food, and insects the size of teaspoons were dive-bombing us in the car park and battering themselves against the store's illuminated windows as we scrambled inside for cover. Meanwhile Rebecca's curly hair was suffering an instant reaction to the humidity, growing bigger and bigger in front of my eyes, and representing an increasingly large insect trap. This was not, I sensed, the best of welcomes to her future home. I thought I could detect some alarm in Rebecca's eyes, and it was no doubt matched by the anxiety in mine.

Still, things got better over those two weeks. The plague of locusts lifted. Rebecca's hair returned to its normal size. Harker Heights, and our proposed new house, was much nicer than the area around it – entirely without broken glass or strippers. And we even managed to get out into the countryside and go camping for a couple of days. Then Rebecca went back to work at Pirbright – still, to my relief, intent on coming back in April – and I promptly got shipped out to Alabama for six months. At least, it was meant to be six months but, for reasons we'll come on to, it turned out to be for longer. Indeed, I would still be there when Rebecca returned.

The plan was that I would be trained to fly Apache Longbow, the newest Apache helicopter and the one which the British Army was on the verge of acquiring. But Longbow was so new at this point that there was no course designed yet which could take a pilot, like me, who hadn't flown an Apache before, directly onto the Longbow. I would need to learn first on the Alpha model Apache and then afterwards do a shorter Longbow conversion course, bringing me up to speed. And the best place for me to do all of that was in the aviation teaching centre at the legendary US Army base, Fort Rucker in Alabama.

So I drove due east from Texas for more than 800 miles, across Mississippi to the Yellowhammer State – the first proper road-test for my still apparently trustworthy Ford Taurus. This, alas, was when the car began to bear out the dire predictions made for it a few weeks earlier. Somewhere on the journey, the engine decided to overheat. I pulled in at a restaurant and went inside for lunch while it cooled down. When I returned, it took me the best part of an hour to get the car restarted. That dissuaded me from switching off the engine unless I really had to. I kept it running even when I stopped off for a drink. Needless to say, before long the temperature gauge was once again indicating heat under the bonnet at volcanic levels. I decided to ignore it. Somehow I made it all the way to Dothan, Alabama, the town closest to Fort Rucker – and just in time. A short way inside the city limits, the car spontaneously combusted.

With clouds of thick smoke billowing from beneath the hood, I did the only thing I could: I got out, taking my bags with me, and let the car burn by the side of the road. Then I called a number for a recovery service and asked them to quote me their best price for carting a charred wreck to the dump. And that was the end of David's reliable Ford Taurus.

Anyway, I now needed another car. I knew Rebecca had always wanted a Jeep Cherokee so, thinking it would make life a little nicer for her when she eventually joined me, I visited the Jeep dealership in Dothan. They were very keen to sell me something on credit, but a little flummoxed by the fact that I was a Brit who had been in the US for less than a month and didn't even have a fully-functioning bank account at this point. But then I showed them my US military ID and the picture completely changed.

'Excellent,' said the dealer. 'We have a deal with the military. Let's get this done.' Next thing I knew, I was a foreign national with an HP credit arrangement on the basis of absolutely no visible financial history whatsoever, and I was driving away in a brand new Cherokee. God bless the USA.

Fort Rucker itself was a sight to see. The US military refer to it, somewhat inevitably, as Mother Rucker. It was only a quarter of the size of Fort Hood but that still meant it was vast by British standards, an entire town with its own grocery store, flower shop, library, golf course and even a vet's clinic, all spread out across 63,000 acres of gently rolling woodland. Meanwhile the quantity of the military hardware on site made your head spin. I would report for class in the morning and there would be anything up to sixty Apaches lined up on the dispersal just to train students – almost as many on that one patch of tarmac than the British Army would be buying in total. Apache had its own dedicated airfield – Hanchey Army Heliport. There was another field further along the road for Blackhawks, another for Kiowa Warriors, another for Chinooks ... It could take thirty minutes in the morning to drive to work from your quarters, weaving through the woods and around the various airfields.

There were ten of us on my course – one other officer and eight warrant officers. We were taught by civilian instructors, ex-military guys who were now on civilian contracts and all of whom really knew their stuff. I went into this course as a twenty-seven-year-old QHI with 1,500 hours of flying behind me. Most of my American classmates were at a different stage in their career – younger and less experienced as pilots, converting to Apache soon after their basic Pilots' Courses with maybe only 200 or 300 hours under their belts. I possibly got a little more respect from the instructors for being an instructor myself and because they knew I wasn't wet behind the ears. At the same time, we were all students, all there to learn a new aircraft and all in it together.

Each morning you would stop off on your way to class to pick up the weather data for the day, and then it would be classroom work for a couple of hours, followed by your flying sorties. Our group was just one among very many, on various courses in various aircraft, all operating on staggered shift times, including at night when you eventually switched to night-flying. It was astonishing to think how many students were passing through Fort Rucker at any one time. It was like a giant, superbly coordinated, pilot-making machine.

Perhaps as an extension of that machine ethos, there was quite a lot of learning by rote. You were taught to recite checklists and procedures verbatim off the page and instructors weren't quite so interested in your understanding of what was going on behind the checklist – what might be happening in the fuel system at certain moments, say, or in the software. This approach applied to practical work, too: so, if you were practising a single-engine failure in an Apache and you didn't hit the prescribed numbers – if you

219

were plus or minus 10 knots or plus or minus 100 feet – irrespective of whether you got the aircraft down successfully, you were downgraded for the exercise and made to repeat it until you were within the defined parameters. I could see and experience for myself how efficient this tight, rote-learning approach was. But I felt more comfortable, overall, with the British attitude which had a bit more space within it to encourage thinking beyond the checklist.

What definitely opened my mind, though, was coming into contact with the Apache helicopter. The Apache was something else. It wasn't subtle, and it wasn't pretty – in fact, it was a big ugly brute of a machine. When you did your pre-flight walk-around, it was robust enough that you could clamber all over it – not something you would have considered doing with a Gazelle or a Squirrel. Neither of those are fragile aircraft but you wouldn't go so far as to climb all over them. In an Apache, the panels would take your body weight and there were steps built into it so that you could climb up and open canopies and properly inspect inside the thing. There were sprung aluminium foot rails in the tail fin which popped out to create a ladder so that you could climb up to the top and check the tail rotor assembly and be sure that everything was OK. And then you could walk up the catwalk and check over the hulking great gearbox, and on either side you could open nacelles and check the engines. So even while you were doing your pre-flight checks, you were manhandling this thing and treading on it and getting a sense of its size, its weight and its solidity. It was easy to see how every part of the aircraft had been designed to withstand a hail of 12.7-millimetre bullets. It was like a tank in that respect – a tank with wheels.

Before this, I had only flown helicopters that were on skids. But the Apache was on wheels, meaning you could pull up a bit of power to lift the aircraft just enough to make it light on its rubber, and then unlock the tail wheel. Then you could steer using the pedals and ground-taxi out to the pad, just like being back in a Chipmunk again, only ten times heavier. And if you had a full load and the aircraft was too heavy for a vertical lift into the air, the Apache could do a running take-off, just like an aeroplane.

So it was a very sturdy animal. But it was also perfectly put together in almost every respect – a pilot's helicopter. You clamber into the cockpits of some aircraft and they may be great flying machines but they have clearly been designed by engineers who have never flown. Nothing is where you want it to be and, in terms of the ergonomics, it's essentially as though someone has just vomited in there after an unwise meal of switches and levers. The Apache, though, had self-evidently been put together by pilots, for pilots, and everything seemed to be exactly where you wanted and expected it to be.

And then the best thing of all: it was absolutely beautiful to fly. It was a tandem seater, with the pilot seated to the rear, sitting high up with great visibility. In a Gazelle you had always felt you could reach out the window and touch the ground, but here you needed to take two big steps up just to get to the pilot's canopy. It was as comfortable a position as any I had experienced in a helicopter – and would remain so, in fact. There was a lot of weight high up, just under the rotor mast where the gearbox was, and with the engines high on either side, too. This meant it was dreadful going across the ground – wallowy and spongy on its narrow undercarriage. But that weight distribution also made it perfect for flying – fantastically responsive and manoeuvrable in the air.

And as for the power ... well, that, too, was a whole new ballgame for me. The Gazelle was a single-engine helicopter with a battery starter. Now I was in something with twin-turboshaft engines, plus a small auxiliary jet engine to get those two engines started. Those twin-turboshafts could easily lift the Apache's ten tonnes when fully laden. That high-pitched whine that I had grown used to from the Gazelle – the scream of the 'screaming chicken leg' – was replaced by a low resonant throb in the Apache.

Perhaps the trickiest thing to get used to at the start was the monocle – the helmet-mounted display. Pilots don't tend to like things near their eyes when they are flying, but the Apache insists on it. Your left eye is unobstructed but your right eye looks through a circular lens, like a Victorian toff's eyeglass, except that it's beaming you the numbers on your airspeed and altitude and every other piece of in-flight data that you might need – your symbology, as we call it. The joke is that Apache pilots eventually turn into chameleons – capable of separating their eyes and looking two ways at once. There's some truth in it. The monocle takes some getting used to, though. In daylight, the image has a slightly pink hue, and it takes a bit of practice to see through it and yet read the symbology at the same time.

At night it becomes even more complicated. In addition to symbology, the monocle displays the image from a forward-looking infrared camera, which is slaved to your head movement. The camera is positioned 10 feet in front of you and 3 feet below you, so the challenge is to marry the two images that are coming into your brain from different eyes looking in different places. You also have to learn to move your head at the right speed: turn too fast and there will be a lag while the camera catches up with you, and then maybe it will overswing before correcting. Do that a few times and you will very quickly start to feel sick.

Now the camera is feeding you an infrared image. That was fine if it was cold or frosty outside, because the camera would have lots of nice thermal variants between buildings, runways and grass, or whatever. If it was a muggy Alabama night, on the other hand – or worse, if a rainstorm came through – the temperatures would equalise and the lovely picture you were trying to fly with would turn green, leaving you saying, 'Crap – I can't see anything' (or words to that effect). And then you were left flying almost entirely on symbology – relying on the numbers to tell you where you were in relation to the ground and other obstacles.

Now, it's a testament to the Apache that the symbology was so good that you could afford to lean on it to that extent. In that Alpha model, it was its saving grace, really. I knew eventually that if I could line myself up on the runway, even though I couldn't see very well, just by using the symbology I could put myself down on the ground. But the first-generation infrared camera was dreadful – really tricky to work with. It got better. Indeed, the next generation infrared camera was immense – three times the magnitude of the first and capable of seeing through cloud and rain. Years later, as a test pilot, I remember having a go with that new camera and zooming in on an object several kilometres away. What was that 'hotspot'? Some kind of moving vehicle? A lorry, maybe? The Boeing test pilot I was with at the time quietly said, 'Tim – come back outta' field of view. You're looking at a cow's leg.' And I was. Compared with the camera I learned on in Alabama, it was chalk and cheese. I used it to fly from Boscombe Down to Yeovilton one time in the thickest fog – fog you could serve with a spoon. The Westland's pilot sitting in front of me spent the whole journey staring glumly into an impenetrable murk. Behind him where I was, using the new camera, I was seeing pylons, telegraph poles, trees, houses ... It might as well have been daylight back there.

But that was off in the future. As an Apache novice in Alabama, most of the time, while flying at night, I would be working 30 per cent on the camera picture and 70 per cent on symbology. Flying by numbers. Accordingly, getting to grips with night-flying in an Apache took a long time – many hours spent in the daytime doing what they called 'the Bag', with the pilot's part of the cockpit blacked out with rubber linings and kitchen roll stuffed in the gaps to simulate total darkness, and with your instructor sitting up front in the daylight while you wrestled as bravely as you could with your infrared image and your numbers.

Of course, there was another all-new element of the Apache for me, over and above the monocle: its weapon system. It had a 30-millimetre cannon, it carried rockets, it had sixteen Hellfire missiles attached to it. Again, such things had been no part of my experience piloting Gazelles and Squirrels. I remember my first time on the range in an Apache, firing the 30-millimetre cannon. The instructor told me to hone in on a target which was laser-ranged at 2,000 metres. I was thinking to myself, 'Really? I'm trying to shoot something 2 kilometres away, with a gun?' The cannon, like the camera, was slaved to my monocle eyepiece; I aimed it merely by looking in the right direction, placing the crosshairs over the target and keeping my gaze steady. I pulled the trigger and there was a low thud-thud-thud as the cannon fired a ten-round burst, rocking the helicopter backwards slightly with the recoil. But even as the helicopter was rocking back it was re-calculating the ballistic algorithms to keep each shot in the right place. After that, nothing happened. Nothing continued to happen for what seemed like an eternity but was probably about six seconds. I'd already given the shot up as a wild beginner's miss. Then, 2 kilometres away, there was a flurry of small explosions and the target was completely obliterated.

When an aircraft is carrying that kind of equipment, its primary purpose is pretty clear. The Apache doesn't come with stretchers for CASEVAC. It doesn't have an underslung hook. It's not there to be a glorified rental van and carry heavy objects around the battlefield. It's built to go out and destroy things. It's an attack helicopter.

As odd as it might sound, this was not something to which I had given an enormous amount of thought before arriving in America. I had gone over there looking at this opportunity very much from a pilot's point of view – the exciting chance to fly and learn about this incredible, complex twin-engine machine, and then to bring that knowledge and experience back to the UK. And straight away I was made to realise, not just how different the aircraft was from anything I had previously flown, but also how different the role of that aircraft's pilot was. The course entirely clarified this, right from the off. It was about the nuts and bolts of flying a new aircraft, obviously. But it was also about instilling a particular mentality – for me, a completely new one: the mentality of a gun pilot.

The American instructors on the course well knew this. The message was constantly reinforced, in ways which ranged from subtle to blatant. I don't think there was a single lesson that went by where the Apache's destructive potency wasn't in some way referenced, underscored or at least alluded to. You might be talking about something ostensibly unaggressive, like, for instance, how to do a sloping ground landing in the Apache – which is especially tricky on account of those spongy wheels and that narrow undercarriage. And there would be some reference to taking extra care of the missiles on the outboard section – how those were the most important thing on this helicopter, so you wouldn't want to go damaging them, now, would you? And it would occur to me that I had been learning about sloping ground landings for years but nobody had ever told me that the point of them was to keep my missiles safe.

I could see what the instructors were doing: they were preparing this group of young American pilots to be completely comfortable in the role that they would be expected to carry out. And if they didn't like it – well, there's a Blackhawk field down the road, a Kiowa Warrior field, a Chinook field ... By all means, go off and fly those aircraft. But you're at the Apache field, and this is what we do.

It's not like the ethos was in any way a secret. Back in Fort Hood, my battalion's motto, after all, was 'Attack', which offers a pretty clear clue. On top of that, each company had its own nickname: Vampires, Reapers, Avengers ... When I eventually became a platoon leader, my call sign was Vampire 16. Nobody seemed to be calling themselves Fluffy Kitten. Absorbing all this, one of my first thoughts was how much of a change this was going to be for the Army Air Corps, when Apache eventually arrived in the fleet. We didn't go around in the Army Air Corps talking about being killers and vampires. We didn't talk about being fluffy kittens, either, but even so. This was very different from being an AAC pilot on attachment in Northern Ireland. This was different from thinking, 'If we get into a skirmish, I might have to fire a weapon.' This piece of kit that I was being taught to operate could destroy an entire village, and that might be your job: not to wait until you were shot at, but to go out there with a target and a missile load, and to be the one who pulled the trigger first. To be a gun pilot.

Not everybody can reconcile themselves to that. Some time later, when I was an instructor on Apache back in the UK, an AAC colleague took himself off the Apache course. He loved the Army and he loved flying helicopters, but now he was being asked to operate Apache and he realised that this was not a role he could commit to. He simply couldn't countenance sitting there on a battlefield surrounded by all that firepower, and having to employ it. And I admired him for standing up and stepping out, because that takes guts, too.

I remember watching the documentary film *Spitfire: The Plane that Saved the World*. It included interviews with pilots who had flown that iconic aircraft in the Battle of Britain in 1940, and many of them spoke about how beautiful it was, how manoeuvrable, how brilliantly designed – but yet how it was also, at heart, a killer. It was designed to shoot down other aircraft. That was its only purpose. And some of those pilots were clearly still, all those years later, conflicted about that, and remorseful. But at least one very openly admitted that he had enjoyed that aspect of it: that he had been eighteen or nineteen at the time, that it had been a thrill. It would be possible, I'm sure, to uncover the same range of feelings in pilots who fly Apaches.

It was something I had to think hard about – a mental transition I had to make. And I'm not saying I was completely comfortable with it, even after thinking it through. But I ended up choosing to embrace it. I thought, 'OK, this is the role of this helicopter, this is serious. And if this is what we're signing up for, then we've got to do it properly, to the best of our abilities, and we've got to know the job inside out.'

As it turned out, push never came to shove for me in that way. I never flew an Apache into combat. In time I would be deployed to Afghanistan, but for that mission I would be flying an altogether different aircraft. I went on to become a test pilot, following my passion for understanding aircraft, how they work and pushing them to their limits. That was the area to which I felt best suited and where I was most fulfilled. At the same time, for these years in America when I was an operational, front-line Apache pilot, I entirely accepted what my job was and what it might mean. I couldn't have been in the role if I hadn't.

I settled into the course, and I settled into the rhythm of life at Fort Rucker. There was a lot of hard work, but also, thankfully, some free time. I would go running in the woods, where I was warned to keep an eye out for the snakes and, near the lakes, the alligators. (I saw the occasional snake, but no alligators. I had to go on an official swamp tour to see them.)

South of Fort Rucker was a small town called Daleville, which styled itself 'Gateway to Fort Rucker', although everybody on the base preferred to think of it as 'Dalevegas' on account of its one tiny strip mall – about as far from high-rolling Nevada as you could imagine. Here there were a couple of dingy sports bars offering scuffed pool tables and cheap Bud Light, an aviation store that sold notebooks, pens and flying jackets, and a twenty-four-hour café which did extremely well from all the pilots finishing their shifts at odd hours and needing eggs, hash browns and weak coffee on tap.

Alternatively you could head to Dothan, once movingly described by somebody or other as 'that map dot town'. Dothan boasted a solitary nightclub. It also boasted in November, not long after I arrived, the annual Peanut Festival. As they say in Dothan, 'They don't call us the peanut capital of the world for nothin'.'

It turned out to be ten days of funfair rides, open-air concerts and a Miss Peanut Festival contest. It was about as far from an English village fete as you could get – I was just grateful that I didn't have a peanut allergy.

In the main, though, people on my course seemed to be making their own entertainment. It was an entirely male environment – there were no women on the Apache course – and my fellow students were mostly good old boys from South Carolina and Georgia whose idea of an evening out was to sling a cool-box filled with beers in the back of the pickup, head out to the deserted airfields, do a few doughnuts in the dust, crack open the beer and watch the sun set from the tailgate. I went with them to do my share of that, too, getting in touch with my inner redneck. It was a lot of fun.

We were approaching Christmas. All was going along smoothly, my understanding of the aircraft was expanding rapidly. And then suddenly the course got halted. There was an accident somewhere in the US Apache fleet that disclosed a problem with the tail rotor drive shaft which required extensive investigation. As had happened with the Lynx fleet in Germany that time, the entire Apache Alpha fleet was temporarily grounded. The hiatus ended up lasting the best part of a month. Early on, a few of us seized the chance to get in a minibus and go skiing in the Blue Ridge Mountains for a few days, hiring a cabin with a hot tub. By the New Year, though, my Alpha Apache course had resumed and at the end of January, having passed my check-ride which, naturally, included firing multiple rockets and a few hundred rounds of 30-millimetre, I became a qualified gun pilot.

Of course, it wasn't long before there was a wedding to plan. In common with so many aspects of our relationship up until that point, this planning happened across a great distance, by phone. And though I would love to claim that I had a crucial input into key decisions on flowers and table decorations, I should further hold my hand up and confess that the bulk of the arranging was done by Rebecca, along with her mother Maddy. Anyway, April 2000 soon rolled around. Since timings were tight, an exotic stag night in a premium European location was out of the question. Instead my mates chose a premium location closer to home: Swindon. Why Swindon? I have no idea. What I do know is that we checked into the King's Hotel, spent the day go-karting and clay pigeon shooting, had a meal back at the hotel and then headed out into town to descend on one of the top-ranking nightclubs for which Swindon may or may not be famous.

Prior to dinner I was ambushed, obviously. The guys thought it would be funny to dye the groom's ginger hair black. I would have found it funny, too, if it had proved possible to wash the dye out completely by the wedding day, which it didn't. I was also made to wear a dress, make-up, heels and fake boobs for the evening. Fortunately those *could* be removed in time for the wedding.

It was brought home to me in the course of that stag night that having a doctor as a best man is both a blessing and a curse. On the one hand, Ian had been there on more than one occasion to stitch me up (quite literally) in the wake of possibly alcohol-related injuries, and no doubt he would have come to my rescue again had it been necessary this time, which mercifully (and actually rather surprisingly) it wasn't. On the other hand, he was also someone with easy access to anaesthetic gel and medical grade laxative. By the time dinner ended, my mouth was so numbed by the anaesthetic smeared on my glass that I was a dribbling mess and wouldn't have known if I was drinking water or rocket fuel. And as for the laxative ... well, let's just say the following morning was a write-off and not just because of the appalling hangover.

I shouldn't complain though – I later discovered that I had got off lightly. One of my best mates, Matt Sills (since tragically killed in a motorbike accident), was an RAF officer. He had been our air traffic controller at Gütersloh. We used to wind him up when he was talking us down on a radar approach by slowing the helicopter down and then flying back up the approach the wrong way. I'm not sure you could get away with that these days but he was remarkably laid-back about it all, and highly tolerant of our juvenile ways. Anyway, I discovered that, for my stag night, Matt had arranged for me to be delivered from that Swindon nightclub, in an alcohol-induced stupor, direct to RAF Brize Norton, where I was to be loaded onto a C130 military transport plane that was due to depart for ... Ascension Island, 4,200 miles away.

This hare-brained plan was dangerously close to being implemented before Ian pointed out that, given the unpredictability of a return flight, there was minimal chance of me actually making it back to the UK in time for the wedding and Rebecca would never speak to him again if I didn't. So as it was, instead of waking up with a pounding head in the middle of the South Atlantic Ocean, I merely woke up with a pounding head in my hotel room in Swindon, not really having much idea what had happened since dinner, but definitely very keen to race to the loo.

We were to be married in the beautiful fifteenth-century St Salvator's chapel at St Andrews University in Scotland – a privilege available to Rebecca as a graduate. My family and many of our friends travelled up from England the day before the ceremony, which happened to be my birthday. That night we all gathered in St Andrews, in a bar called Ma Bells where, despite the occasion, I resisted the temptation to celebrate too much, considering that I might welcome some clarity in the morning. Our mates were under no such restrictions of course and the party roared on long into the night.

It was a military wedding so, come the ceremony, all my military mates were in No. 1 dress (blues) – a colourful mix of Army, RAF, Royal Navy and Royal Marine uniform. No. 1 dress also means full medals, swords with leather scabbards, Sam Browne belt and brown gloves. As groom, I wore a red sash in place of the Sam Browne belt, white gloves, gold epaulettes and a silver scabbard. Most regiments have peaked caps with No. 1 dress but the Army Air Corps is one of the few (along with the SAS, Paras and Intelligence Corps) that always wear berets. Rebecca declined her right to wear officer garb and instead reduced the room to sentimental tears in a traditional and utterly stunning ivory gown with veil and train.

The service passed off without a hitch. We had a lovely Scottish minister who persuaded us that it would be nice to memorise our vows and not do the 'repeat after me' thing. Ever the good student, I had my lines off pat a week before the day, whereas Rebecca hadn't even looked at hers with two days to go. Gallingly, it made no difference to our relative performances: neither of us stumbled and I, Timothy Nigel Peake, successfully took her, Rebecca Jane King, to be my wife forever more. I was, I knew, the luckiest man on Earth, and if anyone was fortunate not to be on Ascension Island that day, it was me. Then we headed out, through the traditional guard of honour with raised swords, into the quadrangle under a (remarkably, for Scotland) dry and bright April sky.

That was the moment Claire Mason, one of Rebecca's girlfriends from Sandhurst, arrived, looking more than a little breathless. She had been at a mess dinner in Wiltshire the night before, had woken late and missed her flight, and had then scrambled to rearrange plane tickets and hire cars to get to St Andrews. She now joined us outside the chapel wearing a damp dress (it had been rapidly washed after the night before) and a crushed hat. It's a good job she did though, because that's where she met my best man, Ian, and three years later they were married. That bright April sky I mentioned was perfect for flying which was convenient because, without telling anybody else, I had arranged with some mates at 7 Regiment (TA) Army Air Corps based at nearby RAF Leuchars to do a flypast with a couple of Gazelle helicopters, since they were training that day anyway. That seemed to go down well, catching everybody by surprise as low-flying helicopters tend to.

During the reception, a food fight broke out – maybe because we had chosen chicken stuffed with haggis. I decided it was time to quash it when a carrot landed in Rebecca's grandpa's glass of wine. This was closely followed by two of my mates who shall remain nameless (Jimmy James and Toby Everitt) leaving the table to do a quick 'naked dash' around the quadrangle. (I don't know why, but it was pretty normal that someone would get naked at some point back then.) We had decided it would be fun to leave a disposable Kodak camera on each table, so that everyone could take pictures during the afternoon and evening for us to enjoy and share later. That was a mistake. The majority of these snaps were not suitable for consumption by anyone, although as collateral to be held against my mates they have proved useful since.

After the reception we headed upstairs to the bar and the dance floor for Scottish reels, with music provided by Hamish Reid and his Ceilidh Band from Rebecca's village. Since Rebecca and I were among the first of our friends to get married, there were plenty of single ladies (mostly Rebecca's Army girlfriends) and gentlemen present, and all could hardly have gone more swimmingly during the Gay Gordons and the Dashing White Sergeants. I had arranged for a vintage Rolls-Royce Silver Shadow to take Rebecca and me to the hotel where we were staying that night. The name of the hotel was a closely guarded secret: Rebecca's mum was the only other person entrusted with the information. This was for a very good reason. I mentioned a little earlier the wedding of David Amlôt, my Army Air Corps mate and predecessor at Fort Hood, to his Northern Irish fiancée Norma. This was when we were all serving together in Northern Ireland at 5 Regiment AAC in Aldergrove. The officers in Aldergrove were a pretty lively bunch who tended to let off steam in fairly prodigious quantities and knew the opportunity for a practical joke when they saw one. On David and Norma's wedding evening, as the reception raged, a few of us somehow managed to persuade the hotel staff that we needed some advance access to the happy couple's bedroom. David was a keen climber and amongst his bags was a climbing rope, carabiners and harnesses which he had brought over for some pre-wedding climbs. In addition to the usual tricks, we used the climbing rope to string up every piece of furniture in the room so they were hanging from the beams of the newly-weds' four-poster bed.

Norma was already a bit pissed off that a few of us had relieved the DJ of his duties earlier in the evening (we were entirely right to do so: he was dreadful) and there may have been some naked dancing to Right Said Fred's 'I'm Too Sexy' (remember *The Crying Game*?) which her relatives had been obliged to witness. Then there was the Dance of the Flaming Drambuies, for which Midge Neil had pulled down his trousers and stuck two glasses of ignited whiskey to his cheeks. Perhaps you are familiar with this distinguished tradition: the flames go out pretty quickly but not before a decent vacuum is caused, sticking the glasses firmly to the dancer's bottom. What was more funny than watching it at the time was the fact that Midge couldn't sit down for about three days afterwards, such were the burns. Anyway, as far as Norma was concerned, the bed was the final insult and I was well aware that there was an outstanding debt yet to be paid. Hence my anxiousness that our hotel location should be guarded as closely as the nuclear codes. Fortunately the word didn't get out. I don't like to think what would have greeted Rebecca and me in there otherwise.

Not that Rebecca would have known much about it. With the ceilidh growing more and more boisterous by the minute, I went downstairs briefly to sort out the Rolls-Royce, leaving my bride on the dance floor, enjoying one last Strip the Willow. On my return I discovered a group of people gathered on the floor, huddled anxiously over a semi-conscious bride. In the process of being spun around by her old school friend Jamie Lang, Rebecca had caught her heel on her wedding dress, tripped and broken the fall with the back of her head. And so it was that everyone formed an arch as I supported Rebecca unsteadily on her feet and escorted her to the waiting Rolls, where she promptly conked out and stayed conked out, holding to her head the ice pack that I had secured from reception on the way past.

The following morning, we had breakfast overlooking the Firth of Forth and then promptly decided to go and find all our mates, waking up for brunch in St Andrews. Here we learned how the reception had continued deep into the night in our absence, and that certain newly formed couples had last been seen leaving 'to get some air' on the fabled St Andrews golf course, specifically in the famous deep hollow on the eighteenth hole, appropriately known as 'the Valley of Sin'. There were also rumours of skinny-dipping at West Sands. Quite shocking and unnecessary behaviour, really.

After the wedding, Rebecca and I had two days at a B&B just outside St Andrews. (No, it didn't count as a honeymoon apparently … and nor did the five subsequent camping trips in the US that I tried to pass off as honeymoons.) We then headed south with a car full of wedding presents to be loaded into boxes and sent out to Texas. In my haste to get everything done in time, I was pulled over (yet again) by Dumfries and Galloway's finest on the M6, which brought the points total on my licence dangerously near to the maximum. Three years in the States was going to be good for my UK driving licence, if nothing else.

The next few days passed in a frenzy of packing and organising and saying farewells. And then, pausing only to squeeze in a fancy dress thirtieth birthday party for our friend Andrea Luck (the same Andrea who helped matchmake us in Gütersloh), Rebecca and I boarded a plane bound for Atlanta and set off for a new life together in America.

CHAPTER FOURTEEN

Apaches and a Cherokee, a trip to the bottom of the
Grand Canyon, and the day the world changed

Rebecca's and my first married home together was in the middle of Fort Rucker – or, as the bumper stickers around the place liked to put it, 'Sweet Home Fort Rucker'. I don't know about 'Sweet Home'; it was certainly 'Small Home' – a tiny twin-roomed officer's quarters with a minuscule galley kitchen. If things had run to schedule, we would have been in Texas, where Dave and Norma's long since vacated and far more comfortable house in Harker Heights awaited us. But my Alpha model Apache course had overshot, so there we were, temporarily, jammed together in Alabama.

To compound things, I had finished doing my stint in 'the Bag' and had moved on to actual night-flying, meaning that almost as soon as we touched down and moved in after the wedding, I was heading out for weeks of night shifts – not the most sociable of starts and a fairly abrupt welcome to her new life for Rebecca. We had the Jeep but a lot of the time I was driving to work in it, so Rebecca was frequently left on her own, trying to get her bearings around that massive military compound on a bike.

When I had a free weekend, the pair of us would gratefully head to the beach at Destin, Florida, which was only a couple of hours away. Yet for me, with my Longbow conversion course to complete, this was a period of manic studying, both in and out of school. I would always be bringing along with me for the weekend a thick Apache AH-64D manual that I was trying to learn by heart. Rebecca would test me on it during the drive – which was absolutely fascinating for her, I'm sure. But I was in the grip of something. I wasn't content simply to know the course flashcards; I wanted to understand the aircraft as completely as I could. That Apache Longbow manual was four inches thick, but even that wasn't enough technical detail for me. I wanted the stuff that came from Boeing, the manufacturers, that went another level deeper. I seemed to be becoming a complete geek in that respect. In the car, beside the bed, next to the loo ... Apache manuals and checklists and data sheets were never far away from me in those days.

Indeed, I was reading nothing else – no novels, no autobiographies, no history books; only things which would expand my comprehension of flight in general and of that one helicopter in particular. And this in turn required me to revisit my education a bit, too. For example, I realised pretty quickly in America that my D in A level maths wasn't going to cut it at the depth I wanted to go. So I bought myself a core maths book to give me some additional grounding and worked outwards from there. I acquired a book called *Aerodynamics for Naval Aviators* which I think was issued as standard to US Navy pilots at the time, and which proved very useful for developing a better understanding of that particular branch of science. And there was a brilliant book called *Understanding Flight* by David Anderson and Scott Eberhardt, which came out in 2000 – more basic but very descriptive and pin-sharp about how and why things fly and, of course, why they don't. Eventually the sophistication of Apache would push me out into other niche areas – radar theory, infrared theory, laser theory, ballistics. I read up on those, too. And I was extremely dogged about it. If I reached the end of a chapter where I hadn't comprehended something completely, I wouldn't turn the page until I had done some more reading and made sure I fully understood what was in front of me. As a schoolboy, I had always been very dutiful about studying, but also a bit lost – unable to quite put it all together. Now that my studying had a definite context, and now that I could apply my learning to something very specific that absolutely enthused me, I seemed to be away.

And the Apache Longbow really did lend itself to study. It was a phenomenal advance on the Alpha model – radically digitalised. In front of you in the cockpit now were two black computer screens, instantly making the Alpha, with its analogue dials and switchgear, look like some kind of relic from an earlier civilisation. The Longbow also now incorporated radar, which the Alpha model didn't have, and specifically a Fire Control Radar, or FCR, with an air-to-air mode capable of presenting you with an indication of any other aircraft within 8 kilometres. Those identified aircraft could then be projected to the pilot's monocle so, if the radar screen showed a target, left at 10 o'clock, say, then you could look left at 10 o'clock through the monocle and an icon representing that aircraft would appear in your sightline. In other words, the radar information was getting translated into the pilot's real-world view, massively enhancing situational awareness. You could also slave your day or night camera to the target, enabling instant visual recognition of friend or foe.

The Longbow also had, as well as its laser-guided missiles, radar-guided missiles, meaning you could use that radar information for targeting. In ground modes, you could be sweeping the territory in front of you, for up to 8 kilometres, and picking up all sorts of targets which you could then lock your weapons and cameras on to. The radar was smart enough to distinguish a tank from other wheeled vehicles, moving targets from stationary ones and to prioritise all the threats. Later generations would detect targets out to 16 kilometres, but even in the form in which I first came upon it, this was a very sophisticated, very clever piece of kit, with massive implications for the helicopter's lethality. If you were coming up to a battlefield, a single scan with the radar would identify multiple targets and you could click through them, visually identify them on the computer screen, link a radar-guided missile to them, dispatch the missile ... It was possible, easily, to destroy ten tanks in sixty seconds using these systems, and all without removing your hands from the helicopter's controls.

Meanwhile, your radar warning system could detect enemy surface-to-air missiles in time for you to evade or deceive them. And the Longbow would know if it were being painted by enemy radar. It would be able to inform you that there was, for instance, a Russian ZSU-23–4 Shilka anti-aircraft gun 7 kilometres away at 4 o'clock with its radar trained on you, and within seconds you would be able to send a missile and destroy it. All of this meant the Longbow was absolutely bristling with complex tech – cutting-edge sights and sensor systems. I lapped this stuff up. I couldn't get enough of it.

While I was on the Longbow course, I was flown out to the vast Boeing campus at Phoenix, Arizona, where the US Apache is built, to spend a fortnight training on the full-motion simulator which they had there. Non-motion simulators were easier to find but, at that point, if you wanted to work in a full-motion simulator, which was an enormously sophisticated piece of kit, this was the only place you could find one. It was a great chance to tour the factory, see Apaches being made, witness those systems being developed, and understand the aircraft a little more.

But there was a break, too. I was working with a couple of other pilots on my course and on the middle weekend of our training we had the Sunday off. We'd had our heads down all week and we figured it would be nice to get out for a bit, so we decided to go and see the Grand Canyon, to which none of us had ever been and which is, of course, on everybody's checklist of the great American sights. It was a three-and-a-half-hour drive from Phoenix to the South Rim, so we left early in the morning, made that fantastic journey through the red rocks and desert, and got to the South Rim of the Grand Canyon at about 11.00 a.m. And it's true what everybody says: nothing quite prepares you for that sight – the intimidating immensity of it and yet, at the same time, the calm and the stillness.

So we stood and drank it in, and we did the touristy thing and took our photographs. But then we kind of thought, 'Well, what now?' I had a suggestion. 'Do we want to go back and say that we've seen the Grand Canyon?' I asked. 'Or do we want to go back and say that we've *done* the Grand Canyon?' My notion was that we should hike down to the floor of the Canyon and take a dip in the Colorado River. The idea was enthusiastically received.

I thought my appetite for under-prepared adventuring had waned a couple of years earlier in the potholes of Gloucestershire, but apparently not. It was the middle of a baking hot summer's day. We were in shorts and T-shirts. One of us was only wearing flip-flops – although we did take the trouble in his case to dive into a gift-shop where he bought himself a pair of Croc-like sandals which were, at least, slightly more suitable for the kind of hike we had in mind. We had no hats, no sun protection and no food. All we had was a water bottle each. And off we went: a 4,500-foot descent down sheer paths through the rocks and dry scrub, a quick swim in the blissfully refreshing Colorado, some photos to prove that we'd done it, and then (the bigger ask) a 4,500-foot ascent, under blistering afternoon sun.

By the time we got back up to the rim and across to our car, it was around 7.00 in the evening. Completely exhausted and utterly famished, we drove to the nearest McDonald's, six and a half miles down the road at Tusayan, piled through the door and ate our body weight in burgers and fries. I rank that as probably the second greatest meal of my lifetime, after the visit to McDonald's which followed the Combat Search and Rescue course.

I realise, in retrospect, that I have been unusually fortunate as a visitor to the Grand Canyon. In 2008, as a test pilot, I would return to Arizona and fly down its tributaries in an Apache during a trial. Flying doesn't get much more spectacular than that. I have also photographed it from the cupola of the Space Station, and I can report that it looks overwhelmingly impressive from up there, too. But I still maintain you haven't really done the Grand Canyon until you've swum in it.

*　　*　　*

When the Longbow conversion course finished, Rebecca and I loaded everything we had, which was practically nothing, into the back of the Jeep Cherokee and, pausing only to savour the delights of a casino hotel in Biloxi, Mississippi, drove west to Texas and our house in Harker Heights. Goodbye, Mother Rucker; hello, Fort Hood. Here, formally back in the bosom of my adopted unit, the 1st Cav, life improved greatly for both of us. The US Army does a very good job of creating a family feel within a battalion, embracing not just the service people but their partners and their kids. There seemed to be social engagements going on all the time – events to draw you in and make you feel a part of it. The US Army does a better job of that than the British Army, if truth be told. The only time I have experienced anything like it was when we were living in Houston with NASA – and that organisation is heavily influenced by its serving US military personnel.

At Fort Hood, we quickly found good friends in Colin Frisbie, one of my fellow company commanders, and his wife Kate, and our American adventure instantly became a lot more fun. Colin and Kate took us to the massive spectacle which is a US college football game and patiently explained what was going on to the Brit novices. I think I got the gist of it although, frankly, an Apache manual seemed simple by comparison. Later Colin and Kate performed the same service during a baseball game. Having Kate as a friend was invaluable to Rebecca, not least when I was off flying, which was frequently.

Meanwhile we had acquired a dog. Rebecca went to 'help out' at a rescue home, fell in love and returned with Foss, a lovely mutt who was part Rhodesian Ridgeback and part other things, altogether less identifiable, but a wholly great companion for both of us. The parameters of Rebecca's Texan life shifted again when she discovered that, if you wanted to teach in America, you didn't need a specialist PGCE qualification, as you did in the UK: it was enough to be a graduate. She started teaching at a primary school in Texas and, in the process, discovered a new talent and something that she wanted to continue doing when we returned to England.

And thus began two busy years in Fort Hood. The immediate region offered less interesting flying than Alabama. You only needed to go five minutes from the base and you were in unbroken miles of flat brown Texan desert with nodding donkeys and tumbleweed – a mind-boggling vastness. (Texas, it always sobered one to remember, is bigger than any country in Europe with the exception of Russia.) But I was flying Apaches, and that more than compensated.

That said, my detail in my first year was as battalion operations officer – what the US Army calls an assistant S3. Although this did include some flying, it was a relatively desk-bound role – my first and only time doing something close to a staff job, which I was always keen to steer myself away from. However, it seemed that I had arrived in the operations team at the perfect moment. Our battalion was assigned to support the 4th Mechanised Infantry Division as part of its transformation into the US Army's first 'digitised' division. This immense feat involved, essentially, networking every armoured vehicle, artillery piece, tank, helicopter and command centre to operate an all-informed digital battlefield. The inner geek in me that had been awakened at Mother Rucker saw an opportunity to go fully ballistic.

If truth be known, I didn't have to fight very hard to get the job. It was coming my way anyway. Digital transformation was one of the US Army's top priorities and the operations team was going to have to make this work for Apache one way or another. We were also up against a tight timeline. There was a major exercise planned the following April, in 2001, at the US Army's vast National Training Centre in the Mojave desert. This was going to be the largest deployment of troops – around 8,000 soldiers – to the NTC and it was the division's opportunity to prove to the US military how well this all-informed network was able to function. I remember a US Army 2-star general briefing us and borrowing that famous NASA line: 'Failure is not an option.'

I spent the best part of that year working alongside defence contractors who were under equal amounts of pressure to make their various digital modems and communications systems work. The Apache Longbows already had full digital communications between themselves, but making this work between every other vehicle in the division was a whole new ballgame. And it wasn't just the US Army involved. I found myself talking with a US Air Force colonel, arranging to link up with JSTARS surveillance aircraft, then sitting on dispersal in an Apache running through checklists until we could send digital information seamlessly, JSTARS to Apache.

I realise now I was essentially doing the work of a test pilot: test, evaluate, learn, innovate – repeat. It may not have carried with it the excitement of a flight trial, but it was pushing my knowledge to new boundaries and, although I may not have known it at the time, it was also laying the groundwork for a whole new career path.

There was some flying to be had, although as an assistant S3 I was low on the priority list – the lowest, in fact. So when it was our battalion's turn to conduct night gunnery, I wasn't surprised to see my name at the bottom of the sheet – the 3.00 a.m. slot. I walked out to meet my pilot, Brian McFadden, who told me to hurry up and jump in. The range was closing and we had precious little time left. Strapping into the gunner's front seat, I heard Brian asking the ground troopers what ammunition they had left.

'Six Hellfires, thirty-eight rockets and three hundred rounds of 30-millimetre,' came the reply. 'How much do you want?'

'All of it,' said Brian.

I silently grinned.

It seems that waiting until last had paid off. We got to clean up. The only trouble was, by the time we got to the firing point we only had five minutes of range time remaining.

'Do you reckon you can get rid of all this in five minutes?' Brian asked.

I gave him the best Texan 'Hell, yeah' I could manage.

Five minutes later, with a red-hot barrel and smoking weapons pylons, we headed back to base a whole lot lighter.

Payback for the assistant S3 job came in August 2001 when I was employed as platoon leader of C Company, the Vampires. There was no shortage of American lieutenants coming into the battalion who were waiting to do that job. The fact that I was given it for a year was a huge show of faith in me, and also a huge show of faith in the whole concept of the US/UK exchange programme, for which I was very grateful to Lieutenant Colonel Danny Ball, my commander.

A couple of weeks later, I was in bed one morning after a night-flying sortie when I was woken by the phone. It was my mum, calling from the UK.

'You need to switch the television on.'

I did so, and saw what practically everybody alive with access to a screen saw that day: smoke pouring from the towers of the World Trade Center in New York and the world changing live on TV.

Not long afterwards, the phone was ringing again. This time it was the battalion's cascade call-out system. We had tested that system a few times, the way you might test a fire alarm. They were fond of staging practices for it at 4.00 a.m. But this was the real thing. The call was going out from Fort Hood to all personnel. 'Report to base immediately.'

I got dressed and jumped in the car, listening to the news on the radio as I drove over: the fall of the second tower, the attack on the Pentagon, another plane downed in a field in Pennsylvania ... It was hard to suppress a feeling of rising alarm. The roads into Ford Hood were bedlam, jammed with traffic as everyone descended en masse on a base that was now abruptly on high-security lockdown. I flashed back to that first morning in Texas with Dave Amlôt, strolling around the Apaches in our T-shirts and shorts. Those days were over and they weren't coming back.

The mood I found among my US military colleagues that morning was of shock and anger, but also of confusion and anxiety. When was the next attack coming and where? Were military bases the next target? America appeared to be at war and there was a bristling desire for retribution – but against whom? In those first hours, nobody even knew who the assailant was. And then, very quickly, all aircraft were grounded, including ours. So now you could add to that mix of feelings at large on the base a sense of impotence. All we could do was sit on the ground and wait.

It put me in a unique position. As a member of the British Army on exchange, there was an understanding that I would live, work, train and patrol with my American colleagues, but not that I would go to war alongside them in the event that the time came. Technically, a full-blown military operation on another territory would likely be my cue to step aside. Yet I was a newly appointed platoon leader. I didn't want to be standing on the dispersal, waving my platoon off the minute there was some serious work to do.

There was one other Brit in the division at this time – Major Charlie Eastwood. Both of us approached the military attaché at the British Embassy and sought permission to deploy. And both of us were granted it, which was an exceptional move, though this was an unprecedented moment in global politics and in the wake of 9/11 the UK was keen to show support for its US allies.

Over the next weeks, my platoon was prepped to go out to Afghanistan. We were trained up, geared up and mentally prepared, all ready to head to the Tora Bora mountain cave complex in Nangarhar Province that December and join the hunt for Osama Bin Laden. And then, at practically the last minute, we were scrubbed and told to stand down. The US decided to send an Alpha model Apache unit instead from the 101st Aviation Regiment.

I could understand why. They were sending aircraft into a region where hand-held surface-to-air missiles and rocket-propelled grenades were rife. They were anticipating the loss of helicopters in this operation, and I suspect they didn't want those helicopters to be brand new, state-of-the-art Longbows. Even so, it was massively deflating. It was the one time I rued being involved at the sharp end of the Longbow programme – the one time I longed to be behind the curve in the development of that helicopter. I was bitterly disappointed to be stood down.

Still, there was other work in which to immerse ourselves. In March 2002, I found myself heading south as part of Joint Task Force 6, Border Patrol, linking with the Drug Enforcement Agency on a counter-narcotics operation on the Mexican border. The US has strict regulations on the use of the military in policing activities which meant that our role was slightly aloof. Essentially, we were allowed to observe and report, liaising with DEA forces on the ground, who were doing the hands-on business. We certainly weren't allowed to engage, but the Apaches made a persuasive deterrent nonetheless, and to see the dramatic impact we were having in reducing the volume of drug-trafficking was rewarding. It was also a fantastic flying opportunity, across sensational mountainous terrain with the occasional suspect-vehicle pursuit thrown in. We flew by day and also by night, sometimes honing in on suspicious orange glows on the hillsides, which turned out on closer inspection to be college kids having bonfires. This was six weeks with an extended-range fuel tank in place of one of the weapons pylons, flying for hours on end up and down the Rio Grande, which is a beautiful place to take a helicopter. It would be a pity if anyone were ever to build a wall along it.

The time soon came to head home. I was excited to be going back to the UK. I knew I was doing so at a key moment for the Army Air Corps, when the first Apaches were arriving and when I would be able to take back three years of Apache experience to help get the pilots trained and the UK fleet of attack helicopters up and running. It was a fantastic prospect.

At the same time, we were genuinely sad to leave Texas – in a way that might not have been entirely predictable that night when Rebecca first arrived in Killeen and we crossed the parking lot to the grocery store through a veil of locusts. But it's never about the place, of course; it's always about the people. Fort Hood and the US military had embraced us, and when we said goodbye in Texas, it wasn't just to a job, it was to a community and a bunch of really great friends and some top-class colleagues. It was a real wrench.

Just before we left, the battalion gave me a farewell dinner. I stood up afterwards and made a leaving speech, at the end of which I presented the 1st-227th with a token of my gratitude. It was a print of the 5th Airborne Brigade, the Paras, securing the Kaçanik Defile on D-Day in Kosovo on 12 June 1999. The picture seemed meaningful and appropriate to the occasion for an obvious reason. In the background of the composition were Apaches from Task Force Hawk, supporting the British Infantry, so the image stood to illustrate the collaboration of US and UK forces, embodied in the exchange programme of which I had been a beneficiary.

But the image had a personal layer of meaning, too – a private resonance that only I would know about. Just a few hours after the depicted scene, Rebecca crossed that same bridge. Her job then, as a lieutenant in the Royal Logistic Corps, was to provide close support to the Infantry and armoured units. She had served fearlessly in a really punchy operational role, and had returned from that mission on a high. Whereupon she walked away from everything: left the Army career that she loved, left her family behind, left her friends behind, and came with me to start forging our new partnership in the middle of the desert. The three amazing years that we had had in America, and all that I had experienced and learned from them, would not have happened without Rebecca's sacrifice, and I loved the thought that there would be a picture hanging somewhere in Fort Hood that quietly marked that.

CHAPTER FIFTEEN

Larkhill rising, effects of controls,
and the helicopter in the garden

So it was back to work at Middle Wallop – but not to live this time. All the married quarters there were full, so Rebecca and I were offered a place 13 miles up the road in the Royal Artillery barracks at Larkhill. This, though inconvenient in some ways, had its advantages. For one thing, those barracks are in a very nice spot, right on the edge of Salisbury Plain. All that open space suited both of us, and also Foss our rescue dog, who had returned with us from America. My morning runs with Foss took me up the approach to the monument at Stonehenge, then out onto the plain and back – a decent backdrop, we seemed to agree.

For another thing, Larkhill gave us a generous detached three-bedroom Army house in a quiet and green cul-de-sac. What Army houses may be said to lack in style they more than make up for in robustness of construction, as I discovered when I went into town to buy a drill in order to put up some pictures. The guy behind the counter in the hardware shop eyed the bog-standard tool I had chosen and asked me where I was living.

'Larkhill,' I told him.

'Hmm,' he replied. 'I know those quarters. Built with double-baked bricks. You'll never get *that* through those walls.'

He advised me to upgrade to something more on the scale of an AK-47 machine gun. Which I did, and then left, wondering if I'd just been had. Not at all. A machine gun was exactly what it took. If I'd used the tool I first chose, I'd still be drilling holes for those pictures now.

As I pointed out to Rebecca, there was another reason to be glad we were in Larkhill. It was well placed for Boscombe Down, the home of the Empire Test Pilots' School and a big military aviation test site – and that might be handy eventually. This was looking ahead a fair while, I knew, but my heart was already set. All that detailed work on Apache in America had entirely persuaded me that what I wanted to become was a test pilot. Flying aircraft every day, seeking to understand what they could do, pushing them to their limits, helping to adapt them and improve them ... It was the dream job as far as I was concerned.

In the Army, of course, you mostly do what you're told, but there are moments also when you can make your preferences clear, and I let it be known what mine were as soon as we got back from the States. To my delight, the intimation from on high was that it could be made to happen. I would be needed for a couple of years of instructing on Apache first, but after that I would be able to go for test pilot selection.

Rebecca started working as a fundraiser for the charity Hope and Homes for Children, which works to get children out of institutionalised care and into families. After three years there, she enrolled at Southampton University to do her PGCE so that she could carry on teaching, as she had been doing in Texas. When she qualified, she would take a job at a primary school in Tidworth, another military town not far from Larkhill. Meanwhile I knuckled down for my own year of teaching as a QHI in the newly formed Attack Helicopter Training Unit at Middle Wallop.

It was September 2002 and, thanks to the British Army's promised investment, the hangers at AAC HQ now had eight brand new Apaches in them. It's always a pleasure to be working with box-fresh equipment, although I can't pretend that there is a 'new helicopter smell', the way that there's a 'new car smell'. All helicopters tend to smell of fuel and oil, whatever their age, and there never seems to be much factory freshness about them even when they are actually fresh out of the factory. Also, no one would ever describe the Apache as a shiny aircraft even at the best of times. It's matte green and black, decidedly un-glossy and even the new ones have a slightly 'previously loved' look about them. Still, it was a thrill to be operating all that recently manufactured, perfectly functioning gear.

At this point, there was just a small team of us working on Apache. Bloo Anderson was the boss, himself having completed an exchange tour in the US on the Apache Alpha model some years earlier. Flying with Bloo was like learning from the Jedi master himself. Then there was Chris Hearn, Bill McPhee, Al Whittle and Jan Ferraro, whose training had coincided with mine at Fort Rucker, along with senior instructors Bill O'Brien, Mark Torpy and Howard Floyd, who had taught me on my pilot's course. Nick Wharmby was an AAC test pilot who had been flying the Apache at Boscombe Down. These guys were all 3,000-hours-plus pilots, some of the most senior and respected pilots in the AAC. I was younger and less experienced than any of them, and it was a real privilege and honour to find myself in this elite group. You learn fast in such company.

Our job was to train up the first group of instructors, but before that we had to develop the course that we wanted those instructors to teach. We had all been through – and owed a great debt to – the American way of teaching how to fly the aircraft. But we now set ourselves the target of doing it even better if we could, with a bit more time spent going into the background – the whys as well as the how-tos. I mentioned before the rote-learning aspect of the Apache tuition that I received in the US. We wanted to get our pilots behind the scenes more – get them deeper into the workings of the helicopter and its systems, not just for its own sake, but in the belief that that deeper understanding would produce a better pilot.

Between us, we developed a lesson flow for the course and went out to practise the sorties on each other to make sure we had got it right. That was a great period, jumping into helicopters and flying off over Hampshire to test our own lessons. We had plenty of aircraft and all the resources we needed, and we were working on something new and exciting that everybody around us wanted to be involved in. It was enormously energising. And that energy only increased when we began instructing the instructors, getting that first wave through and qualified so that they could carry on and teach the first operational pilots.

It was interesting to see how our approach was received by the highly experienced Army instructors who were coming to us for tuition. These would regularly be warrant officers with years of operational service and more than 2,000 hours on Lynx behind them. And there would be me, a captain, with fewer hours and less time in the military, standing up in front of them in a classroom and saying, 'Right, we're going to do lesson one – effects of controls.'

You would see them rolling their eyes and thinking, 'Really? By now, I think you can trust me to know what the effects of the controls are.' But then you would start taking them through a few things – the Apache's autopilot system, for instance, with its various modes (all geared around providing a stable platform for weapons delivery of course), or its fly-by-wire backup control system … And they would realise that, actually, Apache was deeply complex and new and that understanding it from the roots upwards was going to be extremely helpful to them. It was also exciting to see how quickly Apache sold itself to these pilots – how rapidly they would realise, as I had done, that they were dealing with an exceptional machine and become enthused by it. Seeing your enthusiasm shared is the best feeling for an instructor. This was hugely rewarding work.

One of my first students was a warrant officer called Tom O'Malley. He, in common with so many of the students coming through the course, was vastly experienced as a Lynx pilot and had seen service and operations in Bosnia and Northern Ireland. He, too, would have been in a position to wonder what this guy, five years his junior, was doing, telling him to get ready for a starter lesson on effects of controls. But he bore with it, and also with the four months of instruction that followed. In 2007, in Afghanistan during Operation Herrick, Tom, under heavy fire and with two Marines strapped to the outside of the aircraft, landed an Apache within the walls of a Taliban base in order to recover a wounded Royal Marine, Lance Corporal Mathew Ford. Tom was awarded the Distinguished Flying Cross for his flying that day. All of us were extremely proud to think that Tom was an alumnus of our course, and proud, too, of all of the pilots who eventually took our training into front-line service in Afghanistan.

The one cloud on my personal horizon in this period was my looming Junior Command and Staff Course. This was a statutory hurdle for all captains, which you were expected to complete at some point within a two-year window. And it was a prerequisite for promotion to major, which was an incentive to do it, of course. But I can't pretend I wasn't deflated at the prospect. There I was, in the swing of this exciting project, flying to my heart's content; and now I was going to have to set it all aside and spend four months in a classroom.

There was no getting round it, though. In April 2003, I reluctantly left the airfield and enrolled at the Defence Academy of the United Kingdom at Shrivenham in Oxfordshire. Here I spent the first month sitting in a big group of other captains, listening to lectures on military thinking, decision-making and leadership, and being brought up to speed on the latest developments in the wider Army beyond our various regimental bubbles.

I have to confess that I didn't take the course especially seriously. I never saw myself as a staff officer and, although I was keen to be promoted to major, I had no ambition to advance all the way up the career ladder. In fact, to the extent that the traditional career ladder seemed invariably to end up behind a desk, I was keen to keep off it if I could. Most of the people on the course were great, and inevitably there were a few highly career-minded officers – those who were trying to thrust their way through the system and make it to lieutenant general as rapidly as possible. The 'thrusters', as we called them, were easy to spot: they were the super-keen ones doing their best to impress the staff. On this occasion I had no desire to compete. I knew what I wanted to do, which was get back in a cockpit as soon as possible and go on to be a test pilot. The pressure was off me, in a rather nice way.

Ironically, this may have been why I ended up doing quite well on that course. We had to study something called a 'combat estimate'. This was a system which was intended to supply a structured planning process for military action and which took the form, essentially, of a set of logically ordered questions enabling you to analyse the advantages and disadvantages of each of your options while factoring in all the variables. The idea was that you would pass your particular problem through this verbal machine and arrive at the other end with a beautifully ordered plan for how you were going to rescue a bunch of people from an island or do an armoured attack on an enemy position, or whatever the applicable military scenario might be.

Now, maybe I had been away from home too long, but I remember looking at all of this one night and becoming strangely absorbed by it. I guess it chimed with two of my pet interests: logic and structure. Anyway, for a few days there I decided to dig down – first to question it and then to really understand the 'combat estimate' and, very largely because of the work I did on that topic, I finished in the top 10 per cent at the end of the course. That was a big surprise for me – and, I'm sure, an even bigger one for the course tutors, and quite possibly an even bigger one than that for some of the thrusters. Even so, those four months at Shrivenham had cemented my feeling that this was not the career route I wanted to take. I was happy to say cheerio, head home to Larkhill and jump back into my helicopter.

By the time I got back to Middle Wallop in July 2003, our group of qualified Apache instructors had grown to around twenty and the programme was about to move into its second phase. In September, a handful of us were posted up to RAF Dishforth in North Yorkshire to be QHIs on what was rather clumsily entitled the Air Manoeuvre Training Advisory Team, but which could also be more snappily called AMTAT. Our brief was to spend six months preparing a course that would take the first operational pilots and teach them how to fight the aircraft, and then we would teach that course when those pilots arrived. My specialist area was to be communications and mission planning and, along with everyone else, I would also develop and eventually teach battlefield tactics – formation flying, engagement, use of sights and sensors, etc.

All of this meant that for the next fifteen months I would be in Yorkshire from Monday to Friday and home in Wiltshire only for the weekends. That put Rebecca and me back in the old routine of living mostly apart. Still, whole countries had separated us before and this was a mere 250 miles on the M1 – peanuts as far as we were concerned. It was a painful commute all the same, though, and I still shudder to recall the gloom that would descend at 4.00 in the afternoon on Sundays with the weekend over and a five-hour drive ahead.

Still, it was an extremely busy period. The timescale was tight and the six of us QHIs on the AMTAT team worked flat out all week, day and night. We were billeted in some unoccupied married quarters by the airfield where I shared a house with Sean Dufosee, a good friend and Commando pilot in the Royal Navy, on attachment to the Army Air Corps. The house was pretty bare, there was a kitchen which only ever had tea, coffee and the occasional box of cereal in it. We were too busy to do anything other than grab meals in a hurry in the mess.

And you couldn't get much better than Dishforth for the kind of flying training we needed to do. You had the Lake District on one side and the North York moors on the other – a vast real estate of perfect Apache flying territory. Just up the road from us, at RAF Leeming, was the RAF's Air Combat Training Squadron, flying the Hawk T1 fast jets, which was also handy. If we were returning from a sortie, we could call on an agreed frequency to signal that we were game on, and there might be a couple of Hawks returning to the area and willing to play, in which case we could stage an engagement with them. Then we could get to use our radar, practise fighter evasion, and so on. It was useful for the Hawks to see if they could spot us, and it was useful for us to see if we could spot the Hawks.

On top of that, we had access to the 9,600 acres of RAF Spadeadam in Cumbria. Spadeadam can claim to be Europe's only Electronic Warfare Tactics facility. It owns some pieces of captured hardware, including some Russian surface-to-air missile carriers so we could stage mock battles with the real thing, in what was often, if not an aggressive atmosphere, then certainly a highly competitive one. The British military guys on the ground, operating the Russian kit, were as desperate to shoot down an Apache as we were to shoot up a ZSU-23-4. Again, I come back to that theme of making training as real as possible. Spadeadam was great for that.

We also used the Otterburn ranges in Northumberland. While we were putting the course together, I flew up there in an Apache to do a recce, taking along Captain Matt Roberts, our range specialist who was going to check the place over from a safety point of view. On the way up, Matt had told me that he had a friend with a helicopter landing site – an official one, in the HLS directory – so maybe we could use it on the way back. Then he could say hi to his pal, and we could have lunch together. It sounded good to me, so I put that into the schedule as my fuel stop.

What with all the flying that was necessary during our recce at Otterburn, I was getting pretty low on fuel by the time we started descending towards Matt's friend's landing site. I was also slightly disarmed to see how small that site was. Frankly, it was the guy's back garden. It would have been relatively tight in a Robinson R22 or a Gazelle. In an Apache, it was like lowering an elephant into a rabbit hutch. And it didn't help that the site sloped off to one side, either. By flying very carefully, and breathing in very deeply, I could just get the front wheels on the very forward edge of the grass landing area, with the tail wheel squeezed on at the aft end.

Anyway, one way and another I got the aircraft on. I shut down and we got out and said hello. On the way into the house for lunch, I casually mentioned to our host that I would be needing some fuel – that I was almost out, in fact.

'How much?' he asked.

'About 1,000 kilograms,' I replied.

'Ah,' he said. 'I'd better start rolling out the drums, then.'

A good joke. I laughed. The idea of filling an Apache the old-school method, by hand from the barrel, was highly amusing if you were a pilot.

Then I realised the guy was entirely serious. No high-pressure fuel-pumping system here. Just a stack of barrels and a little hand-held Zenith pump.

Gravity refuelling? I was flashing back to Kenya and the Gazelle days, tilting the barrels to try and persuade the Zenith pump to suck up the dregs. But this was an Apache. Nobody does gravity refuelling on an Apache.

Actually it now occurred to me that, in my case, I had never done *any* kind of refuelling on an Apache. It might sound strange in the circumstances, but as aircrew, we don't tend to get too involved in that side of things. We're usually strapped in for the refuelling, monitoring from the cockpit and letting the professional refuellers do it for us. Anyway, this was obviously going to be an excellent time to learn.

I opened the fuel tank caps at the front and back. We rolled out a barrel, stuck the Zenith's nozzle into the Apache and began to pump. And immediately a backwash of fuel welled up in the opening and slopped onto the ground. It turns out that the tube leading to the fuel tank in an Apache has a rollover flap – just like in a car – that you need to poke the nozzle past. But in the Apache that flap is quite far down the tube – too far, certainly, to be opened by the short nozzle on a Zenith pump.

This, I was obliged to realise as I stood there with my hands on my hips, was quite the quandary. I'd parked an Apache with an empty tank in someone's back garden and now I couldn't get any fuel into it. Moreover, there was no room for another aircraft to land next to us and help us out – not unless we took the owner's house down first.

The word 'stranded' was beginning to take shape in my head. Basically, I appeared to have condemned an extremely expensive attack helicopter to a new life as a garden ornament. I wasn't sure the Army were going to be too pleased about that. I wasn't sure the owner of the garden was going to be all that happy about it, either.

All was not entirely lost, though. Grubbing around in a state of thinly disguised despair in our host's garage, I found a 3-foot length of steel rebar. Could this be my saviour? I took the metal rod into the kitchen and gave it a thorough wash under the tap, then took care to wipe it completely dry. Bingo: I had just patented my own unorthodox Apache-compatible pump-extender. Back out beside the helicopter, I thrust the metal pole down past the fuel flap and stood there while our host squeezed the Zenith's feeble nozzle in alongside it. At last, some fuel was reaching the tank.

It was a painfully slow process. Eventually the first barrel drained and we moved on to the second, then the third, then the fourth, filling the forward tank and then moving down to fill the aft. It took about an hour and a half and six barrels, all in all, which is no one's definition of a quick refuel. Also, stabbing the filler-flap with a metal pole didn't feel like quite the kind of thing you should be doing to a new Apache's fuel system, so throughout the entire caper I continued to flush hot and cold with the thought that I was doing some irreparable damage to the aircraft, or that I might let go of the rod and watch, in horror, as it disappeared into the tank. But at least the Apache now had gas. And the lunch afterwards was good.

Otterburn proved very useful to us. It was nothing like as massive as the ranges in Texas, but it was at least big enough for us to teach the use of the Apache's cannon and rockets. As for the Hellfire missiles, the Apache had its own built-in simulator, meaning you could practise missile-firing with no missiles on board. This is the point, I suppose, where computer gaming and flying the actual aircraft overlap entirely. In sim mode, the weapons page on the Apache's screen could behave as if the aircraft were fully loaded. You would be using the real laser to guide the virtual missiles down, you would be seeing the time of flight, you would even be getting audio through your headphones – the whoosh of the missile leaving the rail. Yet nothing was being fired. It was an extremely smart training tool – and highly cost-effective, of course, not least in the context of the British Army's extremely limited Hellfire missile stocks.

At Otterburn we also taught running firing techniques and diving firing techniques because we knew they would be applicable to the mountains of Afghanistan. We wanted to give the pilots a broad spectrum and teach them to fight the aircraft in all terrains, but we knew that Afghanistan was coming up. That focused our minds and it focused the minds of the students who were coming to us, who could see straight away that Apaches would be going out to Afghanistan and what their role was likely to be.

Our job at AMTAT was to turn pilots into gun pilots. We weren't as gung-ho about it as some of my American instructors had been. We weren't shouting 'Attack!' at every opportunity. But we were mindful of the shift in mentality that our students were having to make, just as we had had to make that shift ourselves. You would see a distinction – sometimes very early on, during exercises in the simulator – between those who were ready to engage and those who were more reluctant. We had one student who was softly spoken – an extremely nice person, and a highly skilled pilot, but who showed no obvious instinct for being an attack pilot. As the course wore on, we had started to have a few quiet conversations among ourselves about whether this person was ever likely to make the transition. Then it came to his final check-ride during which he was set an exercise in which he was required to engage with a convoy of vehicles. In a devastating display of targeted aggression, our mild-mannered student took out the leading vehicle to block the route, then took out the last vehicle to block the escape route. Then in swift succession he took out every vehicle in between – ding, ding, ding, ding – pausing only to switch to rockets when the missiles ran out. We instructors all looked at each other and nodded as if to say, 'OK, he's ready.' Sometimes it's the quiet ones you need to watch.

* * *

'Right,' said someone on the panel. 'The Lynx rotor blade.'

And off I went, explaining the shape of the BERP blade with its special tip, and how that shape is designed to prevent retreating blade stall and to improve forward speed performance ...

I reckon I did as much preparation for my interview at the Empire Test Pilots' School in 2004 as I did for astronaut selection four years later – even though astronaut selection was in several stages over the space of a year and test pilot selection lasted only one day. I knew that, as an Army pilot, I would be expected to know everything about Army aircraft – Gazelle, Squirrel and Apache, which I had flown, and Lynx which, at this stage, I hadn't. I also knew that I would have to show an understanding far beyond those aircraft – of flight dynamics, engines and how they work, electrical systems, hydraulic systems. All of this I did my best to revise. Plus I knew I would be expected to demonstrate a strong knowledge of the aviation industry generally – the commercial side, trends within it, and so on. In the months beforehand, in any spare time I could find while working on AMTAT, I steeped myself in aviation and space magazines. This must have been the first time I had consciously applied myself to thinking about the space industry and noticing some of the overlaps between the technologies used in spaceflight and the technology used in helicopters. I would be doing a lot more thinking about all of that later, of course.

Something else I knew about the ETPS selection process: how tough it was to get accepted. The Empire Test Pilots' School only takes one candidate onto its helicopter test pilot course from the Army each year (and one from the RAF, and one from the Royal Navy, plus some students from other militaries around the world, making an intake of about seven or eight each time). The prestige of the place goes ahead of it. There are only a handful of these kinds of schools in the world – America has Edwards AFB for Air Force and Pax River for Navy; France has EPNER – and ETPS was the first, founded in 1943. To graduate from here is to reach one of the aviation world's undisputed pinnacles.

Oh, and the one thing you won't be doing during the selection process? Flying. It's understood that you can fly to a high standard. The ETPS has written references from your bosses and reports from your annual check-rides to tell them that. You wouldn't get through the door at all if you didn't already have the 'outstanding pilot' box ticked.

And so, undeniably feeling the pressure, I drove myself to the hallowed Boscombe Down Airfield, where an English Electric Lightning T4 stood guard at the gate and where the school motto was emblazoned on the wall: '*Learn to test: test to learn.*'

First up was an interview with the boss – Commander Charlie Brown, a Navy pilot. That turned out to be quite sociable. He knew, of course, that a thorough grilling was coming my way later in the day, so he was more interested in finding out about me and getting me to explain what I had been up to in my career. I was happy to tell him.

Then came the panel. Across the table from me were a couple of ETPS tutors – one rotary wing, one fast jet – and two civilian guys who were experts in aerodynamics and flight control systems. They peppered me with questions – quite lightly at first, and then gradually more thickly – and then I had to give a prepared presentation. No props – no PowerPoint slides or indoor fireworks; just a whiteboard and some pens, while these experts looked on with folded arms. Naturally I chose to do something on Apache. At the same time, my specialist areas – communications and mission planning – didn't link too easily to a broader discussion of how things fly. So I chose instead to talk about the Apache autopilot system, which took me into the realm of new technology but also allowed me to show some knowledge of aerodynamics.

The point of that panel was to dig down to the bottom of your knowledge bank – to find out something about what you knew but also to get to the point where your understanding ended. So it was pretty taxing, all in all. I suddenly realised how an engine must feel when it gets stripped down. But I had a good tip, going in, from someone who had been through the process, who told me that just about the best thing you can say in front of that interview panel when the going gets tough is: 'I don't know. That's beyond my knowledge. But this is what I would hazard a guess at.' That way, you've been upfront, you haven't opted to bullshit, which is never a great idea, and you've also shown yourself happy to try and think outside your constraints a little. It would be a good tip for any kind of interview, really.

Anyway, the panel seemed to go OK. The only cold moment came when I had a complete mental blank when asked the simple question, 'What unit do we measure power in?' To which I panicked and, knowing that power is equal to a force applied over a distance in a specific time, eventually replied, 'Power is Newton-metres per second.' Which is true, up to a point, although 'watts or kilowatts' was actually the far simpler answer they were looking for. But maybe they thought, 'Well, at least he can convert the formula in his head. He can't be *completely* stupid.'

An agonising three-week wait followed. This was going to be the fulfilment of a dream if it came off, but I still had to wonder whether the Army would let me go. The Apache force was so valuable and so small that if you were an Apache QHI you were like gold dust to the Army Air Corps at this time. So I felt a flood of relief when my military commander at AMTAT, David Meyer, eventually told me that I had come through and had been selected. Perhaps ultimately my Apache connection had worked in my favour. The Army wanted an Apache test pilot at Boscombe Down. They knew that a potential Apache deployment to Afghanistan was coming up and they saw the value in having an experienced Apache pilot on the testing side when that happened. Right place, right time, right helicopter.

I saw out the year in North Yorkshire and began 2005 where I longed to be: at the Empire Test Pilots' School.

CHAPTER SIXTEEN

Embracing the avoid curve, high speed R&D,
and a secret mission

The button was marked '*Regime 21*' and I knew it had been mentioned in passing a couple of days earlier by my instructor, Kostya. But he had been speaking in Russian, which was always a bit of a struggle for me, and I couldn't for the life of me remember what he'd said about that button, and what it did.

One way to find out, of course. So I pressed it.

And that was the point after which all the instructors on the Soyuz simulator at Star City knew who I was. I was the one who pressed the '*Regime 21*' button – known widely as 'that button you should never press' – thereby crashing not just the Soyuz I was in, but shutting down the entire simulator complex, putting it out of action for about two hours.

Actually, I don't think the instructors minded. It was lunchtime, so they got an extended break. The programmers who had to reset the simulator and restore its entire navigational profile might have felt differently.

All this was during astronaut training, of course. I was being instructed to fly in the right-hand seat of the Soyuz capsule that would take us up to the ISS and bring us home. But Kostya knew I was a test pilot and he recognised the value of having crew who understood everything they could about flying the spacecraft, and not just their own role. In spare moments, if Tim Kopra and Yuri Malenchenko weren't around, Kostya would have me climb into the left-hand seat or the commander's seat and teach me about the controls in those positions, too. He ended up giving me loads of extra tuition, including on that day when, after we had completed a sortie, he left me alone to familiarise myself with the screens and I dug down deep enough into the system to unearth the dreaded 'Regime 21'.

If you are a test pilot, it probably goes without saying that you are an incorrigible presser of buttons – not to mention also a switcher of switches and a dialler of dials. That kind of curiosity goes with the territory – the urge to play around and see what happens. Test pilots and little kids who get in cars and annoy you by fiddling about with everything on the dashboard have a fair bit in common, all in all. We'll also be the ones adjusting the position of our car's steering wheels (sometimes while driving, which our partners may not appreciate us doing) and thinking rather more intently than most people about our seating positions at our desks and our eyelines in relation to our computer screens. I seem to have had that instinct, to some degree, all of my life. But even if I hadn't, it would have been instilled in me forever by my year at the Empire Test Pilots' School.

That was a very fulfilling time for me. My six colleagues at Boscombe Down on the No. 43 Rotary Wing Test Pilot Course were a great group of elite pilots from the Royal Navy, the Royal Air Force, the French Navy, the Canadian Armed Forces, the Italian Army Air Cavalry and the United States Navy. (There were similar-sized, and similarly international, groups on the Fixed Wing and Flight Engineer courses which formed the other two thirds of the year's intake.) It seemed to me that there could be little to beat the thrill of striding out in your flight suit in the morning for your first sortie as an aspiring test pilot. Plus the course offered a perfect balance between the academic and the practical. I would get to fly more than thirty different types of aircraft in those twelve months, undertake some extremely exciting test sorties, and take both my understanding of flight and my ability as a pilot to whole new levels.

We were all made to start with our feet firmly on the ground, though. In fact, a large part of the initial week was spent sitting in a helicopter in a hangar, going absolutely nowhere. Our first topic was Cockpit Assessment: how to go about it and, equally importantly, how to write it up. Obviously it's all very well being able to critique an aircraft, but if you can't come back and communicate what you've discovered effectively then there's not much point. It would be like being a Formula 1 racing driver who couldn't tell the pit engineers what he was experiencing in the car.

My assigned cockpit was in a Sea King HU Mk5, the classic Search and Rescue helicopter, one of which was parked in a Boscombe Down hangar. It was my job to give the inside of this aircraft a thorough going over, uncovering the good, the bad, the satisfactory, the unsatisfactory, the deficiencies that might be safety hazards. How well could the pilot see? I did a mathematical analysis of the field of view from the pilot's seat. How easy was it to get into the pilot's seat while wearing full flying kit? Not easy at all, I discovered. How quickly could you get out of this aircraft if you needed to? I timed myself repeatedly doing so in order to find out.

I scrutinised the instruments and controls, honing in on the bad low-height warning system – a dim light in a remote area of the panel – and the unhelpfully small and cluttered compass. And I raised my eyebrow at the lack of lumbar support and back padding offered by the distinctly uncomfortable pilot's seat. I then made my recommendations, ranked according to whether they were merely 'desirable' or 'highly desirable'. My finished report ran to twenty-one pages in total, with appendices and detailed supporting material, as instructed, right down to a full list of the clothes I had worn while forming the impression that the pilot's seat was difficult to clamber into. ('Boots, 1965 Pattern; Socks, Terryloop, green Mk 3; Drawers, long, cotton-ribbed.')

When the work came back, I anxiously scanned my tutor's remarks. 'You have grasped the 7-part paragraph without problem and avoided nearly all the common snags,' he wrote. 'My one point would be in the tendency to include data that isn't relevant to the deficiency.' Was there such a thing as too much data? My student self seemed to struggle with the idea, but clearly so.

Once that exercise was completed, we finally got off the ground. And once we were up there, I was continually surprised and impressed by what our tutors were prepared to show us. For instance, what's the lowest point from which you can do an engine-off landing? From 500 feet? OK, now try and do one from 400 feet. OK, let's see if you can do one from 300 feet. With helicopters, we work with something called the 'avoid curve' – also known more grimly as the 'dead man's curve'. It's a graph, specific to each type of helicopter, which plots height against speed and reveals the danger area within which the pilot would be too low and too slow to execute an emergency autorotation and safely land the aircraft in the event of a mechanical failure. Pilots are well advised to take the avoid curve at its word and avoid it. Here, though, the tutors would get us to creep into that curve and touch the edges of it so we could understand how the test pilots had drawn up that graph in the first place.

And then there were engine relights – how low could you be, with an engine failure, and still attempt to relight it? As a Gazelle pilot, I was used to flying within the parameters of that aircraft's Flight Reference Cards or FRCs. There it states that, if you're below 2,300 feet, you shouldn't attempt an engine relight but should commit instead to an engine-off landing. But here at ETPS, on one memorable sortie, I remember my instructor saying, 'Well, let's have a look at that, shall we?'

So we descended to just below 2,300 feet and he switched the engine fully down and said, 'I want you to relight the engine and fly away from this.'

Now, I should say that we were coming into position over the airfield at this point so, in the event that I didn't manage to get the engine going in time, we would still have had somewhere to do an engine-off landing. Nevertheless, I was now firmly in the hot seat, having to maintain autorotation and watch my rotor speed while at the same time doing all the button switches that are required to start the engine and simultaneously turning towards the airfield in case we needed to put the aircraft down. And meanwhile the ground was getting horribly close. I got the engine up to speed and pulled away with about 50 feet to go. As we surged back up into the air, the instructor was laughing. 'So, you see, there's not much leeway,' he said. I certainly did see that.

But perhaps most memorable of all was the day we explored vortex ring. Vortex ring is a highly dangerous state that helicopters can get into when their rate of descent is too high and their speed is too slow. It can happen if the pilot messes up an approach, coming down too fast with not enough forward speed. The emergency comes when the pilot starts to pull in more power but only then succeeds in forming a bubble of air around the descending aircraft – the vortex ring bubble – which causes it simply to drop like a stone. This nasty little trick of aerodynamics is why every helicopter pilot is taught that, below 30 knots, your rate of descent should never be more than 500 feet per minute. Otherwise the vortex ring bubble will form and it's very possible that there will be no room to recover from that.

So, considering all the obvious dangers, when my instructor said we were going to explore vortex ring, I assumed we would be going out to inspect the phenomenon in its incipient stages – maybe witness a bit of vibration in the cockpit and notice the rate of descent starting to increase and then snap out of it, before the real trouble started.

Not a bit of it. I was flying that day in a Lynx with an Army instructor, Eric Fitzpatrick, both of us wearing parachutes as we did for all test pilot exercises above 3,000 feet. We climbed up to 10,000 feet and Eric told me to put the aircraft into a hover.

'OK,' he said, 'now let's autorotate.'

So I dropped the lever and the Lynx began to fall vertically in a clean air autorotation. At this point, I was still in complete control of the descent.

'Now pull in power,' said Eric.

'How much?' I asked.

'Everything,' he said. '100 per cent torque.'

I pulled in full power and ... holy smokes. It was suddenly as though the Lynx was a solid block of metal falling through the sky. The aircraft simply plummeted, the rate of descent shooting right off the clock. Meanwhile I had lost tail rotor control. The tail rotor was now yawing and the whole aircraft was pitching and rolling slightly from side to side, none of which was particularly encouraging to witness from the pilot's seat.

'Right,' said Eric, as we continued our rapid drop in the direction of Wiltshire. 'Try and fly out of it.'

I went full forward on the cyclic, utilising the Lynx's considerable control power. Nothing happened. We just remained in the bubble, falling. Adrenaline was now flooding through me at quite a rate. This helicopter was dropping like a brick and I appeared to have no means to stop it.

'That's it,' I said to Eric, trying to prevent my voice from rising an octave. 'It's not responding.'

Eric looked at me calmly with a smile and said, 'Yeah, this is vortex ring.'

To get out of it, which we had briefed earlier, I had to lower the lever back down to restore normal autorotation and then go forward on the cyclic again and hold it there. Then I had to hold it there some more, and some more after that until, gradually and reluctantly, the aircraft flew out of the vortex ring zone. After that I was able to recover it from the autorotation. I started the recovery at about 4,000 feet, and completed it by 2,000 feet. This was the reason we had gone up so high before attempting this.

But what had happened to those feet between 10,000 and 4,000? They had gone by in the blink of an eye, flashing before me along with my life. I never for a minute thought that I would find myself intentionally putting an aircraft into full vortex ring. It seemed utterly brilliant to me that we would be taken up into the air and actually shown this phenomenon in order to understand it. Again, it's one thing to read about something like that in a book or see a warning on a checklist, and quite another to experience it from the pilot's seat.

Another fantastic thing about the Test Pilot Course was how rapidly it moved you from one type of aircraft to another. By the time I started at Boscombe Down, I had flown Gazelle, Squirrel and Apache, and as an operational pilot I was used to doing long conversion courses between one type of aircraft and the next – weeks, or, in the case of Apache, months. In the test pilot world, in stark contrast, a conversion course was a solitary check-ride with somebody who was qualified on type. When it was time for me to fly a Sea King, for instance, rather than just write about its cockpit, I had one hour-long trip with a QHI and that was it.

Next time in the air, I was on my own at the controls and accompanied by an Australian flight test engineer, who hadn't flown in a Sea King before and wasn't even a qualified pilot. Same thing with Lynx: a quick check-ride, and off I went.

It was a massive amount of trust in your abilities as a pilot, and a huge responsibility. But it also meant you could tick off new types of aircraft with enormous speed. We flew in Eurocopters, Chinooks, Agustas. We were taken to the airfield at Upavon in Wiltshire to spend a day trying out different types of glider. And along the way, the rotary pilots would get to fly fast jets – a Hawk, an Alpha Jet, a Tornado – and the pilots on the Fixed Wing Course would find themselves in helicopters, which is always interesting, and a great leveller. A helicopter pilot can be flying a fast jet in five minutes, more or less, but it's going to take the best fixed-wing pilot in the world an hour at least to learn how to get a helicopter into the air and keep it there. Fast jet pilots (without wishing to generalise) tend to have, shall we say, a bit of swagger about them. As the old joke goes: How do you know if a fast jet pilot has just entered the room? Answer: They'll tell you. I'd be lying if I said we rotary pilots didn't enjoy a quiet snigger at the hovering attempts of our fast jet brethren.

You could write whole papers on the differences in mentality and character between rotary pilots and fixed-wing pilots. Back in the early 1970s, the US journalist Harry Reasoner made some remarks, much quoted in discussions of this topic ever since, about the basic distinction between planes and helicopters – the former inclined to fly naturally, the latter kept in the air by effort. And he suggested that this explained why '*airline pilots are open, clear-eyed, buoyant extroverts, and helicopter pilots are brooders, introspective anticipators of trouble. They know if anything bad has not happened, it is about to.*' There is a lot of truth in that, although I have met plenty of helicopter pilots who were open, clear-eyed, buoyant extroverts, too. Still, it was nice, during those phases of the Test Pilot Course, to offer the fixed-wing guys a welcome to our broody world.

What we all shared as aspiring test pilots, though (over and above an instinct for button pushing), was that broader desire to probe the limits – those of the aircraft and our own limits as pilots. All of us, in our various disciplines, had come up against the rules and regulations that bind operational flight. And at some point along the way it had clearly occurred to each of us to think, 'Who put that rule there? And why? And how did they arrive at that rule? There must be someone out there pushing at the boundaries in order to establish these regulations. There must be someone flying to the limit in order to come back and tell us what that limit is. So why can't that person be me?' It didn't matter whether your background was rotary or fixed-wing: this yearning we all had in common.

And the ETPS course backed us all the way in our ambition. We flew heavy transport planes, including a C-130. And one day a DC-3 arrived at Boscombe Down, a beautiful old Second World War aircraft, and we took that up. When we were at altitude, the female pilot I was flying with invited me to stall it. I was happy to do so, but I suppose I was being a little timid with it. After all, at the time this was the heaviest aircraft I had ever flown. Plus it was a slice of history. I didn't want to be the person responsible for breaking a museum piece. So I was easing back on the control column gently to lift the nose and the stall was coming on slowly.

'No, no, no,' said the pilot. 'If you're going to stall it, stall it.'

And with that she grabbed the column and yanked it right back. The DC-3 instantly went nose up, pitching high into the sky until it stalled and then did a huge great wing drop which I pulled us out of by dropping the nose, putting in opposite yaw and applying full power. It was quite a ride, so we did it a couple more times. I was thinking, 'This is nuts, throwing around a big old aircraft like this. This shouldn't really qualify as work.' Yet it did.

Of course, other aspects of the course were more readily recognisable as labour. A lot of hard study had to be done out of the cockpit. Indeed, I think of this as the year when my weekends entirely disappeared. I thought I had been immersed in books and manuals while I was learning Apache in America, but Empire Test Pilots' School was on another level in terms of the amount of studying that was required. And just to crowd the schedule even more, I had decided to take the option to work towards a degree. Boscombe Down had a partnership with Portsmouth University which enabled students to enrol for a BSc Honours qualification in Flight Dynamics and Evaluation. I hadn't gone to university at eighteen, but I could, in effect, do so now, at thirty-three. It was too good an opportunity to miss. My Sandhurst training and my various Army qualifications, including the ongoing ETPS course, counted as points towards the degree and then there were further modules to complete on top of that. That meant additional tuition and extra tasks on top of my regular coursework, and the pile of books by the bed grew even higher. It was worth the push, though. I came out of it with an expanded understanding and a BSc. Not that I had any thought of this at the time, but it would prove enormously valuable to be able to put that Honours degree on my CV when I eventually applied to the European Space Agency.

In November it was time for Preview. That was the course's final exam where we were put into groups of three and sent off to evaluate an aircraft that none of us had flown before. (Despite the ETPS's best efforts through the year, there were still some of these left.) My team included a Canadian and an Italian student test pilot and for our preview aircraft we were allotted the Kiowa Warrior, which is a small, armed reconnaissance helicopter used by the US military. And that meant spending two weeks at good old Fort Rucker in Alabama. It was fun to be back there, hook up with some old acquaintances and show the other two guys Mother Rucker and all the places I knew and loved from when I had been there before.

We performed a complete evaluation of the aircraft in all the categories that we had been taught – cockpit assessment, engine performance, stability and control, sights and sensors, and so on. Then we wrote a report which we had to present back at Boscombe Down. Our grades for that were combined with the grades for all our sorties and pieces of written work across the year to form our final qualification grading.

It was enough for me that, when my marks were added up, I had passed and graduated, but actually it got even better. In December, the Empire Test Pilots' School held the McKenna Dinner, the formal post-graduation presentation evening: supreme of chicken with watercress sauce, strawberry bavarois, and music from the saxophone quartet of the Central Band of the Royal Air Force. No expense spared, clearly. After the dinner, when the prizes were given out, I found that I had been awarded the Westland Trophy, which is given each year to the best rotary wing pilot on the course. Given the company I was in, that was a lofty distinction, and most definitely a career high. I found it hard to think that things could get any higher.

* * *

It was a short walk to my next job. The Rotary Wing Test Squadron was just across the runway from the ETPS at Boscombe Down. I went over a couple of times in December to do some preliminary sorties and to refresh myself on Apache, which I hadn't flown at all during my year on the Test Pilots' Course. And then in January

2006 I reported for the official start of my three-year tour as a fully operational military test pilot.

I was the senior test pilot for Apache and a primary test pilot for Lynx and Gazelle. Shortly after that I became qualified on Puma and Chinook as well. But Apache remained my main concern. My arrival on RWTS – and this was no fluke, clearly; the Army had thought this through – coincided with my AAC friend Andy Cash and his squadron finishing their time with AMTAT, the course I had helped design in Yorkshire, and then becoming the first British Army Apache Squadron deployed out to Afghanistan. So this was a phenomenal time to be an Apache test pilot at Boscombe Down. We had that operational squadron's needs to service and we must have made about twenty-five mission-specific modifications to the aircraft in that first year alone. Though I worked with other aircraft too, I was to remain busy on Apache for the rest of the tour.

When I arrived in the squadron, though, Apache was still going through some of its initial trials and testing. It still didn't have clearance to fly in icing conditions – a basic service clearance for a helicopter, and nothing to do with fighting it in an operational environment. That January, two of my colleagues, Andy Ozanne and Mark Chivers, were out in Nova Scotia in eastern Canada doing an Apache icing trial and I flew out to join them for a couple of weeks.

Nova Scotia was a good place to go for this test because you need warm, moist air which gets raised up in the clouds to form lots of supercooled water droplets. The idea is that you then put the aircraft into this supercooled water, let it freeze onto the airframe and see what happens. You're monitoring how much ice the aircraft is accumulating and whether you're able to exit those conditions if the ice accumulates too quickly and gets too heavy. You put a gauge by the pilot's window – essentially a metal ruler

– which shows you how much ice you're picking up and then you note what the accumulation is doing to the power you need for straight and level flight and to the vibration levels in the aircraft's frame.

This was where we discovered something rather fortuitous about the Apache: when it got heavy with ice, the vibration would build up, as expected. But then the vibration would get to such a point that it would shed the ice off the blades and the air-craft would start flying smoothly again. It would do this far more efficiently than the blade de-ice system that was actually installed to do the job. It was almost as though the Apache had its own instinctive de-ice system, or was capable of shaking itself out, like a wet dog.

Nova Scotia was the last baseline test, though. After that, everything we did with Apache was Afghanistan-focused, and fell under what's termed Urgent Operational Requirement, or UOR. That, too, was lucky from my point of view because UORs tap into a different and deeper funding stream than standard testing, so money was pretty much always available to make sure that Apache was able to go and do its job. Also (and, again, the clue is in the name), UORs are deemed urgent, so there was a much faster and less laborious decision-making process at the procure-ment end of things, and a greater energy and intensity all the way down the line to get things done. Again, these were dream condi-tions for a test pilot. Your work was relevant, it was operational and it was fast.

Once the guys were on the ground over there, we were respond-ing to reports from Afghanistan about what they wanted. For instance, pretty soon after arrival, the aircrew realised that there wasn't much call for Hellfire missiles; what they needed more than missiles was a lot of 30-millimetre ammunition and a greater fuel capacity. So back at Boscombe we fell to exploring the possibility of hanging a fuel tank in place of one of the missile pods. That was a fine idea in principle, but you don't really want large amounts of fuel on the outside of a helicopter where small-arms fire could strike it and cause it to explode. So that in turn meant a smaller, self-sealing and crashworthy fuel pod had to be developed. A whole new system was designed, built by Robertson over in the US and then flown over, and we launched, tested and implemented it, all in a highly compressed time period. It was high-speed R&D.

We had to think hard about surfaces for landing and taxiing, as well. Our Apaches had only really landed on tarmac up to now – unless, of course, you count the time I landed that one in somebody's back garden. But in Afghanistan they would be landing and taxiing on dust tracks and semi-prepared surfaces. What kind of dust cloud does a landing Apache generate? What is the degree of brown-out? How suitable were the wheels and suspension for landing on tough, gritty surfaces? What kind of damage would sand do to the aircraft in the short and long terms, and what were the implications for the engine filter system? All of this had to be analysed and adjustments made accordingly.

And then how would the aircraft operate in 'hot and high' conditions, such as above the Afghan mountains where the heat makes the air very thin? We couldn't simulate that climate above Boscombe Down, obviously, but Arizona, in the screaming dry heat of the summer, was perfect. So we flew out there for three weeks in September 2006 to conduct high-altitude trials. We would wear oxygen – which was itself an unusual thing to be doing in a helicopter – and climb up to 12,000 feet pressure altitude, which in the heat was the equivalent of about 18,000 feet density altitude. This was a high-risk trial. The air was so thin that the helicopter became harder to control, and the vibrations in the frame extreme – so extreme at some points that it was impossible to sit in the forward seat and use a pen to make notes. All this had massive implications for the stability of the aircraft, which we needed to be right on top of. We would fly the aircraft heavy and do aggressive manoeuvring, taking it to its limits, well beyond the point that test pilots from Boeing, the manufacturer, had explored, defining what it could and couldn't do in those conditions. It was a revealing and highly productive flight trial with real implications for the eventual operation of the aircraft in theatre.

We had been steaming through the UOR Apache trials at Boscombe, one after the other. Then one afternoon Andy Ozanne and I were told to report to the boss's office. The commanding officer at the time was Ian Burton, an Army lieutenant colonel. Ian explained that what he was about to tell us was highly confidential.

'We need a couple of test pilots to go out to Afghanistan and check out an aircraft.'

Andy and I automatically assumed that this must be Apache-related. Apaches were in Afghanistan, Apache was what we had been doing, our bread and butter. If this was something to do with Chinook, or with Merlin, other people would have been summoned to Ian's office, not us.

'It's not Apache,' said Ian. 'It's Mi-17.'

That threw us. Mi-17 is a Russian-built, multi-role, military transport helicopter. So far as we knew, nobody in the NATO operation in Afghanistan was using Mi-17s. We would know more, Ian said, after we had been briefed by Joint Helicopter Command at RAF Benson in Oxfordshire.

Andy and I were thoroughly mystified. But at the briefing we learned (again, in strict confidence at the time) that a very small pool of pilots had been flying Ukrainian Mi-17s out in Afghanistan for Special Forces as part of something called Operation

Emperor – a counter-narcotics operation that was attempting to disrupt the opium trade and its funding stream to the terrorist organisations, Al-Qaeda and ISIS. Unfortunately, the Mi-17, although a brilliant hot-and-high workhorse, had notoriously poor tail rotor authority, not to mention a dreadful cockpit design with regards to instruments, dials and switches. There had been three crashes. No one, thankfully, had been injured thus far. But clearly Joint Helicopter Command now thought it would be prudent to get someone to go out and provide a formal assessment of the aircraft. That 'someone' was going to be me and Andy.

So in November 2006, under a thick shroud of secrecy, the pair of us packed for a three-week trip to Afghanistan. We were allowed to tell close family where we were going, but nothing about the operation or the purpose of our visit. All the same, I decided to tell my parents a white lie and informed them I was off to Arizona instead. I just knew I'd be giving my mum a month of sleepless nights if I mentioned Afghanistan in any shape or form. Clearly, the old 'off to Arizona' ploy wasn't going to wash a few years later when I boarded a Soyuz rocket headed to space, but for now, at least, I felt I could spare Mum some worry. I said farewell to Rebecca who, as ever, took an eminently practical view of the circumstances. Then I drove to RAF Brize Norton and boarded a TriStar bound for Kabul, via Cyprus.

By military standards, a TriStar is a comfortable way to fly. It's not like being strapped up against the walls of a Hercules. It feels like any commercial flight, really. At one point, the aircrew came down the aisle and handed Andy and myself our RAF lunchboxes. There was even an in-flight entertainment system, and you certainly don't get that on a Hercules. At the same time, on no commercial jet will you hear the announcement: 'Ten minutes to landing. Body armour on, helmets on.' That was how we arrived in Kabul: helmets attached, bulletproof jackets zipped up, entering an operational environment.

I had been on the verge of going to Afghanistan with the 1st Cav, after 9/11, only to be stood down at the last minute. Now I was finally there, and in extraordinary and highly secret circumstances. It was winter, and I was struck by the beauty of the place – stark desert and mountain. It reminded me strongly of Arizona, in fact, though far colder at this point. Not that we had much time for sightseeing. Kabul was an extremely high-risk zone and we were obliged to hunker down on the base and get on with our work of assessing the aircraft.

I had never flown an Mi-17. However, during the Test Pilots' Course, I had gone out to Poland and flown one of the Polish Air Force's Mi-2s. That was my first time in an Eastern Bloc helicopter, and it was an experience I hadn't forgotten. During the sortie, I was flying from the left-hand seat and trying to absorb the very strange cockpit layout when my colleague on the flight, a large, tough-looking Polish Air Force pilot, decided to switch off the hydraulics.

Now, this was something we did all the time while testing aircraft, and for lighter helicopters, like the Gazelle, it's a little like driving a car without the power steering; you can manage it, but it's going to take some extra muscle. So I began pushing at the Mi-2's controls, expecting them to yield eventually. They didn't, though. I was heaving away and nothing was happening. At the same time, I didn't want the big Polish Airman thinking that I was some kind of weed, so I continued to shove at the controls with gritted teeth. Still no movement. I realised I would have to concede defeat.

'I can't get it to respond,' I said, trying to sound nonchalant about it. 'Is there a button I need to push or something?'

The Polish Airman shook his head.

'No,' he said. 'We can't fly this one without hydraulics. Just showing you what it's like.'

And then he switched the hydraulics back on again.

Well, thanks for the demo, I suppose. At the same time, to switch off a helicopter's single hydraulics system mid-flight when you know that the aircraft is unflyable without it ... it probably goes without saying that that's not the wisest course of action, especially if you haven't bothered to warn the person who's actually flying the thing at the time. Anyway, that was my introduction to Mi helicopters.

I learned much more on the ground in Kabul, where there were a couple of Ukrainian engineers to explain the Mi-17 to us, albeit in pidgin English, and where an RAF flight sergeant was also in attendance. The first thing Andy and I did was conduct a cockpit assessment – a task complicated by the fact that, as far as we could tell, no two Mi-17 cockpit layouts were the same. Certainly none of them conformed to the standard T-pattern, which teaches pilots where to expect to find their aircraft instruments. The airspeed indicator in one aircraft looked like it had been taken from a MiG-29 jet fighter and it didn't even register anything until about 100 kmh. That might have been fine in a fast jet, but in a helicopter you certainly like to know what's happening at low speed, for reasons I think I've already explained. Furthermore, nothing was NVG-compatible for night-flying. There were no five-point harnesses for the pilots – just a lap strap, which would hardly have improved survivability. It occurred to us relatively quickly that, for very little money, you could make some meaningful modifications to that cockpit with significant consequences for the aircrew's safety.

Flying was necessarily restricted, because of where we were, but despite this we were able to conduct a fairly comprehensive flight trial over a couple of sorties. It was just like doing Preview at the end of ETPS. We covered everything: stability and control, engine performance, running take-offs and landings and, of course, we did some tail rotor authority testing on it, to establish just how robust that notorious tail rotor was. On the back of that we wrote a full report and were able to make several recommendations, not just aircraft modifications but also handling advice to aircrew. Our work made the Mi-17 a safer proposition for the operational pilots and, as a test pilot, there is nothing more satisfying than that.

When our assessment was finished, we had a two-day wait for the TriStar that was going to take us home. During that time the Special Forces guys asked us if we would like to join them on a night op. We were delighted. We had picked all over this aircraft – it would be good to see it in action. Andy and I climbed in the back with the SF boys, who were getting dropped off on an operation somewhere, and, once they had disembarked, we went back up in the air to loiter and patrol. We were waiting to be called back in to collect the unit on the ground when the flight sergeant on board mentioned that we were close to a range area that Special Forces often used for practice.

He said, 'Would you boys like a go on the GPMGs?'

Us? On the General Purpose Machine Guns? Well, if you insist ...

So we flew to the range. The Mi-17 was fitted with two GPMGs, one up front on the left-hand side and one at the back on the right. Andy took left, I took right. Down below us lay a field offering old vehicle hulls for targets which we could see through our night vision goggles. Our aircraft was generously stocked with belts of 7.62-millimetre ammunition mixed with tracer rounds, and the only instruction from our RAF friends was, 'Fill your boots.'

I had spent many years flying helicopters, but I have to say, right up there in my Top Five All Time Helicopter-related experiences would be the half-hour I spent down the back of that Mi-17 firing a machine gun out of the door.

Two days after that, the winter snow came driving in and our TriStar out of Kabul was cancelled. But planes were still flying out of Kandahar, an hour away, and a C-130 crew was prepared to defy the conditions and take us there, so we headed south. That was a bonus because in Kandahar we were able to go and see our Apache friends on the ground. They were a little mystified as to why we were there, and why we had come from Kabul, and, of course, we weren't able to tell them. It was good, though, to see how they were coping in the conditions and to be able to get a first-hand debrief which we could then feed back into the work that we were doing in the UK. It's an interesting thing about the mentality of front-line troops that a certain kind of fatality can descend on them – an acceptance of their lot, whether that lot is good or bad. So it was nice to be in a position to invite the guys to compile a kind of Christmas list of things they would like to see improved on the aircraft. We could assure them that, even though they might not get it all, we were in a position to bring about some of it, which would be better than nothing.

Shortly after this, QinetiQ, the private company that manages Boscombe Down on behalf of the Ministry of Defence, acquired a couple of Bulgarian Air Force Mi-17s, enabling us to do some more in-depth evaluation and adaptation of that aircraft and also, in due course, to invite over and train some pilots from the Afghan National Army Air Corps to fly it, both at Boscombe and on Salisbury Plain. All of that came on the back of the work that we had done in Kabul, so it clearly proved a useful interlude from our ongoing labours on Apache.

PART 3: SPACE

CHAPTER SEVENTEEN

Situation vacant, forms to fill,
and a boy called Makunudu

How do you become an astronaut? I get asked that question a fair bit. And the answer is that, in at least one respect, it's the same as with any other job. You answer the ad.

Rebecca saw it first. 'You should have a look at this,' she said. My friend Charlie Howard-Higgins sent me the link, too. 'This might be for you, Tim.' It was an announcement by the European Space Agency. For the first time in a decade, they were looking to recruit new astronauts – four of them, to be precise. And, for the first time under ESA's auspices, Brits could apply.

In previous astronaut selections member states which were paying to be involved in ESA's non-mandatory Human Spaceflight programme had put their own candidates forward, and ESA had taken their pick: we'll have this German astronaut, that French one, this Spanish one, and so on. This time, however, ESA had decided to run its own selection process from the get-go, and to make it open to all member states of the European Space Agency, regardless of whether or not they had signed up for the Human Spaceflight programme.

That meant the UK was included and the announcement of ESA's European Astronaut Selection process created quite a buzz in the British media. Not since Helen Sharman became the first

British astronaut, with the Soviet–British co-sponsored Project Juno in 1989, had the UK been in a position to send anyone to space. Suddenly the door was open again.

I didn't know a huge amount about the European Space Agency. I had read aviation and space magazines, so I knew something, very generally, about what we were doing there. But I knew very little about human spaceflight. I did know that there was a long-standing and distinguished connection between astronauts and military test pilots, extending all the way back to the Mercury Seven, and on through Gemini and Apollo. But I had never considered making the connection myself. Like a lot of people, I had followed the Shuttle missions and the ISS missions. And, as a seventeen-year-old, I had shared the thrill and the sense of national pride when Helen Sharman arrived at the Mir Space Station. But I regarded myself as an interested observer of human spaceflight, and certainly not as somebody who was considering it as a potential career.

Now, though, with a nudge from my wife, and another from one of my best mates, I was looking at the ESA website and starting to read the application material … and finding myself intrigued.

'The exploration of the Solar System will be one of humanity's most exciting adventures in the near future,' ESA's advertisement said. 'All of the world's spacefaring nations are preparing for this huge enterprise, and an astronaut corps is essential for Europe … Now is the time for ESA to seek new talents to reinforce its astronaut team, to prepare for missions to the ISS, the Moon and beyond.'

Two people close to me clearly seemed to think this kind of work would suit me. But what did I think? Did I even qualify? I began to examine the small print. First things first: the preferred age range was twenty-seven to thirty-seven. I was thirty-six, so within the scope. Secondly, the successful candidates would '*be able to speak and write English*': I could do that. (Russian, the second language on the ISS, would also be handy, but it wasn't a requirement and would be taught, apparently.) Thirdly, candidates would be expected to have a degree in Natural Sciences, Engineering or Medicine and at least three years of related postgraduate experience or flying experience as a pilot. Well, it was a good job that I had taken the option at Boscombe Down to get that BSc under my belt; I might have qualified on flying experience alone but, with so many candidates to choose from, it's more likely I would have been a faller at this hurdle otherwise.

What about physique, though? I kept myself fit; but did I keep myself astronaut fit? Well, there was no indication in any of this publicity material that they were looking for superhumans. Before you were issued with a password to access the application form, you needed to submit a Class 2 medical examination certificate from a certified Aviation Medical Examiner. No problem for me, of course, and a smart move by ESA, clearly: without that stipulation they would have been inundated with people just giving it a punt. Nevertheless the job description made it clear that they weren't seeking '*top-level athletes*'. It was enough, it seemed, to be '*healthy, with an age-adequate fitness level*'. You needed to have '*the normal range of motion and functionality in all joints*' and 20/20 vision, '*either uncorrected or corrected with lenses*'. And you needed to be free from disease, dependencies on drugs, alcohol and tobacco and '*psychiatric disorders*'. I could tick yes to all those.

Digging down deeper, I read the following: '*The applicant must demonstrate cognitive, mental and personality capabilities to enable him/her to work efficiently in an intellectually and socially highly demanding environment.*' There was more detail about what that might mean a little further on: '*General characteristics expected of applicants include but are not limited to: good reasoning capability, the*

289

ability to work under stress, memory and concentration skills, psycho-motor coordination and manual dexterity. An applicant's personality should be characterised by high motivation, flexibility, gregariousness, empathy with fellow workers, low levels of aggression, and emotional stability.'

I read all this and absorbed it. There was no doubt that there were things here which broadly tallied with skills I had acquired and the training I had undergone during my career. And there was no question that the timing was good. It was April 2008 when ESA announced that they were hiring. My three-year tour at RWTS had a few more months to run but it was time to start thinking about what I should do next. I knew that there would be future Army roles for me to consider in due course; I was less sure how I would feel about accepting them. In the meantime, astro-naut? Well, why not? Surely it couldn't hurt to apply.

Looking back now, it's interesting how the selection material gave you a flavour of what the job would be like, but didn't really make clear what shape your life in the broader sense might be likely to take. Where were you going to live, for instance? Were you going to be travelling around a lot? How long were you likely to be away for? The first year of basic astronaut training seemed to be set to take place at the European Astronaut Centre in Cologne, Germany. But what would be happening during the subsequent two stages that the material outlined – the advanced training and the mission-specific training? I don't think ESA themselves really knew how it was going to work for the new intake at that point. So just by applying you were announcing your willingness, at some level, subconsciously or otherwise, to cede control over the stability of your home life. Of course, Rebecca and I were hardly strangers to that. It was the military mindset: this year Germany, next year, we'll see. But the act of applying was still a leap of faith in that regard.

And it was clearly a leap I was prepared to take. I submitted my medical certificate and then received the magic password which unlocked the application form. As I scrolled through it on my laptop at home one evening, I was struck by how basic it was. It seemed to be no more than a long list of questions demanding straightforward answers about yourself and your qualifications. Can you swim? If you have a degree, in what subject is it? If you're a pilot, what have you flown, and how many hours? What are your hobbies? Would you say you were good at working with tools, and if so, what experience have you had? There really wasn't a huge amount of scope along the way to expand and express yourself. Only at the very end were there questions which invited you to break free and open up a bit.

Why do you want to become an astronaut? (750 characters)

In your opinion, what are the main tasks that should be performed by an astronaut? (750 characters)

Write a candid description of yourself as a person (including assets and limitations). (750 characters)

It was obvious that the answers on this part of the form were going to be pretty much the only way in which ESA would be able to differentiate between thousands (no doubt) of similarly well-qualified candidates and decide who was going to make it through to the next round. So I thought to myself: 'These three answers had better be bloody good.' I spent ages thinking about what I was going to put in those boxes, mulling over it on the drive to and from work and at home in the evenings over the next few days, making notes, whittling away at my responses, drafting and re-drafting those 750-character clusters. I didn't want there to be an ounce of excess flesh on them. I ran them past my dad, too. He had been a journalist most of his life: if anyone could spot the flab on a piece of prose, he could. Ditto any grammatical and spelling mistakes. I wanted those statements to be as powerfully persuasive as they could be. I felt they might be the only hope I had.

Eventually, I was ready to enter my words onto the form. So, why did I want to become an astronaut? '*I wish to participate in a programme that seeks to further the knowledge and understanding of humankind, expand frontiers and have an enduring positive impact on society and our planet as a whole,*' I wrote. '*My experience as a military experimental test pilot has given me the opportunity to explore boundaries, overcome challenges, manage risk and cope with unforeseen events. It is within this domain that both my strengths and passion lie. I believe that the skills and personal qualities that I possess, which enable me to do this job well, will also enable me to make a valuable contribution as an astronaut within the European Space Agency.*'

What did I think I would be doing? '*An astronaut should be trained and prepared to perform any task required to support ESA's mission to shape the development of Europe's space capability and ensure that investment in space continues to deliver benefits to the citizens of Europe and the world,*' I wrote. '*Specific tasks may include scientific experiments and research, systems operation and maintenance and mission support tasks. During a space mission an implied task would be to ensure that the integrity and well-being of the crew and spacecraft are maintained to the highest possible standards at all times.*'

And what assets did I suppose I could bring to this job? '*I am a highly motivated, extremely loyal team member with strong powers of initiative and good communication skills,*' I wrote. '*I have proven leadership skills as an officer and as captain of a crew in peacetime and combat environments. I am a quietly confident and unassuming person, who is able to speak with conviction and authority to varied audiences. I have to guard against assuming an excessive workload. I pay meticulous attention to detail and have a sharp analytical mind. I enjoy holding positions of responsibility and being challenged both physically and mentally. I work well under pressure, maintaining a calm and level approach in all situations and have a well-known sense of humour.'*

It was the most convincing case for myself that I could make in a short paragraph. And whatever the people at ESA ended up making of it, in the course of writing it I had at least gone some way towards convincing myself. Filling out the form had inevitably caused me to look back over my life and experiences: my work as a test pilot; my time as a flight commander; Kenya, Bosnia, Gütersloh; that Resistance to Interrogation course; the parachuting, the potholing; my training at Sandhurst; Operation Raleigh, even those young escapades in a scratchy shirt with the Cadets and campfires with the Cub Scouts – all of the things I have written about here in this book. Maybe all of it together had brought me to this point. Maybe I had been preparing to become an astronaut all my life. I just hadn't realised.

I filed my application. Meanwhile, around Europe, 8,413 other people were doing the same. And then I went back to work at Boscombe Down and did my best to forget about it. After all, seriously: over eight thousand people for four jobs? What were the odds?

Soon after this, my mum – whose son had thus far put her through two active military service campaigns in Northern Ireland, another in Bosnia and one (covert) in Afghanistan, and whose day job was testing helicopters to the limits of their physical capacity – was off to have coffee with her best friend Val.

'Tim's applying to be an astronaut,' she said, with all the enthusiasm, I imagine, of a parent whose offspring has just indicated that his future might lie in strapping himself to a 300-tonne rocket.

'Well, he's going to get that, isn't he?' said Val.

'Yes,' said my mum, resignedly. 'He is.'

I only wish I'd shared their confidence.

*   *   *

Had Rebecca and I known that having children wasn't going to be easy for us, we might have started trying a bit earlier. Instead we had happily indulged our child-free days, travelling in the States for three years and then returning to the UK to busy ourselves with our respective careers for a couple of years more. By the time we started thinking seriously about starting a family, both our sets of parents were probably beginning to wonder whether they would ever see grandchildren from us.

And then, when we finally did get round to thinking about it, we couldn't seem to make it happen. We spent two years trying, without success. Our best chance, it seemed, was going to be IVF. We had gone right along that route and were ready to embark on the first treatment, when, in the early spring of 2008, Rebecca discovered she was pregnant. Her grandmother did some sums and became firmly attached to the theory that conception must have taken place during a wonderful romantic holiday we had recently taken in the Maldives, on a remote island called Makunudu where we had blissfully dived and eaten shellfish and sunk cocktails at sunset. In fact, such was Nana's conviction on this matter that she felt the child should be christened Makunudu.

The reality, alas, was a little more prosaic. Conception must have happened before that holiday, meaning Rebecca had clearly spent the early weeks of her pregnancy doing things that you shouldn't really do, such as diving, eating shellfish and sinking cocktails at sunset. And as far as names were concerned, Larkhill didn't really have the same ring as Makunudu. As it happened, Rebecca's due date was 25 December, so we gave the infant the working title 'Baby Jesus', with a commitment to review it nearer the big day.

Come the summer, Rebecca was now visibly pregnant. And I had just had an email from ESA. My application form had done the trick. I had been invited to attend round one of the selection process in Hamburg in August. Over eight thousand applicants had been reduced to around nine hundred after inspection of the application forms. And I was one of them.

Flights and accommodation were all covered, and my portion of the nine hundred was instructed to arrive the night before at a standard conference hotel in the city. We all gathered in the bar that evening, mingling in our polo shirts and chinos in a perfectly friendly atmosphere, yet very obviously checking each other out as rivals. I began to find it all a bit intimidating. There seemed to be an awful lot of people there with PhDs, and an awful lot of highly intelligent people in the room in general. It was all quite deflating. I had gone there, quietly thinking I was bringing at least something to the table, and the next minute I was talking to a physicist with a doctorate who had spent the last two years working at CERN. I was thinking, 'This is a tough crowd. Do I even have a hope in hell here? An Apache pilot with 3,000 hours? That can't be what they're looking for.' My confidence was badly chipped that evening. One person particularly stood out from the throng: a tall, bald test pilot from Italy – very gregarious, very confident, very happy to tell his stories. His name was Luca Parmitano. I had a feeling he might do quite well.

The next morning, I woke up feeling terrible. Nothing to do with the previous night in the bar. Nobody was going to get carried away the night before an astronaut selection test. Like everybody else, I'd had one beer, just to show that I was willing to socialise. But now I felt awful – hot and achy. Something was clearly up. Great timing. But I just had to ignore it, get myself together and press on with the day.

The tasks were fairly broad spectrum: a memory retention test, a science and engineering test, tests for spatial awareness and concentration. There was a fast mental arithmetic test, which I found a struggle. And there was a multiple-choice psychological test, where you had to respond to loads of quick questions in a short space of time. Some of the questions were repeated several times during the test, albeit in a subtly different way, but you were too overwhelmed to remember what your response had been the first time, so you couldn't falsify your answers.

There were a couple of points in the day when I thought my flying experience was helping me out, such as the coordination and reaction tests, but there were plenty of others when I thought, 'This is just plain tough.' It didn't help that I kept wondering what the policy was: were they after perfection here? It was a Space Agency, after all. Would they rather I got to the end of the test, and answered all the questions, but got a few wrong on the way through? Or would they prefer that I only completed 80 per cent of the paper, but did that much perfectly? All that second-guessing was going on in my mind, rather distractingly.

I battled on and eventually the day completed. I returned to the airport and flew back to the UK, quite dispirited and also, by now, feeling properly ill. Rebecca was away and I put myself to bed with what had become, by the time I got through the front door, a raging fever. The next day, I dragged myself to the doctor and was diagnosed with a urinary tract infection. The gods didn't seem to have smiled on me, exactly, in Hamburg. The field was huge and impressive; the tests had merely gone OK, I thought; and I had felt like death warmed up throughout.

Forget it and move on, I told myself.

At least I had a big work trip coming up to take my mind off it. In September, along with a handful of my RWTS colleagues, I was off to Arizona to conduct an asymmetric trial on the Apache. We based ourselves at Show Low Regional Airport in Navajo County, a very high-altitude airfield not far from the Grand Canyon. There our task was to load up the aircraft's wing pylons to shift its lateral centre of gravity and then take it up to see how that imbalance affected the Apache's performance. Our findings would have important implications for the loading of the aircraft during operational use. Imagine you have a heavy external fuel tank on one side of the aircraft and missiles on the other. You might well be balanced on take-off, but if you fire all those missiles early on in a sortie the aircraft could become dangerously unbalanced. Our role was to assess how far you could push it. It was another occasion when I had cause to love my job: doing some amazing and challenging flying in the most beautiful scenery. We managed to have some fun at ground level, too. All of us were motorbike nuts and on our free weekend we hired Harley Davidson hogs and rode through the canyons in full *Easy Rider* mode. It was a sensational trip altogether.

On top of that, Rebecca had some time off and we tagged a holiday on to the trial, taking a 4x4 around the stunning canyons of Arizona for a few days – our last chance, we knew, to grab a far-flung vacation before the arrival of Baby Jesus. One evening in our hotel room in Phoenix, I logged on to my email, and there in the inbox was a message from ESA: my rejection letter, presumably. Oh, well. I wasn't going to let it spoil the mood. I opened it up.

'*Congratulations: you're through to the next round.*'

I was elated – and also stunned. I genuinely hadn't expected this at all. Hamburg had been tough and at times I had found myself well outside my comfort zone. Yet I was through. And now, for the first time, I allowed myself to wonder. If I could survive that … well, maybe I actually had a chance here.

Then again, eight thousand had become nine hundred. And nine hundred had now become about two hundred. And that was still a lot of people, going for only four places. All those doctors and CERN scientists, and that bald Italian guy with the stories ... They'd all still be in play, you could bet. It continued to feel like the longest shot.

Anyway, the second round of the process was going to be in Cologne at the end of October and the beginning of November. '*Would you be available?*' the email wondered. I couldn't think of any reason why not.

* * *

Back in the UK, I now had to do some serious thinking about what to do when my tour with Rotary Wing Test Squadron concluded at the end of the year. For a major, at the stage I was at in my Army career, one of the natural steps would have been to apply for a role as squadron commander. That's a prestigious position that a lot of people aim for and not everyone gets the opportunity to fill. There appeared to be a couple of options for me in that line. I'd been told I had a good chance of commanding 657 Squadron, which was then part of the Joint Special Forces Aviation Wing, based at RAF Odiham in Hampshire. That was considered to be one of the best squadron command jobs you could do. Or I could perhaps go to Hereford and command 8 Flight, which was working on counter-terrorism with the SAS and flying Agusta 109 helicopters – a smaller flight than at Odiham, clearly, but a unique and elite role in a highly active area.

My dilemma was that I wasn't simply weighing up the merits of the two postings. I was trying to work out whether squadron command was actually a job that I wanted to do at all. Obviously I knew that it would be interesting and exciting work and valuable experience. But it also had one important shortcoming, as far as I was concerned: it wouldn't put me in the cockpit of a helicopter anywhere near as much as I had grown used to as a test pilot. The squadron commander shouldn't be in the cockpit, after all: they're there to command their squadron, to oversee and provide guidance and direction, not to do all the flying. Did I really want to be doing less flying at this point? Wasn't being a test pilot the thing that I really loved?

Also, there was the question of the future. Beyond squadron commander, the next job you would be looking at, if everything went smoothly, would most likely be a staff job of some kind – again, a challenging, interesting and privileged position, but one taking place even further away from a cockpit. My concern was that, by accepting a squadron command, I would be starting myself down an Army career path that would lead me inexorably towards the thing that I had been doing my best to avoid the whole time – a desk job.

I struggled with the decision for a while. I had been in the Army nearly eighteen years. I had made it through to Major and I had worked myself into a position where, as my seniors made clear to me, further promotions and greater responsibilities quite likely lay ahead for me if I continued to apply myself. Yet, for all that, my heart was still where it had always been, from the very beginning: in aviation. Flying was the soul of it, for me. Flying was what had drawn me here in the first place and flying was what I had devoted the best part of my energies to. I knew that the best way for me to carry on flying in the way that I most enjoyed would be to become a test pilot on the civilian side of the aviation industry. I could possibly have gone further up the Army ladder and then made my way across to civilian aviation a little later, but it would have been a harder transition to manage. The optimum time for me to get a job as a civilian test pilot was clearly now, when I was at the top of my game as a military test pilot. As hard as it was to contemplate, this was tactically the best moment for me to leave the Army.

I began to explore my options. Our work at Boscombe inevitably put us in close touch with people on the civilian side at Westland and at Boeing. I knew Rich Lee, a Boeing test pilot who frequently came over from Arizona to work with us, and I talked to him about the possibility of going to work with Boeing in the States. But there were rules and regulations regarding foreign nationals working on military test programmes and Boeing, although I think they would have been happy to take me in some capacity, wouldn't have been able to employ me on the development of Apache.

I also interviewed for a test pilot job in Brisbane, Australia – which is a long way to go for an interview. Conveniently for me, during your three years as a test pilot at Boscombe Down, you were allowed to attend one overseas conference of professional interest. People would often go to Los Angeles, for instance, for the annual test pilot conference there, or to helicopter-related conferences in all sorts of exotic places. But I was coming to the end of my three years and I hadn't cashed in that ticket. I now did so, flying out to the Heli-Pacific Conference in Brisbane, and fitting in an interview at a company where an ex-AAC guy was already employed and where they were hiring test pilots to work on Tiger and MRH-90, which the Australian armed forces were just bringing into service.

But then Richard Morton, a test pilot at Westland, told me in confidence that he was thinking of leaving the company shortly. And soon after that, Don Maclaine, Westland's chief test pilot, whom I also knew well, casually asked me if Richard's job was something I might be interested in doing. It certainly was, and the timing couldn't have been more fortuitous. Richard was planning on going in early 2009, exactly as my tour was ending. I went and had a more formal chat about everything with Don in his office at Westland's Yeovil headquarters. It sounded as though it would be a virtually seamless transition. I would be flying Apache, Agusta 109, EH101, and something called Future Lynx, which was a brand new Lynx helicopter, still in the process of being built at that point. That was going to be really exciting. In the military, you don't get to do first flights of production line aircraft: you take production helicopters and modify them, push them to go further, higher and faster. But Future Lynx would be entirely unflown. It would be a question of jumping in, seeing if the thing flew straight and taking it from there. All of this was extremely alluring.

However, during that conversation, I did feel that I needed to make a confession to Don.

'Look,' I said. 'I've applied to be an astronaut.'

On reflection, it was an odd conversation to be having in an office in Somerset. It felt a bit like telling someone: 'I've applied to be a cowboy.' Lord knows what was in Don's mind at this point. He was probably thinking, 'Yeah, an astronaut. Right. Well, come back later and tell us how you get on.'

But I persevered. I explained that I was a little way into the selection process, that there were still hundreds of candidates in the running, that they were only going to take four, that they were highly unlikely to choose a Brit on account of the background politics, and that, accordingly, my chances were worse than vanishingly slim. But I at least wanted to be upfront with Don about it, and I wanted to state that I was really keen to see the process through to its conclusion, whatever that might be, because, when all was said and done, it was a fascinating process to be involved in.

Don had no problem with that. 'Just keep me posted,' he said, even managing to stop himself from laughing as he said it. I could start at Westland Helicopters in Yeovil in February.

With my future seemingly in good shape, I flew to Cologne for round two of the ESA process. Again, the travel, the hotel and expenses were all smoothly sorted. Gerhard Thiele, a German astronaut, was in charge of managing that part of the selection campaign and it was formidably slick. You felt like you were dealing with a finely tuned organisation, which wouldn't always be the case further down the line, but for now it was extremely impressive.

This time the focus of the testing seemed to be on teamwork, communication and leadership. We were put into small groups and given initiative tests in the form of problem scenarios: you've got to get off an island, but this side is mountains, and that side is crocodile-infested water, and you only have these pieces of equipment ... You would have twenty minutes to talk about it between yourselves and then you would have to present your plan. In another phase you were split into pairs and given a traffic management problem – controlling a city's traffic-light system on computer screens, with one of you operating the east–west roads and the other operating north–south. And there was a team goal (to get as many cars through as possible), but there were also individual goals, and from time to time a piece of information would pop up on your screen about the developing situation, handing you a dilemma: should you pass it on, to the benefit of the group goal, or intentionally withhold it, to the benefit of your individual goal? Or, given the complexity and frenetic pace of the test, did you even have the capacity to notice the message in the first place?

It was amazing how much pressure that test seemed to put people under. The key to it, I guess, was to try and relax about it: to accept that you were going to get overwhelmed with information and that there was no way you would be able to absorb all of it, and simply to try and take in as much as possible, help your teammate as much as possible and try and achieve your own goal as much as possible. But some people came out afterwards and you just knew that they were blown: they were fuming with anger or utterly stressed out. It was really interesting to watch and all the time during these tests, there would be a psychologist standing over your shoulder with a notepad and pen. I don't suppose that helped to diminish the tension, either.

There was also an interview at this stage, which meant I met my first astronaut – the Frenchman Jean-François Clervoy, known to the world as Billy-Bob. That's because when he reported to NASA, prior to flying the Shuttle on his first mission, and introduced himself, a Texan sparked up and said: 'What? We can't say that. From now on, you're Billy-Bob.' He seemed to love that and embraced it. Jean-François was part of a four-person interview panel, along with two ESA instructors and a psychologist. I had about fifteen minutes with them in a relatively genial atmosphere – a little bit of probing, but nothing too painful or searching at this stage and no obvious trick questions. Years later, I was chatting to Jean-François and I asked him what his criteria had been when he was conducting those interviews.

'Quite simply,' he said, 'I would come out of that room and if I thought I wanted to fly to space with that person, then that was a plus. If I didn't want to fly to space with them, then no.'

For all the refinement of the selection process, that one question did seem to nail it, really, or certainly from a fellow astronaut's point of view. 'Could I live with this person in a tin can for six months?' I had to hope that I had come across as a viable colleague.

I left Cologne in pretty good spirits – definitely better than I had felt after Hamburg. The tests had gone OK, in so far as I could tell – no obvious disasters, certainly. I had felt altogether less intimidated by the other candidates than I had in round one. I still had no real concept that this venture could yet end with me becoming an astronaut. That still felt like some sort of mad dream. But even if it stopped here, it had been an experience, something you don't get to do every day. And at least I had got to *meet* an astronaut.

A couple of weeks passed. We were getting on towards December 2008 when another ESA email dropped into my inbox. It appeared I was through to round three. The axe had fallen again and two hundred had now become forty-five. Next stage: Cologne, again, for a week of medicals.

Maybe this would be a good moment for Rebecca and me to have a little chat.

From the application form to the two selection rounds thus far, I had fully committed to the ESA process. But I had fully committed to it in the context of a genuine belief that I didn't actually stand a cat's chance in hell of succeeding. So up to this point, Rebecca and I had never felt we needed to have a serious discussion about what me becoming an astronaut might mean for the future direction of our life together. The prospect seemed so outlandish, it would have been hard to have the conversation without laughing.

But now things were looking a bit more serious. True, forty-five candidates remained, going for just four places, so the odds were still slight. And the political background hadn't changed: the likelihood of a Brit getting picked over candidates from countries who had historically invested in the relevant ESA programmes still seemed remote. And yet we did appear to have reached a crunch moment. I knew that a week of medical testing for forty-five people would be an extremely expensive thing for ESA to organise – a decent slice from the €6 million budget that was being spent on this selection process. From that point of view, it wouldn't really seem right to go through the medicals and then ultimately say no. Agreeing to participate in the medical round seemed to me the equivalent of agreeing to say yes, in the (unlikely) event of being offered the job at the end of the process.

But did we want to say yes?

On the face of it, it seems such an obvious answer: who wouldn't want to be an astronaut? There are many people out there who would give their right arm to be an astronaut, under any circumstances. But there was a lot at stake here and it wasn't a decision I could take alone. Incidentally, the people who would give their right arm to be an astronaut aren't the people the Space Agency is looking for. It wants people who have taken a well-informed, considered approach and used good judgement in making their decision to commit.

Rebecca and I sat down and talked it all through – the pros and cons, the whys and the wherefores. There was, as I mentioned before, a lot of built-in vagueness about where in the world this job would take us, how much time I might have to spend away from home and the extent of the upheaval that might lie ahead. My test pilot career had just taken a new direction that promised both excitement and stability for the family that we were now starting. We could buy a house in Yeovil and settle down after all the moving around. With ESA, there was no guarantee of a mission to space. Being selected as an astronaut is just the first step on a long and difficult path to getting a spaceflight. It felt like there were a million uncertainties and this needed to be a decision that the pair of us made together.

And then Rebecca said what we were both thinking. 'I can't see us turning this down and forever thinking, "What if?"'

Sometimes you just need someone to verbalise something and everything clarifies. The sacrifice and the separation and the disruption that you might be letting yourself in for by accepting a challenge could prove hard. But if you didn't accept it, if you walked away and all you had to live with was 'What if?' – well, wouldn't that be harder still? Better to make the leap.

I signed up for Cologne.

*   *   *

Christmas came and went. Still no sign of Baby Jesus. New Year came and went, too, and still no indication that this character had any intention of making an appearance. So much for due dates.

The thought occurred: Was it actually a phantom pregnancy? The size of Rebecca's tummy suggested otherwise, and yet …

It was 1.00 a.m. on 5 January 2009, the coldest night of the year. We had retired to bed, following a Sunday-night curry and a movie, and I was just dozing off when Rebecca suddenly announced that her contractions had started. I sleepily suggested that it was probably just the curry and asked if maybe we could chat about it some more in the morning. I then opened one eye wide enough to see Rebecca's expression and realised that this was not the response she had been hoping to get.

I stirred myself and began carefully timing the contractions, as briefed. Despite my reluctance to face reality, I had to concede that they were coming less than four minutes apart and seemed to be causing Rebecca more distress than any curry had previously managed. Now operating on autopilot, I grabbed the prepacked 'go-bag', helped Rebecca into the back of our VW Golf and set off for Salisbury hospital, 15 miles away. Thankfully there was little traffic and, crucially, no police cars around at that time of the morning to witness Rebecca balanced on all fours on the back seat of the car while I negotiated the snow and ice en route to Odstock Hospital.

On arrival, I expected things to move pretty quickly. It seemed to me that we had got there in the nick of time and I was going to be happy to take the credit for some pretty heroic driving, if I say so myself. However, a less than impressed midwife informed us without ceremony that we shouldn't really have bothered. She was obliged to agree that the contractions were indeed as frequent and as regular as we had reported half an hour earlier on the phone, but nevertheless we were going to be in for a long wait.

She wasn't wrong. Twelve and a half hours after my highly skilled mercy dash through the Wiltshire snow, and with no pain relief taken by the mother at any point (unless shouting counts as pain relief), Rebecca's labour began to reach its climax. Those were tense moments, as it happened. The baby had decided to arrive with the cord wrapped around his neck, and his heart rate dropped dramatically in the final moments. A bed had been prepared in the operating theatre for an emergency Caesarean and they were going to let Rebecca have one last push before they whisked her away. So Rebecca pushed. And there was Thomas.

There is life before children and life after children – every parent knows this – and it's fair to say that I was not prepared for the dramatic extent of the difference. I'm not talking about the nappies, the sleepless nights and the shirts speckled with baby vomit, although those were something. It was the shift in outlook that took me by surprise. Suddenly a tiny new human had arrived who was completely dependent on Rebecca and myself and who had instantly and effortlessly taken up a position at the centre of our universe.

Another question I am often asked: Did spaceflight change my perspective? Absolutely it did. You can't look down on the planet from up there and not see the world differently forever afterwards. But did it change my perspective as much as becoming a father did? Not even close.

CHAPTER EIGHTEEN

*The enema within, Civvy Street,
and a big day in Paris*

I had a new-born son, I was about to leave the Army after nearly eighteen years and embark on a new job, and I was on my way to Cologne for a week of medical tests which could possibly lead to me advancing to the sharp end of an ongoing selection process for European astronauts. 2009 was already shaping up to be one of my life's more eventful years. And it was only January.

Again, the slick ESA machine seemed to have clicked into action. I had a plane ticket and a reservation at the Hotel Zur Quelle, a place I would come to know very well in subsequent months, a business hotel situated close to the entrance to DLR, the German Aerospace Centre where ESA is based. I was in a group with ten other candidates – one of two batches going through the process in Cologne in consecutive weeks. The rest of the forty-five were despatched to Toulouse to undergo the same procedures there.

And what procedures. Everything you have read about astronaut medicals in books on the Apollo programme in the sixties and seventies turns out to be true. It's no holds barred. You might as well hang up your dignity on the way in and pick up what little of it is left on the way out. Over the course of that week, every part of me got prodded, poked, squeezed and measured, including some parts of me that I had only been dimly aware I had. If it would take a probe, they probed it. It wasn't that they were looking for super-fit athletes. They were looking, in as much as they could ascertain, for people who were medically low-risk – people who weren't going to cause an inconvenience by going down with something while 400 kilometres away from Earth. And, boy, were they thorough about it.

There were eyesight tests – basic and advanced. There were ultrasounds for your major organs, your arteries and veins. There were DEXA scans for bone density. Every aspect of your plumbing got a thorough plunging, and fortunately there were no repercussions from my infection earlier in the year. And there were so many blood tests, I stopped counting. Blood seemed to be getting taken from me at every turn. I spent one whole morning walking around with a cannula in my arm, waiting for the next person to stick a test tube in and help themselves.

Then there was the VO_2 Max test, the cardiovascular fitness assessment where they hook you up to an oxygen mask and a blood pressure cuff and set you off running on a treadmill. This was the first time I had ever done one of those and I found it extremely strange and uncomfortable, not least because as I ran, every two minutes or so someone was clipping at my ear to get a blood sample to analyse for lactic acid. I love running – but running with wires all over the place while someone periodically punctures my earlobe … that's less enjoyable.

Reporting one day for my sigmoidoscopy – an internal examination of the lining of the lower colon – I rode in the hospital lift with three German women. We exchanged polite nods and smiles. Then I got out and found the necessary department where I was handed a gown and directed to a changing room down the corridor. The gown provided an all-over body covering except for the large hole halfway down at the back. Wrapped in this unglamorously revealing garment, I waddled self-consciously back to the clinic, where I was now sent to a treatment room, instructed to adopt the prone position on the examination bed and to wait for the surgeon's assistant, who would be administering an enema before the procedure.

I had never had an enema before and my mild levels of anxiety weren't helped when the surgeon's assistant arrived dressed in the usual gown and rubber gloves but also wearing a pair of wellies. Wellies? Holy smokes, what were they about to do to me that could possibly warrant wearing wellies?

As all this was running through my mind, the assistant smiled apologetically and proceeded to inject what felt like a startlingly large volume of liquid into my lower intestine.

'*Fünfzehn Minuten bitte*,' he said, and then pointed towards the loo, adjacent to the treatment room.

My German didn't amount to much, but I got the picture. After fifteen minutes of mounting anticipation, there followed a sprint to the loo and an explosion of almost volcanic intensity.

Well, that would explain the wellies, I thought as I headed back to the examination bed, adopted the required position of reclining on my side and waited for the surgeon to do his worst.

'Do you have any objections to some medical students observing the procedure?' the surgeon asked.

'Not at all,' I said, trying very hard to sound as if I meant it. And in walked the three women from the lift.

'Hello, again,' I had to say, smiling more weakly this time as I lay there with my bare butt in the air.

Some of us gathered in the hotel for a drink that evening and told our war stories about what we'd just been through. This was one of the nicest upshots of that medical week. During the previous stages, we had all been looking at each other slightly askance as competitors and being, perhaps, a little guarded. But here there was no competitive element and it released the tension between us. The eleven of us talked and bonded – the pilots, the scientists and the engineers all opening up to one another.

I struck up a friendship in particular with a French candidate called Remi Canton. Remi happily confessed to us all that, during the sigmoidoscopy procedure, he had somehow missed the instruction about going to the loo after the enema had been administered. He had simply lain there, growing more and more desperate. Eventually, when the surgeon arrived, he noticed his pained expression.

'You have emptied your bowels, haven't you?' said the surgeon.

'No,' said Remi. 'And I need to go really badly.'

The surgeon, looking somewhat alarmed, hurried him on his way as Remi, hunched over and not daring to stand fully upright, made a beeline for the loo. He calculated that he had held that enema in place for twenty-three minutes. We all decided that this was the absolute record for enema retention and unlikely to be surpassed. He deserved to get through the medical stage for that feat of self-control alone. Remi, incidentally, would end up doing me an act of enormous kindness further down the line.

When the week was up and I had no more parts to probe, I returned to the UK and to Rebecca and our four-week-old son. ESA would use the results of the medical testing to cut the candidates from forty-five to twenty-two. Those twenty-two would then face the final hurdle – the interview process from which four trainee astronauts would finally be picked. It all seemed tantalisingly close – yet at the same time light years away. After all, the politics were still the same. The tests might reveal my blood to be in good nick, but it was still British blood. There wasn't much I could do about that, except, again, try not to raise my hopes too high.

LIMITLESS: THE AUTOBIOGRAPHY

The problem was, the longer this process went on, the harder it was getting to control those hopes.

*   *   *

Some people find leaving the Army hard. So much of military life happens in a world-within-a-world: a self-contained place where you eat, drink, sleep and work. Breaking out of the bubble isn't always simple, especially for those who have been in there for a long time. I was very lucky in that respect. My own transition, even after almost eighteen years, couldn't really have been much softer. Without really planning it, I had been slowly withdrawing from mainstream Army life for some time. On reflection, going to Boscombe Down was really the start of me unplugging from the military. By attending the Empire Test Pilot School and then joining Rotary Wing Test Squadron, I was already working with Royal Navy, RAF and civilian test pilots and with the aviation industry more broadly. So the whole process had been, unwittingly, a slow and gentle exit. I didn't go through that jarring transition that some military personnel experience: years in the forces and then the next day out on civvy street in a suit and tie, looking around, feeling bewildered.

I even carried on living in an Army quarter. Because my official Army end-of-service date wasn't until April, Rebecca and I were allowed to continue renting at Larkhill while we looked for something else. We began house-hunting around Yeovil, near to the Westland factory.

My new job wasn't exactly a jolt, either. OK, I now had an hour-long commute every day. But I would often be flying aircraft into Boscombe Down, or the pilots that I'd been working with on Rotary Wing Test Squadron would be coming down to Yeovil, and we were all still dealing with each other on a daily basis. The

313

difference was that I had now switched sides. At Boscombe, I was working for the Ministry of Defence and doing my best to ensure the MoD got the best product from Westland, holding their feet against the fire. Now, I was a company man, trying to sell the product to Charlie Pickup, who had taken over from me as the senior Apache pilot on RWTS. But in the context of transactions on Apache, even that wasn't as oppositional as it sounds. We were all effectively working on the same team, all wanting the aircraft to be the best it could.

Anyhow, pretty soon I was having to go to my boss, Don Maclaine, and say: 'You know that astronaut thing I told you about? Well …'

I had got through medicals week. I was now one of the twenty-two. That meant I was into the endgame – interviews at the European Space Research and Technology Centre, ESA's base in the Netherlands, on the coast at Noordwijk. Did Don see where this was heading? I doubt it, any more than I did. But we had our prior arrangement and he kindly gave me the time off to attend.

I went out to the Netherlands that spring knowing nothing about the twenty-one other candidates. ESA didn't give out any information on that. They wouldn't have been keen on us communicating with each other – and frankly, I wouldn't have been keen on it either. In that situation, rumours abound and people tend to wind each other up needlessly. That said, groups had inevitably formed between some of the candidates during the process. There were those with whom you exchanged email addresses and stayed in touch. But by this stage all the people I had been chatting to along the way were out. They hadn't made it, including, sadly, my French friend Remi Canton, the enema hero, who wasn't in the end rewarded for his titanic struggles that day and was cut after that round.

* * *

But Remi did something I will always be grateful for. He sent me an email saying: '*Congratulations on the medicals – fantastic. Wish I could have got through that, but I didn't make it. Here are all my notes that I had prepared for the interview phase, in case they are any use to you.*'

Now Remi had been working as a contractor on the peripheries of ESA, so he had a perspective on the organisation that I, as a total outsider, could only aspire to. That was a generous gesture, then, and those few pages of A4 were some welcome background material for me.

However, with the restriction on chatting to each other about these things, it wasn't until I got to Holland that I learned that there were four other Brits in the twenty-two. Three of them were pilots – one an RAF fast jet pilot and two RAF test pilots – and there was a British-born doctor who was living and working in Australia. Five Brits in total! It only confirmed, really, how the open selection process had galvanised people in the UK and convinced them to give it a go – and also how high-calibre those Brit candidates had turned out to be.

So the day of the big finale duly dawned. I arrived at ESTEC more than an hour before my interview. ESTEC is a vast, impressive modern campus, the largest of ESA's R&D sites and the place they call 'the incubator of the European space effort'. I spent my spare hour walking around the Erasmus Space Exhibition Centre, a museum on the site which showcases ESA projects. Looking for things to take my mind off the interview, I was particularly distracted by the section devoted to the Cassini-Huygens mission, an ESA/NASA project which sent a probe, *Cassini*, to Saturn, and put a lander on Titan, one of Saturn's moons. I remembered reading a little bit about that mission in aviation magazines, but I didn't know much detail about it. The museum had a striking picture of the lander on Titan, sitting next to a lake which you might assume to be water but which was, in fact, liquid methane. It was a photograph, but it could have been a sci-fi poster – a stunning image. I studied it for some time.

At the given hour, I left and reported for my interview. There were eight people in the panel across the table from me, some of them ESA directors, including Simonetta Di Pippo, the director of Human Spaceflight. Also there was ESA's head of HR and Thomas Reiter, the German astronaut who was a veteran of missions to the Mir Space Station and the ISS. I remember being asked: 'Why do you think we should spend money on human spaceflight?' Also: 'What's the one quality above all that it's important for an astronaut to have?' And also: 'What makes you better equipped than any of the other candidates to be an astronaut?' They were quite open questions – the sort with which you could easily talk yourself into trouble by over-answering. I recall trying to say the right thing while, at the same time, trying not to say too much of the right thing.

And then came my big break. Somebody asked: 'What do you think has been ESA's greatest achievement to date?'

'The Cassini-Huygens mission,' I replied promptly. 'Landing the probe on Titan.'

I was then able to speak with conviction, and with some freshly acquired facts to hand, about that incredible feat of science and engineering, so far from Earth. I could tell the answer had gone down well. At the start, when I nominated Cassini-Huygens, Simonetta Di Pippo had laughed and turned, smiling, to one of the other guys on the panel who, unbeknown to me, had been project leader on that mission. That was an extremely helpful trip to the Exhibition Centre.

Soon after that we all shook hands and I left. The selection process was finally over.

Except that it wasn't. I got home from Noordwijk assuming that the next thing I would receive from ESA would be the big letter – acceptance or rejection. In fact, just a few days later, an email dropped into my inbox.

'*Congratulations, you've made it into the final ten.*'

Would I now attend a further interview, this time in Paris, with ESA's director general, Jean-Jacques Dordain, alongside Simonetta Di Pippo?

This was intriguing, to say the least. It was the first time anyone had mentioned a 'final ten'. The plan, as we all understood it, had been to select the successful four from the last twenty-two. But ESA had inserted an extra round into the process.

Now my mind was whirring. I was in a group of just ten candidates, going for those four places. The odds had come down again. Did I have a shot here? I rang my friend Remi to see if he knew anything about the nationalities of the other candidates. He had heard a couple of things, including that I was the only remaining Brit. Was that to my advantage, my disadvantage, or nothing at all? On the face of it, I knew, the maths wasn't promising. Italy needed to be regarded as a special case. More than 50 per cent of the pressurised modules on the Space Station were Italian-built through Thales Alenia, the French–Italian aerospace manufacturer. Those deals had been done directly with NASA, and not via ESA, for which, in return, Italy had received the guarantee of spaceflights for their astronauts. So at this point, in 2009, Italy had a 2013 spaceflight and a 2015 spaceflight already in the diary that were Italy's to give away, not ESA's. That seemed to indicate very strongly that two of the chosen four astronauts would be Italian.

What was also indisputable was that Germany were the biggest contributors to ESA's Human Spaceflight programme and France were the second biggest. So it stood to reason that the third and fourth slots would go to German and French astronauts. It was hard for me to consider all that and not conclude that my participation in this abruptly organised extra round of ten was just a formality.

On the other hand, why had they added it? If everything really was as preordained as it seemed to be, why wouldn't they just make their announcement and get on with it? And hadn't ESA indicated all along that this genuinely was a transparent selection process, open to all member states, independent of funding issues and background politics?

Whatever, I was excited to be getting another chance to prove myself worthy of the job. Having begged for yet more time off from my long-suffering boss, I flew to Paris, took a taxi to a small hotel near ESA's headquarters in the Saint-Germain district, and willingly submitted myself for the final grilling. Unsurprisingly, I arrived early for the interview the next morning. I wore a blue pinstriped suit and my Army Air Corps regimental tie. I'd already enjoyed a nice walk and a coffee in the sunshine, ESA's HQ not being far from the Eiffel Tower and the beautiful Champ de Mars. After a short wait, I was escorted towards a door from which a man of similar age to me, and also wearing a suit, had just emerged. It was clear that we weren't expected to converse with the other candidates, but we managed to chat generally to each other a bit before I was invited into the room. He struck me as a really nice guy.

That man turned out to be Matthias Maurer, from Germany. Matthias didn't make astronaut selection in 2009, but he did join ESA a year later and worked at the European Astronaut Centre on a variety of different projects. I was delighted when he was finally selected to join the astronaut corps in 2015, one of those outcomes which demonstrates how perseverance and determination really can pay off.

This interview felt quite different from the one in Holland. That had been an intimidating experience – tough questions being fired at you from a large panel of some of ESA's most senior and experienced staff. This one was no less intimidating, but much more intimate. This time I was presenting myself to the man in charge, Jean-Jacques Dordain, who with his friendly demeanour and quietly spoken questions seemed to be searching your soul through his intelligent eyes, which lurked behind a pair of comically small, round glasses.

Having already got to know us in Holland, Simonetta Di Pippo seemed content to leave most of the questions to the director general. The purpose of this interview was clearly for the DG to make his final cut. I answered his questions as thoroughly and professionally as I was able, and I couldn't help feeling that what he was really trying to determine, in addition to whether I would make the grade, was if I knew what I was letting myself in for.

I gave it the best I could. Now – unless there was another unscheduled interview stage that we had yet to hear about – it was all in ESA's lap.

There were three weeks between that interview and the date in May, which ESA had already set and told us about, of the announcement. As those days ticked by and no word came, my optimism could only diminish. Rumours had gone around how, in the previous rounds, ESA's communications with the candidates they weren't putting through hadn't been that great. Apparently some people simply never heard anything at all. That was how they knew they were out: the process simply advanced without them.

True or otherwise, it was now Monday night. On the Wednesday morning, in just thirty-six hours' time, ESA would be presenting their four new astronauts to the press at a big launch event in Paris. If I hadn't heard by now, it could only mean one thing, couldn't it?

Rebecca and I were at home in Larkhill, eating supper on our laps in front of the television. My phone rang on the sofa beside me. I glanced down at the screen and saw '+33' – a Paris number.

Here we go, I thought: it's the rejection call. It was too late in the day for it to be anything else. 'Sorry that you didn't make it, but thanks for your participation in the process and good luck with whatever you choose to do in the future …'

And that would be fine, by the way. I'd be a test pilot. We'd move to Yeovil, buy our first home together, settle down there and raise our family in Somerset. And I would always know that I had put up a good fight. I'd always have the satisfaction of having come close to being chosen as an astronaut.

I took the phone through to the kitchen, away from the television, before answering it.

It was the director general's deputy, Ludwig Kronthaler.

'Timothy, I'm delighted to let you know that you've been successful. Would you like to join the European Astronaut Corps?'

'Do you know what?' I replied. 'I think I'll pass. It's been fun and everything, but I think, at the end of the day, it's probably not for me, thanks all the same.'

No, of course I didn't say that. I said: 'Absolutely I would, yes.'

It emerged that, following the round of ten (and presumably in deliberations leading up to that additional round), they had decided to recruit six astronauts rather than four. I was in.

During the year gone by, as the process unfolded, I had occasionally dared to fantasise about the moment when I might tell Rebecca that this whole protracted and strangely implausible venture had come off, and that I had actually been picked to become an astronaut. I pictured the two of us together, somewhere nice, me saying, coolly, casually, with exquisite timing: 'By the way, I've got some job-related news …'

You can plot these things, just as you can plan romantic New Year's Eve proposals in Austrian ski resorts. But life lets you down. And so it was that, having followed me out because she sensed something was up, Rebecca now found me in the kitchen with the phone clamped to my ear, beaming all over my face like a stupid schoolkid, and knew right away.

There were so many feelings charging through us both at that point – elation, excitement, trepidation and, on top of that, an unnerving sense of all the uncertainty that we were about to unleash in our life. Clearly, we were off to Germany for a while, in the autumn. But then, after that, who knew?

'This is it,' I said. 'Last chance. Any doubts? Are we doing this?'

There were no doubts. We were doing this.

I called my parents who were, of course, enormously pleased and excited, despite, obviously, my mum having correctly predicted the outcome from minute one. I think some of our conversations about the politics of it all had undermined even her ironclad sense of conviction along the way.

And then I had to call Don Maclaine at Westland and break the news to him.

'Don,' I said, 'I'm really sorry, but I've been selected.'

He congratulated me and was clearly very happy for me, but at the same time I knew I was putting him in a difficult position. When my predecessor, Richard Morton, left, Don had had to work very hard to keep that job open, against a lot of pressure coming from elsewhere in the company to close it down. And with that battle won, now his brand new test pilot, having been there just long enough to get trained up and qualified on all the relevant aircraft, was buggering off to the Space Agency.

'And also, Don, could I have tomorrow afternoon off to go to Paris?'

This was not the easiest call to a boss I have ever made.

Having told the people who needed to know, I now realised there were some other arrangements I had to make. All ESA had given me over the phone was the address (at the ESA headquarters again) and the time for the press conference in Paris. A flight? A hotel? None of those things had been mentioned. I got on with booking myself a plane ticket. Then I realised that I didn't even know how long I was expected to be in Paris for.

I rang the director general's assistant back.

'Do you need me to stay the night on Wednesday, after the press conference in the day? Are we having, perhaps, a social or a meeting in the evening or something?'

The assistant thought about it. 'Yes, we'll probably have a social after the announcement. Or maybe we'll go out for dinner. Why don't you plan on flying back the next morning? That's probably the best thing to do.'

I hung up and went back to my booking sites, a little bemused. Cracks had finally appeared in the magnificently slick ESA process. Up to now, everything had been so well organised – the emails, the travel arrangements, the briefings, the crisp punctuality of it all. You thought to yourself, 'Yes this is space. This is the kind of organisation I want to be involved in.' Even coming from a military background, where things tend to move pretty smoothly, I was impressed. But in the last three weeks, with the interview that hadn't been scheduled, and the silence until the last minute, and the hasty call on a Monday night with no clear directions, the well-oiled machinery seemed to have dried out a bit. In the heat of the moment, and as I sat there that Monday night, Googling myself a cheap Paris hotel, it was a little disconcerting. What had happened to the super-slick space organisation that I thought I was joining?

At least things seemed OK when I went into work the next day, on the Tuesday morning. I felt a bit sheepish, seeking out Don, but I didn't need to. Westland was owned at the time by Finmeccanica, an Italian company, and Don had already phoned Head Office in Italy to report what was going on. Apparently Head Office was delighted.

'Really? We have an employee who's been picked to be an astronaut? What could be better?'

It was when they used the term 'we' that Don felt his shoulders relax. I think the bosses had sensed the potential value to the company in PR terms of having an astronaut on their books. Don wished me well in Paris. (When I eventually left in the summer, he was also able to keep that golden test pilot job open and appoint my successor.)

I left work after lunch and flew out to France that afternoon. I had been instructed to keep my selection as quiet as possible, to avoid leaks to the press which might have spoiled the big unveiling. But in the taxi from Charles de Gaulle airport to the centre of Paris, I thought of someone I really needed to tell: my friend Andy Ozanne, my accomplice on the trip to Afghanistan, the buddy with whom I had done so much flying over the past three years.

'I'm in Paris, mate,' I said. 'I got through.'

He was delighted for me and we laughed a lot about the implausibility of the whole thing. But after I'd hung up, I had a quiet moment in the taxi, staring out of the window. That was the point at which it really came home to me that I was leaving behind my life as I knew it and stepping out into a whole new phase.

The following morning I arrived, as instructed, at a hotel for breakfast with Simonetta Di Pippo. It was my first meeting with my new colleagues. There was Samantha Cristoforetti from Italy, Alexander Gerst from Germany, Thomas Pesquet from France and Andreas Mogensen, who was Danish. And there was a tall, bald, outgoing Italian test pilot, very familiar to me from the first round of the process back in Hamburg a whole year ago – Luca Parmitano. I told you he would do well. It was a very relaxed, warm, informal gathering, with excited talk of adventures to come, and it was clear we were going to get along well as a group.

Then we were taken across the road to ESA headquarters. Twenty-five minutes before the press conference, we gathered round a table to be told how the announcement was going to work. The director general was going to give an introduction, then each of us would have a moment to say a few words before it was thrown open to questions from the press. During this meeting, the head of HR arrived and handed us each a contract to sign – the terms of our employment. It was a 10-page document and there was no time to read it properly, which felt a little odd. I skim-read it, established, so far as I could, that there was nothing irregular in it, and signed it. But there was a problem for Luca and Samantha because the contract demanded 100 per cent fealty to ESA, and no other employer, and both of them were hoping to maintain their positions with the Italian Air Force throughout their ESA contract, as some of their predecessors had. I think there was a special agreement in place that allowed serving Italian astronauts to do this. It made for an awkward moment at the table, with those two very uncomfortable about being made to sign, and all of us a little put out to be doing this in a hurry and thinking, 'Is this really how it should be done?'

And then we were out and into the glare of the press conference, all six of us lined up behind the vast oval table in ESA's main briefing room – our first taste of the media glare. I was startled by how many journalists were there. The room was heaving. After the director general's words about the rigour of the application process and how we were the future of European spaceflight, we each did our own bit in English. And then we were asked if we would like to say something in our native languages. Luca and

Samantha set off in Italian, Thomas made a little speech in French and Alex spoke in German. Then it reached me.

'Well, here I am in English again ...'

It was Andy I felt most sorry for. He was Danish, but he had spent so much of his life in the States, where he also went to university, followed by the UK, where he had been working at the Surrey Space Centre, that his Danish was extremely rusty. He was in a sweat when it got to his turn, because he thought everyone in Denmark would be laughing at his dodgy pronunciation.

After that, the press were invited to ask their questions. The first person with his hand in the air was Jonathan Amos of the BBC, who directed his question to the director general.

'Why have you given us a British astronaut?'

Now, at this point Jean-Jacques Dordain very graciously blew me some smoke rings, saying something to the effect that I had been such a good candidate that they couldn't possibly say no. Well, I'm sure I did OK, but I'm also sure any one of the ten people in that last interview round would have made an equally good European astronaut. I had been the only Brit in that final ten. So were there politics at work? Was this a clever move by ESA to encourage the UK to start contributing to the Human Spaceflight programme by giving them an astronaut to root for?

Whatever, Jonathan Amos's question had summed up the mood. The mood was shock and surprise: a British astronaut? Seriously? They've picked a Brit? Well, that's a turn-up. Here was a story to get behind, and especially for the British press who duly piled in.

A social on the Wednesday evening? A dinner? My flight home on the Thursday morning? Forget all that. After the press conference, I was separated from my new colleagues and whisked away to the Gare du Nord, where I boarded an afternoon Eurostar to London in the company of Simonetta Di Pippo, her assistant and

Clare Mattock, ESA's UK head of PR. As the train pulled out, Clare handed me her phone.

'This is the Press Association.'

That was my first interview – to ten newspapers at once, a conversation in which I could barely hear a thing on a phone that cut out in every tunnel. I was just happy they spelled my name correctly in the morning papers. And they kept coming, right through the three-hour journey, with Clare taking her phone back, then handing it over again as we picked off the news outlets one by one. Then in London, we were scooped up and driven round the news channels – Broadcasting House, Millbank Studios, ITV, Channel 4. It was 11.00 at night by the time we'd finished and I was drained. Clare told me to expect an early start. 'Up at 05.30, please – we're on GMTV at 06.50 followed by BBC Breakfast.'

The transformation was abrupt, and really quite bewildering. People talk about overnight fame, but this was more like under-the-Channel fame. Whether I liked it or not, by the time that Eurostar pulled into St Pancras I had just about said goodbye to being the Timothy Peake who could wander into Costa Coffee without anybody noticing, and was well on my way to becoming an altogether more public item: 'Britain's Major Tim'.

A few days later, with another round of PR kicking off, I was sitting with a coffee in my hand in the green room at *This Morning*, waiting to go on and talk to Eamonn Holmes. And who should walk in but my boyhood musical heroes, Madness. Suggs wandered over, shook my hand and wished me good luck while I tried to think of something vaguely normal to say.

Madness, indeed. Suggs knew who I was? What was going on here?

CHAPTER NINETEEN

Back to school, business class muffins,
and all aboard the vomit comet

I'm sure you can imagine the stress involved in moving house from the UK to Germany, taking all your possessions with you along with an eight-month-old baby and a dog. I can certainly imagine it – and, indeed, I have imagined it, many times since, with a massive amount of guilt. That's because while my wife was actually doing that, I was on a beach in Florida playing volleyball.

Let me say extremely quickly that this had not been the plan. In the August of 2009, Rebecca and I took a family holiday in Scotland – the last quiet summer of our lives, really. And then we returned to Larkhill to pack up the house ahead of the move, ready for me to start with ESA in Cologne at the beginning of September.

In the meantime, Simonetta Di Pippo at ESA had arranged for the six new European astronauts to fly out to Florida and watch the launch of the Shuttle from Kennedy Space Center, scheduled for early in the morning on 25 August. Christer Fuglesang, the Swedish ESA astronaut, was part of the crew heading for the ISS on that mission, so this was going to be a big event for ESA and it would also be a nice way to welcome us newcomers to the world of spaceflight.

Well, I could hardly say no, could I? In any case, we would be back in time for me to pitch in with the move.

Except that the launch got delayed right at the last moment when a storm blew in – a massive anticlimax, although the storm itself was quite impressive to witness. That moved everything back twenty-four hours, to the Wednesday. The six of us hung around, with an unforeseen empty day tacked onto the schedule and with nothing much else to do apart from take a ball to the beach. Again, it would have been impolite not to.

And then late that afternoon, with the excitement once again beginning to mount, the Shuttle's launch was scrubbed again, on account of a technical fault. Once more the time was pushed back – until Friday night now. That was too late for nearly all of us, with our first official day at work in Cologne nearing. We flew back on the Thursday night, leaving only Alex, who lived in Germany already, to witness the launch.

A somewhat frustrating trip, then – although we had met several NASA astronauts, enjoyed an impressive tour of Kennedy Space Center, and if you like Florida sunshine and beach volleyball ... I got off the plane in Germany on Friday morning and walked into the new house where Rebecca, assisted by my parents who helped look after Thomas, had just finished installing all the furniture. 'But enough about *my* week – now tell me about *your* week.' It was some time before I mentioned the volleyball.

Our new rented home, which had been found for us by a relocation agent used by ESA, was in a small village called Uckendorf, on the edge of Cologne, about fifteen minutes' drive from the European Astronaut Centre. It was a nice enough detached place, with a big garden, and its kitchen was intact – a bonus because in Germany people seem to like to take their kitchens with them when they go, leaving a big hole where you thought the cooker and units were going to be.

An emergency trip to IKEA was still necessary, though, because the previous occupiers had stripped out all the light fittings. Indeed, my astronaut journey nearly ended prematurely that very weekend, up a stepladder in a bedroom, when I touched two bare wires together and got knocked backwards across the room by the spark. (Top tip: check that the house you are hanging lights in didn't use to be two flats, with separate isolation switches for the circuits on each floor.)

That apart, the house was freshly decorated and looked very shiny when we arrived – at least until the damp rose up from the perpetually flooded cellar and the new paint simply fell off the walls. As for the location, I don't think even Uckendorf's biggest fan would describe it as a place where a lot was happening. Rebecca would soon be feeling pretty isolated there. ESA had no kind of family programme and there was certainly no 'Astronauts' Wives Club'. Luca was the only other one of us who was married, and Rebecca and Luca's American wife Kathy became great friends. But Samantha was single and Andy, Alex and Thomas had girlfriends at home who hadn't joined them yet.

Just before we arrived, a Brit working with ESA called Craig Thomson, a former Royal Marine, emailed to introduce himself and offered to show me the local ropes if I needed. I was hugely grateful to him for that. But it wasn't until several months after moving in that Rebecca bumped into a local mum, out walking with her toddler. Christine and her husband Gerd, who had impressive BBQ skills and a profound knowledge of German beer, quickly became good friends of ours. And that was pretty much the limit of our socialising. That was a tough spell for Rebecca: getting by on GCSE German in a quiet village, with a small baby to care for, a dog to walk and a cellar full of water, and with me frequently off training and out of the picture for whole weeks at a time. It would only get better when we eventually moved to Bonn where life had more to offer, including an un-damp house.

At the beginning of September, I began my daily commute to the European Astronaut Centre in Cologne, commencing the 'basic training' phase, which would last just over a year. It was impossible to walk into the building – with its grand foyer and fantastic pictures on the walls and models of spacecraft, and to peer into the training hall, with its life-sized Space Station modules and cargo vehicles and Soyuz simulator – and not feel the hair go up on the back of your neck. So this really was happening: I was going to be trained to go into space.

Those first months of study were exciting and challenging – but also a bit glitchy at times. Bear in mind that this was the first time ESA had ever trained their own astronauts. European astronauts had always gone through NASA training in Houston. That meant there was a lot of pride at stake, and ESA knew that NASA would be watching carefully. They didn't just want their first astronauts to match NASA's standards, they wanted the training to be a cut above. We were brand new recruits, going through a brand new training scheme, with brand new instructors. ESA were feeling their way at this stage, just as the students were, and there was a lot of pressure on the training team charged with delivering this programme and all its bold ambitions. Sometimes we could sense that they were only a week ahead of us. Metaphorically speaking, if not literally, the smell of wet paint was often in the air.

Inevitably, there were one or two misfires. The low point came in an early 'trust-building' exercise in which five of us were required to don blindfolds and then form a line with our hands on the shoulders of the person ahead, while the un-blindfolded person led us around the building, alerting us verbally to hazards. Up and down the stairs we went, in and out of rooms. Heaven only knows what the rest of the ESA staff must have thought as we tottered blindly up and down the corridors in an underpowered conga, but I know that those of us doing the conga felt pretty foolish about it.

I thought back to one of the questions Jean-François Clervoy, the French astronaut, had asked me during that interview in the second round of the selection process. He said, 'You're thirty-six, you've had years of military experience, flown in combat, been a QHI and a test pilot. How are you going to feel about going back into the classroom and being treated like a schoolboy?'

I replied that I had been studying and learning all my life and that I didn't expect I'd find it hard. But I think we were all surprised by the extent to which we felt like schoolchildren in Cologne in those first months. It wasn't just that the course, right up to Christmas, was initially so classroom based. It was also that ESA seemed very keen to protect us and keep us in a bubble. They didn't want us bothered and distracted by other people around the building. We worked together and with our instructors, and had no other interaction. There were 120 people working in the European Astronaut Centre and I would have loved to have wandered around and been introduced and got to know people in their various departments. I was really keen to begin understanding how the organisation fitted together as a whole. But it seemed that everybody had been issued with the instruction not to bother the new astronauts. We were to be left to our studies. Eventually, and gradually, we began to feel our way around a bit more. But at first we were almost completely cloistered.

Still, that had the positive effect of bringing the six of us very closely together, completing the work begun, I guess, by the blindfolded conga-walk. After a month in the classroom, the bonding accelerated when we were sent to Bochum for a four-week intensive Russian course in a residential language institute. Samantha didn't come on this trip because she already spoke excellent

Russian, having studied at the Mendeleev University of Chemical Technology in Moscow. I, meanwhile, had a C in French at GCSE and a less than sound understanding of English grammar, and now I was attempting to take on a language with an alien alphabet, six cases, gender nouns … I struggled and, although I eventually made it to the required standard and got all the passes I needed, I'm not sure that the intensive approach suited me. After a week my brain was full, and after a second week I was frazzled and it was all starting to swim around on the page. A little, and often, might have been the better strategy for me.

Still, I wasn't alone. Thomas had already learned a bit of Russian, so he was OK. Luca was a natural linguist and he flew off with it. Alex wasn't quite so good, but the leap from German to Russian is arguably a little simpler to make than the leap from English, and he was getting to grips with it. But Andy, at least, was like me. It was a great relief to cast an eye around the table and see him looking as puzzled as I was. You know what they say: if you look around the classroom and you can't spot within a minute who the idiot is, it must be you. At least there were two of us.

An hour north of Cologne, Bochum wasn't exactly Las Vegas. The city had more than its fair share of soulless modern concrete, 38 per cent of it having been destroyed by bombing in the war. But we had a good time. Pranks abounded – or, as we learned to call them, 'shenanigans', a word which those who didn't have English as a first language particularly seemed to enjoy using. Shenanigans happened in our extremely rudimentary student rooms, where Gütersloh-style upside-downings were not unknown. Shenanigans could break out at quiet moments in the classroom, too. Alex liked to wear some very Germanic, slipper-type shoes that he would kick off during classes in order to perch cross-legged in his socks. One day, sitting opposite him, I stretched my leg under the table and used my foot to drag his discarded slippers towards me. Then I bent down to pick them up, filled each of them with the drinking water that we all had on our tables and used my foot to prod them back to where they had come from. At the end of the lesson, I had the pleasure of watching Alex slide his feet back into the cold, squishy mess that used to be his footwear. So childish, I know. But it passed a quiet moment in a lesson on verbs.

In due course we decided that we needed to design ourselves an unofficial group patch. We based it on the shape of the ISS's Cupola window and incorporated the flags of our five nations, the astronaut corps values and, inevitably, the word 'Shenanigans'. Eventually, in 2013, the patch would have its maiden trip to space, carried by Luca and, in due course, by all the 'Shenanigans'.

From Bochum it was back to the classroom in Cologne. But after Christmas, things began to open up. The five of us were despatched to St Petersburg for another four weeks of immersive Russian learning. In advance of this trip, there had been some discussion at ESA about the wisdom or otherwise of allowing five sixths of the astronaut class of 2009 to travel together, in case disaster should strike and wipe out ESA's investment almost in its entirety. I'm not sure, however, how thorough that discussion had been. I can only report that, although we were made to travel in two groups, on two separate trains from Cologne to Frankfurt, we all then climbed onto the same plane bound for St Petersburg. Confusing. Maybe ESA knew something about German trains that we didn't.

Anyway, the separate trains ploy almost caused a minor disruption when the service carrying Alex and Andy broke down en route, forcing them to get out, jump in a taxi and bribe the driver to take them as fast as he could push it up the autobahn to Frankfurt Airport. So much for ESA's super-cautious approach to astronaut safety. Alex and Andy dashed to the desk just in time to check in – and also just in time to learn that the plane had been delayed an hour in any case.

Luca, Thomas and myself, meanwhile, were upstairs in the business class lounge, greatly enjoying the comfy seating and Lufthansa's generous provision of complimentary coffee and snacks. We had been down to the gate once, to urge them to keep it open for the late-arriving Alex and Andy. But then the delay had been announced, so we had returned to the lounge. None of us was used to having business class tickets and we were determined to enjoy the privilege while we could. Andy and Alex joined us and the free snacking continued. At one point a trolley of blueberry muffins came out, and I love a blueberry muffin so, naturally, I had to sample one of those, as did all the others, along with another cup of coffee.

Somewhere in the middle of all this, time seemed to become elastic. It was one of those 'group mentality' situations, where nobody was actually in charge so everybody was assuming someone else was. They would call us for our flight, wouldn't they? We had already been down to the gate once, so the people there knew where we were. They would look after us, surely. And Thomas had been an airline pilot, so he knew how the system worked, and he seemed pretty relaxed about the delay.

There was still no call for our plane, though, so eventually, when the blueberry muffins were finished, we slowly gathered our things and sauntered down to the gate to see what was going on.

Oddly, the sign above the gate now read 'Algiers'.

'Hello, it's us again,' we said. 'Did you change the gate?'

'No,' said the woman from Lufthansa. 'The flight to St Petersburg left five minutes ago.'

They don't give out PA announcements in Lufthansa's business lounge – or so we now learned. Apparently, they disturb the customers who prefer to eat their blueberry muffins in peace. The

Landing from an Apache flight to be told that I had been assigned to a long-duration mission to the ISS . . . now THAT was a good day.

Below: Explaining to my boys what Daddy was off to do – I felt it was important to help them understand what was going on; how I would travel to space and maybe do a spacewalk.

I'll always be so grateful for the amazing support from friends and family throughout my career.

Right: I'm stuck behind the glass in quarantine – Team Peake is on the other side. They, along with my fellow astronaut Tim Kopra's high-spirited guests, fully embraced the Kazakh hospitality at Baikonur Cosmodrome.

Above: There is nothing that can truly prepare you for launch. It's a visceral experience – 9 million horsepower accelerating you to twenty-five times the speed of sound. It's not for the faint-hearted!

Below: My home for six months – the International Space Station. To my mind the greatest engineering accomplishment of humankind, and a shining example of what we can achieve when many nations work together.

Expedition 46 (from left to right) Tim Kopra, Yuri Malenchenko, me, Sergey Volkov, Misha Kornienko and our commander, Scott Kelly. Misha and Scott were soon to finish their year-long stay on the ISS.

Right: One of the many blood donations during my time in space – all in the name of science.

Below left: Jeff Williams (left), me and Tim Kopra celebrate the successful capture of a Cygnus spacecraft – a high-pressure moment for any astronaut.

Below right: And with fresh cargo, comes fresh fruit – a real treat to smell and taste something grown on Earth.

The US lab – where, along with many science experiments, I also practiced capturing the Dragon spacecraft using this simulator.

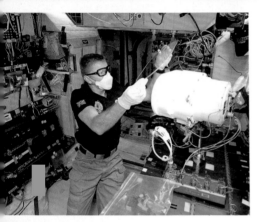

Left: Repairing one of the Space Station's external lights that Tim Kopra and I would exchange on our spacewalk.

Below left: Running the 2016 London marathon from space – a hugely memorable experience, albeit somewhat painful, given the harness I was wearing.

Below: Sealed in the airlock for a complex experiment to investigate airway inflammation. Many asthma suffers will benefit from this important research.

Above: The Sinai Peninsula squeezed between our sturdy Soyuz spacecraft and the Cygnus cargo vehicle. I would check our spacecraft daily from the Cupola window, inspecting every panel in case of damage from space debris.

Below: The lights of Italy, Sicily and Malta on a clear night, shortly before sunrise. The weather wasn't so kind when I flew across the Adriatic in 1996 in a Gazelle, from Split to Venice.

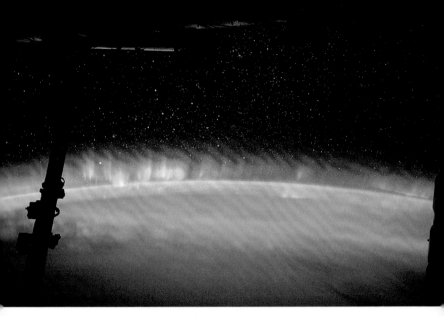

Above: During our mission we had particularly high levels of solar activity, enabling us to witness the awe and wonder of the aurora borealis and aurora australis.

Below: With no light pollution in space, more than 100 billion stars of the Milky Way shine brightly when seen from the Cupola window.

Scott Kelly helping me suit up ahead of my EVA.

The first Union Flag to be worn in the vacuum of space – an incredibly proud moment for me.

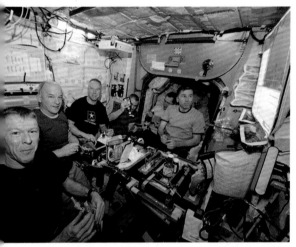

Left: Friday nights were always sociable occasions on the ISS. Here, the Expedition 47 crew: me, Jeff Williams, Tim Kopra, Aleksey Ovchinin, Oleg Skripochka and Yuri Malenchenko enjoying a meal together.

Below left: After a fiery re-entry, enjoying the final minutes of calm before hitting the ground. Life on Earth would never be the same again.

Below right: 18 June 2016 – that cold towel was pure bliss.

Above left: Enjoying a cuppa with mum and dad, following a calf muscle biopsy, just hours after returning from space.

Above right: My first run in the fresh air back on Earth – so good to be surrounded by nature again.

Left: I'm so inspired by young people's passion for science and exploration. Maybe I have already met the first person who will set foot on Mars.

My wonderful family – with me every step of the way, embracing each new adventure head-on and always keeping my feet on the ground. © Michael Cockerham.

airline people had gone to the trouble of taking our bags off the plane, but they hadn't gone to the trouble of looking for us in the most obvious place. We were quite irritated about that. We were also pretty sheepish, though. It was the first time we had been trusted to look after ourselves outside the ESA schoolroom bubble, and look what had happened: five astronauts who couldn't even catch a plane on time.

Fortunately Alex managed to sweet-talk the Lufthansa desk into putting us on the next flight out, sparing us an embarrassing call to Cologne. However, instead of arriving at 4.00 in the afternoon, we landed at 10.30 at night. That meant it was past midnight when most of us reached the homes of the various St Petersburg families who would be hosting us for the next month.

Well, I say 'families'. In fact only Thomas ended up with an actual Russian family – an upper-middle-class one in a plush district of St Petersburg where he had his own spacious room and convivial meals round the dinner table every evening. The rest of us found ourselves billeted in less salubrious circumstances, with widowed Russian grandmothers offering B&B in communist apartment blocks. Andy and I were on different floors of the same enormous building. My babushka turned out to be a lovely woman, but also one of very few words. I was embarrassed, when she let me into her home at that ungodly hour on the first night, to discover that she had given me the one bed in her tiny flat and was herself going to be sleeping on the sofa in the very small living room. I spent a long time trying to convince her, in my very sketchy Russian and with many accompanying gestures, to have her bed back and let me sleep on the sofa. But she would have none of it. This was the arrangement: the room had been rented, I was to sleep in the bed. I felt terrible.

I also felt hot. Outside in the street it was minus 15° C most nights – a typical St Petersburg February. Inside the flat, though, where the radiators were pumping out the heat, it was upwards of 30 degrees and sweltering like Turkey in August. I slept with the window open, just to get some air in the room.

Every morning my landlady would give me a cup of tea and a bowl of porridge and the two of us would exchange a few stilted pleasantries before falling silent. Then I would meet Andy and leave for school – which was actually great fun, as it turned out, with terrific teachers. Thomas and Luca, who were surging ahead, were in one class; Alex, Andy and I were in another, but even the slower among us came on in leaps and bounds over that month. And we all got to know St Petersburg, which is a fantastically beautiful place.

Was it there that we finally started to exude the calm and quiet competence associated with astronauts as they go about their daily business? No, I don't think it was, actually. One Saturday morning I was at a cashpoint in the centre of the city with Alex and Luca. I put in my card, took my cash and then stood waiting.

'What's up?' said Alex.

'My card hasn't come out,' I said. 'I've got the cash, but my card is still in the machine.'

I was really keen not to go through the inconvenience of getting my bank card replaced while out in Russia. The bank was shutting but we managed to talk our way inside and explain my dilemma. Nobody seemed especially keen to help, but we were as insistent as we could possibly be, pidgin Russian allowing. Eventually the manager came down and he in turn summoned an engineer who turned up about five minutes later with a toolbox, while everyone else stood around muttering because they wanted to lock up and get off home.

The engineer had practically finished using his screwdriver to remove a long and fairly involved series of panels from the back of the cash machine when Alex casually mentioned to me a thought he had just had: namely that it was quite unusual for a machine to distribute cash before returning the user's card. Was this a Russian thing, maybe? Didn't the card normally come out before the cash?

I took his point, but I checked my wallet again and showed him: no card.

Then I slipped my hand into my pocket and felt something – a hard plastic shape …

Wincing inside, I furtively flashed Luca the card that I must have unthinkingly withdrawn and stowed in my pocket before my cash appeared. The engineer was still hard at work exposing the intestines of the machine as the two of us sidled to the door like cartoon characters, slipped out and then bolted up the street, leaving Alex to explain as diplomatically as possible that the situation appeared to have resolved itself.

Spatial awareness: massively important for an astronaut. But hey, I was still a student. This episode was known forever after as 'the self-shenanigan'.

When the time came to leave St Petersburg, we were driven to the airport altogether in a minibus. Clearly the notion that, like the royal family, ESA astronauts should never travel together had by now properly bitten the dust. And yet, ironically, that journey was possibly the closest we came to disaster during our whole training.

It was late February now and, after the hard Russian winter, the thaw had begun. Icicles and slabs of ice were falling into the streets all over the place and the roads were a slushy ice rink. At the wheel of the minibus, our Russian driver, who seemed to be enjoying himself a little bit too much, had been pulling off a succession of barely controlled slalom moves around the city's arteries, and was now brought to a sliding halt by a red light. As we sat and waited, there was an almighty 'crump!' and the car parked on the pavement beside us simply imploded in a white cloud. A slab of ice the size of a waste skip had dropped off the adjacent building and, narrowly missing our minibus, smashed down onto this car's roof, almost certainly totalling it.

The lights turned green and our slippery journey continued.

* * *

One morning at the European Astronaut Centre, I went to the office to use the printer and found a document that someone had left in the tray. It wasn't meant for my eyes, and perhaps I shouldn't have read it. But it's quite hard not to be curious when you go to move a piece of paper out of the way and your name jumps off the page at you and you realise the subject is your future.

The document was a summary of the options available to ESA management as it attempted to resolve the quandary it had created by taking six astronauts onto its new graduate programme rather than four – i.e. too many astronauts, too few scheduled missions. ESA had secured three missions to the space station through NASA and, along with the two missions that Italy had separately bartered, that meant that five astronauts would fly, with the last of those missions scheduled for 2019. The document observed that the UK was still not contributing to the ISS programme. And it drew the following conclusion: '*For the moment it is most probable that Timothy Peake from the UK may be the one with the lowest chance to fly.*' It was suggested that I now be regarded as '*the reserve astronaut*'.

Now, nothing here was especially news to me. I knew my nationality currently put me at a disadvantage to the other five astronauts, and it was no mystery why. But it was still hard to see it written down like that in black and white. Hard, too, to reflect that these things were being discussed openly at management level but not with me in person. At no stage during the selection process nor thereafter had it been suggested that I would be on a different footing from my colleagues. The public line had always been that mission assignment was done on merit, when clearly the truth was that it came down to funding and politics.

Soon after this I had a conversation with one of my superiors who finally conceded frankly that my chances of getting a mission were dependent on the UK changing its position on the funding of human spaceflight. It was nice to hear it openly stated. It was also empowering, and inspired me to begin doing some lobbying of my own. Human spaceflight had been off the British agenda since Margaret Thatcher turned against it in 1986. But the value of the twenty-first-century UK space industry (supporting 68,000 jobs and worth about £6 billion to the economy in 2009, and expanding rapidly) was already changing minds on that score, and I would find myself pushing against an already opening door. I began contacting science ministers, talking to UK space industry experts, consulting with people at the new UK Space Agency, the government body which replaced the British National Space Centre in April 2010. My efforts seemed to be met with enthusiasm wherever I went, so I banged the drum when I could over the ensuing months.

Nevertheless, I knew it could be several years before my place on a mission, assuming one could be acquired, might be confirmed. These would be years of doubt and uncertainty – and I realised that I needed to form an attitude to protect myself against that, as best I could. At home I talked it through with Rebecca. Clearly, there was a strong chance that I would never get a flight. So I needed to be open-minded about it. I had to be happy to potentially leave ESA after many years, un-flown and without so much as a single regret. Because the last thing I wanted was to end up a bitter and twisted un-flown astronaut, who walked away from the test flying career that he loved for a big dream that came to nothing. The way to avoid that, I felt, was to fling myself into these years with ESA – to see the training itself as the opportunity and the challenge, and to grab everything it offered with all the energy I had.

After all, things might open up somewhere along the way. And even if they didn't, at least I would have buckled up, enjoyed the ride and contributed in some small way to the human spaceflight programme.

Happily for me, this was exactly the point at which our training began to get serious and properly exciting. The slightly stiff early classroom sessions of the winter were now well behind us. It was spring and we were off to Bordeaux for a week to do the famous 'vomit comet'.

The vomit comet is designed to give astronauts their first taste of weightlessness. It was actually an adapted Airbus 300, with a few passenger seats at the back but with most of the wide-bodied cabin stripped out to provide an open space, framed by netting, and with the walls, floor and ceiling all lined with padded material. When you first step inside, it has the forlorn feeling of a slightly unimaginative children's soft-play area, but once the plane gets up in the air and starts pulling its crazy parabolas, it's an altogether different story.

I'll admit I was a little apprehensive the night before our first parabolic flight. The reduced-gravity aircraft's reputation precedes it. There's the fact that it's called the vomit comet, for one thing. I was also wondering what the sensation of weightlessness was actually going to feel like. Seven months into an astronaut course would be a slightly awkward time to find out that it didn't agree with you at all.

Before we flew, we were all given a small shot of scopolamine, an anti-nausea drug. Clearly it was in nobody's interest that any of us should spend the three and a half hours of useful and important (and expensive) training at the back of the plane with our heads in a sick bag.

They broke us in gently. On the first parabola, the pilot was pulling about 2 g on the pull up and 2 g on the pull out. G-forces I had plenty experience of as a pilot. They're only a problem if they take you by surprise. If you're ready at the onset, you can cope with them by tensing your stomach muscles, locking your core in place and riding them out. It's also a good idea not to turn your head when pulling g until you become used to it. They had us lying on the floor staring at the ceiling. As the plane went over the top of the parabola you felt yourself lift off a little bit.

We worked upwards from there. They gave us Mars, which is about a third of the Earth's gravity. You can stand up, but you still feel that you're in a gravity environment. Then they gave us the Moon. That's a sixth of the Earth's gravity, and fantastic fun. You can do those big bounding steps and realise what it felt like for the Apollo astronauts. It's impossible to walk normally in those conditions. The bounce is the only way forward.

And then we were into the full zero gravity parabolas, where you go from a 2 g pull up to weightlessness in a second or two. This is the point in the exercise where people are most likely to need to empty their stomachs, though I'm happy to report that nobody on our flight ended up requiring a quiet moment alone. Nothing can prepare you for that weightless sensation. As a pilot, I had experienced zero-g for short bursts at a time. But it's very different when you're strapped tight into a seat and when you're free to float around in the back of an airliner. You get about twenty-two seconds of weightlessness within the minute of flight through the parabola. As you begin to go into weightlessness, your stomach lifts slightly and there's a moment of slight discomfort. But then comes a feeling of release and the next twenty seconds are an absolute joy – the closest thing I'd felt to an out-of-body experience.

We did about thirty zero-g parabolas on that flight, acclimatising ourselves and then completing some exercises: some spacewalk techniques (also known as EVA – extravehicular activity), tethering and using tools; passing objects around to get a sense of the way things with momentum handle in zero gravity; running on a treadmill to begin to get used to how fitness activities felt in those conditions. By the end, thoroughly emboldened, we were curling up in balls and getting spun around in the air by the instructors just to see how much provocation and disorientation we could take. I think we all ended up enjoying our time in the vomit comet and it was another important box ticked.

Next thing, we were being flown to Sardinia, driven for three hours down dusty tracks into the hills, and abandoned there. This was Survival Training, a truly gold-plated package, organised by ESA's Loredana Bessone and run by Italian ex-Special Forces, and probably my favourite part of the entire basic training course. We pitched up initially at a set of dilapidated old farm buildings in the middle of nowhere – our base for the next few days. Looking inside we found extremely rudimentary bedrooms – sleeping bags on thin mattresses. There was a shared bathroom, again very sparsely equipped, and with the water heated by solar panels. Then we found the kitchen where, somewhat out of keeping with the whole deserted rustic vibe, there was a €3,000 coffee machine. But this was Italy, after all, and some things are essential. The kitchen, unsurprisingly, became the focal point during our stay.

A barn with a high ceiling and sturdy beams had been made into a classroom where we practised rope work with our instructors, ahead of the rock climbing and abseiling that we would be doing as part of the main survival exercise. Why did astronauts require survival training? Because your capsule – returning to

Earth in a hurry, say, after an emergency departure from the ISS – might end up landing somewhere remote, and it might be a couple of days before Search and Rescue reached you, leaving you grateful that you knew a few things about how to look after yourself in the wild.

To this end, the six of us were going to be spending two days and two nights in a clearing in a remote valley. But first we were directed to an old wreck of a car, handed a couple of sledgehammers and a hacksaw and given twenty minutes to strip it for anything that might be of use to us during our adventure. We solemnly set about trashing this car for all it was worth, whether it was the rubber seals from around the windows (handy for tying bits of wood together to build shelters) or the fabric off the seats (re-workable as hats to protect us from the sun) or the electrical wiring that we could use as cord.

Then we set off on the long journey to our remote clearing, where we built our shelters and made a fire. All of this inevitably took me back to the Army in 1995 and that Combat Search and Rescue course – except that this was Sardinia in the summer, rather than Dartmoor in late autumn, and there were no leery Paras tracking us down through the undergrowth, so I suppose it was relatively civilised by comparison.

It did contain its hardships, though. We had to survive without food for the first two days. Late on that second day, when we were all really beginning to feel it, we were supplied with a live chicken, which we killed, plucked and cooked. Spatchcocked and grilled on an open fire in the warm Sardinian evening, we agreed that it was the most delicious chicken we had ever tasted, though that might have had something to do with how desperately hungry we were.

On the last day, now at least fortified by a meal, we had to abseil and climb down a steep canyon and then hike along a gorge, the latter stages of which was flowing river, which meant donning immersion suits. We eventually emerged intact and trudged up a hill, where a Sardinian Police helicopter was primed to collect us. Rescued at last? Not quite, because the function of the helicopter was to drop us straight into the Mediterranean, along with an inflatable dinghy, for the 'sea survival' phase of the exercise.

Alex got dropped first, and the raft was tossed in after him. He had to pop the raft and climb in and then the rest of us had to swim up and join him. Our spell of bobbing on the Med went on through the night, and was not without its mishaps and moments of mild chaos. An instructor was there to brief us on various procedures, including the correct operation of the emergency flares. Not quite getting to grips with it, Thomas managed to set one off at the wrong time and also (even more remarkably) in the wrong direction – 'down' rather than the always recommended 'up'. This burning flare lay smoking on the floor of the dinghy until we managed to scoop it out and extinguish it in the sea.

The instructor also explained crew overboard drills and warned us that, at some point in the night, when we least expected it, he would be throwing himself out of the dinghy and he would like to see us perform the crew overboard drill and 'rescue' him, just to prove that we had been listening.

It must have been about two or three in the morning. I was 'on stag', keeping watch while the others slept, when I heard some rustling and detected movement among the slumbering forms on the other side of the dinghy. It was the instructor. I watched him in the dark as he rose and very quietly began to get into a kneeling position on the dinghy's rubber edge. 'Rumbled,' I thought. 'It's crew overboard time.'

I shouted to Luca and Thomas who were closest to him.

'Luca! Thomas! Grab him! He's going to jump!'

Luca and Thomas were instantly awake, grappling with the instructor and pulling him backwards into the dinghy. It was only then that we noticed he had an important part of his anatomy in his hand.

Moral of the story: be extremely careful about when you get up to take a night-time pee during a sea survival exercise.

The instructor seemed to forget the idea of trying the crew overboard ruse after that and the rest of the night passed off relatively peacefully. Food-wise, our at-sea rations had proved as miserable as ever (those nutritious firelighters again), but there was a bright moment when a passing fishing boat took pity on us and threw us a freshly caught fish. I had never eaten sushi before, but I certainly ate it then, and I was impressed. Who could have imagined raw fish was so mouth-watering?

Anyway, we had cause to be thankful that we'd had some solid food on the final morning when a proposed heli-winching exercise turned into something more like a swimming marathon. It was the same Sardinian Police helicopter that had dropped us off the day before – and, to be fair to them, it was a training exercise for those guys, too. But, at the risk of sounding picky, I have to say that I had seen heli-winching more skilfully performed.

The helicopter only had a short winch, so it was obliged to drop in low over us, and every time it did, the downwash would blow our little rubber dinghy off to the side, like a beach toy in a gale. Then the helicopter would lift and drop again, and the same thing would happen: the dinghy would pop out from under. We were getting buffeted to high hell down there, the dinghy's canopy collapsing in on us. Eventually the police realised that the most prudent approach was to do the winching at a distance from the dinghy and get us to swim out to the hook. The helicopter then took up a hovering position at least 500 metres away – playing it safe, presumably. Alex was first off, and it's a good job he's a strong swimmer because that was quite a stretch in a choppy sea, not least after three exhausting nights living rough. Once Alex was on board, the helicopter seemed to pluck up the nerve to get in a little closer. The rest of us followed in turn. The helicopter then flew us directly back to the farm headquarters where, without pausing to change out of our damp immersion suits, we went straight into the kitchen for a nice coffee from the expensive machine.

Back in Cologne, the training experiences kept coming. We began diving sessions in the pool, wearing full face masks and bulky spacesuit gloves – the first stages of learning to spacewalk. We did robotics training, beginning to understand how to work the robotic arm which is a key aspect of life on the ISS. And we made what would turn out to be the first of many trips to Star City, the Yuri A. Gagarin Cosmonaut Training Center near Moscow, to be introduced to the workings of the Russian modules of the ISS.

It was an extraordinary fourteen months, all in all. We had covered so much ground: chemistry, biology, solar physics, space propulsion systems, space law, electronics, computer technology … It had been challenging, intense in places, mind-expanding and a real privilege to have been part of. At the end of it, all six of us graduated and could now officially regard ourselves as astronauts, albeit un-flown ones. And we had the certificate to prove it. ESA held a graduation ceremony for us. In fact, they held two. The date of the original got changed at the last minute, causing us to point out that we'd all got friends and family booked to come out. Luca and Samantha in particular had VIPs from the Italian Air Force flying in whom they wouldn't have been looking to inconvenience. So the original ceremony went ahead as planned, and so did the second one for ESA's benefit a few days later.

But I suppose you can't have too many astronaut graduation ceremonies. My parents and sister came out to attend. And so did Don Maclaine, my long-suffering Westland's boss. I was pleased to be able to show him that I hadn't simply been winding him up with all the astronaut talk and the requests for days off.

So I was an astronaut. Now all I needed was a mission.

CHAPTER TWENTY

Heavy metal surgery, life on the ocean floor,
and another important arrival

About a month before graduation, in September 2010, Luca left us to go to Houston. He was already assigned to the Italian mission, due to launch in May 2013, and needed to start the two-and-a-half-year mission-specific training phase. That was a big moment for ESA: Luca was formally assigned to fly ahead of any of the American, Canadian or Japanese astronauts who had been selected at the same time as us and were going through NASA's training programme. It was an emphatic statement: the European training programme was up and running. And it was exciting for the rest of our class, too. The faster Luca got to fly, the sooner the next of us would be up, and so on down the line.

However, it was already clear that Alex would be next, a year further on, in line with Germany's commitment to ESA, and with Samantha after that, occupying the second pre-bartered Italian seat in 2015. Assuming, quite reasonably, that Thomas and Andy would be going ahead of me, my hopes at this stage were pinned to an as yet entirely notional mission in 2021. All was shrouded in uncertainty and the future was very hard to predict.

All I could do was to keep working hard in Cologne, and in all the other places my training was sending me, and continue agitating as much as I could for government support in the UK. In the wake of my selection, I had had a meeting with Lord Drayson, the Labour Minister of State for Science and Innovation in the government of Gordon Brown, discussing the UK's future involvement in the ISS space programme. He had been very positive about it all. But then, in May 2010, the government changed. The Conservatives were taking over under David Cameron, and I wondered if I was back to square one. Not a bit of it, in fact. David Willetts became what was now known as the Minister of State for Universities and Sciences. He too was enthusiastic about the British space industry, and he too saw the potential value to the industry, and beyond, of sending a British astronaut to the Space Station. Influential British figures seemed to be coming on board but there was still a long way to go. I continued to lobby in the background.

Luca was gone. Alex's assignment was only a year away and Samantha's only two years, so their paths were set for the immediate future. The question for senior management was, 'What are we going to do with Tim, Thomas and Andy?' And the answer seemed to be, 'You know what? We're not entirely sure.' Again, this whole situation was new and unprecedented for ESA. There was talk of splitting us up – one of us going to Russia, one of us heading to Houston, and one of us remaining in Cologne as reserve astronaut, and then perhaps rotating those positions. But that came to nothing and we all ended up being based in Cologne. My official designation as reserve astronaut would enable me to keep going with my skills training. Otherwise, in this slightly structureless interim period – the 'pre-assignment' phase – we could carve out our own roles, up to a point, by opting for what were known as 'collateral duties'.

During the first months of the year, Thomas, Andy and I went to the ISS Mission Control Centre in Munich, and qualified as Eurocoms. The Eurocom is the person at Mission Control who talks to the astronauts in space. Think Charlie Duke speaking with the Apollo 11 crew during the Moon landing. There are all sorts of people operating in Mission Control, but they like to have an astronaut on the ground pushing the button and communicating with the crew. As Eurocom, you are the flight director's voice, but sometimes what you communicate to the crew will involve condensing five minutes of heated exchanges between many different parties in the control room into one short, clear and concise sentence of instruction. It's a challenging role.

Learning how to fulfil that function in Munich involved understanding how all the elements of Mission Control work. It was truly eye-opening. I saw at first hand how the entire support structure on the ground works and came to appreciate the extraordinary planning cycle that goes into forming just one day of activity on the Space Station. Everything that happens up there is planned and honed and constantly reviewed from whole months out. Later, when I was waking up on the ISS, having my morning cup of tea and looking at my daily schedule, I would be profoundly aware that hundreds of hours of work had already gone into the structuring of that one day.

Qualifying as a Eurocom also allowed me to see what the knock-on effect could be of even the most casual, throwaway remark by a crew member on the ISS. If you let Mission Control know that some piece of equipment is playing up, or that something is proving a struggle, you unleash a flurry of concerned activity on the ground, possibly leading to a whole day of backroom meetings where the problem you have raised is being addressed and resolved. Having witnessed the consequences, I was more careful, as an astronaut, about choosing the moments when I really needed to bring everyone's attention to a problem I was having, and times when I just needed to suck it up.

When I was finished in Munich, I seized the opportunity to work with Loredana Bessone, who had organised our survival training and handled the human behaviour and performance elements of our basic training, and who was in charge of developing ESA's CAVES training programme. CAVES stands for Co-operative Adventure for Valuing and Exercising Human Performance and Behaviour Skills, which is a very baggy title for a very smart concept. Loredana had come up with the idea of using a cave as a space analogue. Caves are an excellent extreme environment in which to isolate people, put them under pressure, set them to work in confined spaces and hone all the skills – teamwork, communication, leadership, followership – that an astronaut is going to need to apply. You can also do genuine scientific research in cave networks – taking samples, storing them, analysing them – while using ropes and tethers, operating technical equipment, even learning photography, all of it applicable to life on the ISS.

My role was to help Loredana put the course together, trial it and be the consultant astronaut on the project. We flew out to Sardinia on a recce and hooked up with some of the Italian Special Forces team who had organised our astronauts' survival course, along with a couple of local caving experts and Jo De Waele, a professor of Geography and Speleology. So it was back to the remote farm complex and the top quality coffee machine. In the vicinity of those farm buildings was a 15-kilometre network of uncharted caves, which was perfect for our purposes. We did a lot of caving, climbing and exploring to check the locations that would best suit us, work out how the course should run, and test it all out.

With my test pilot mindset reasserting itself, I was often pushing Loredana to increase the jeopardy, which she didn't always agree with. To my mind, it was all about risk management and if you wanted people to go and do a risky job, you needed to let them handle some actual risk along the way during the training. Did we really need instructors helping us on and off the climbing ropes, for instance? Once taught, couldn't we be trusted to take responsibility for that? After all, nobody was going to be holding our hands on a spacewalk. But I understood why Loredana was reluctant to indulge me in some of this. This was going to be an international programme, with astronauts joining from various different space agencies, and she must have had a whole viper's nest of health and safety initiatives to comply with.

Anyway, I returned to Sardinia eventually to be part of the first group of students going through the course. I thought the finished product worked brilliantly. It was demanding, physically and psychologically, and it compressed a vast amount of relevant learning into just two weeks. The caves themselves were astonishing: tight sections opening into massive caverns, wet sections requiring dinghies and immersion suits. After seven days of working by the light of head-torches, commonly on their dimmest settings to save the batteries, our eyes had completely accustomed themselves to the dark. Accordingly, the sensation when you emerged at the end of the exercise was amazing. People often talk about that supposedly overwhelming moment when a capsule returns after a space mission and the hatch is prised open and the astronauts finally re-encounter the sights, sounds and smells of Earth after their time away. To be honest, my sense of smell is pretty appalling anyway, and, in any case, my initial impressions on landing back from the ISS were a little less romantic, as we shall discover.

Coming out of those caves, though, was something else. My senses were in absolute overdrive. I could smell the moss on the ground and the sky was a crazily intense blue, as if someone had switched the contrast to maximum on the world's television set.

The effect only lasted a few minutes, but I wish it could have gone on longer because for that short spell you felt superhuman.

Back in Cologne, one of my further 'collateral duties' was working with the medical division, looking at how to manage fitness training and medical issues in space, considering the life sciences that we do on the Space Station and examining rehabilitation for astronauts, post-mission. Again, my role was to give an astronaut's point of view on the regulation of these matters where it was helpful. It was a massively interesting area, and it also gave me an opportunity to open up a dialogue with the people in the UK's space biomedical community and discuss with them ways to encourage the government to commit to ESA's life sciences programme, another vital part of my background lobbying push.

Astronaut training has a sizeable medical skills component – certainly several stages beyond the rudimentary first aid that I learned in the Army, which was basically tourniquets, slings and a bit of CPR. On the Space Station, any of us would be able to give injections, draw blood, administer a local anaesthetic, suture a wound and remove a tooth.

True, major surgeries would be considered beyond our remit. But that didn't mean it wasn't considered a good idea, and valuably educational, for all of us at some stage to go into a hospital and witness the operating theatre in action. To that end, Samantha and I attended the main hospital in Cologne during this period of training for three days of close observation. I believe the NASA astronaut course allows its students to get quite hands-on during their medical training but, due to tighter regulations in Germany, our brief was simply to watch.

And what a fascinating spectacle it turned out to be. Samantha and I separated and I found myself witnessing a hip replacement, the removal of a brain tumour, the repair of a detached retina and an open heart surgery. In the case of the last of those, it was apparently the third time this patient had been through this procedure and his poor battered sternum had fused and was clearly resisting the attempts of the German surgeon to saw through it. Eventually the circular saw broke – to the surgeon's evident fury. A nurse was despatched without ceremony to fetch a replacement.

I was standing at the head, talking to the anaesthetist when the nurse returned with the vital part. Still clearly frustrated, and not waiting for his assistant to use the suction tube to clear the area of blood and debris that had pooled, the surgeon grabbed the new saw and ploughed back in. The consequences were spectacular. Perhaps you are familiar with the Kenny Everett character, Reg Prescott, the haplessly bungling DIY-er. There was a considerable overlap between a Reg Prescott sketch and the scene in that operating theatre. The surgeon was spattered with blood and bone fragments, the floor was spattered with blood and bone fragments and all of us in the vicinity were spattered with blood and bone fragments.

At one point Samantha and I switched rooms. In an atmosphere of complete calm, and to a soundtrack of classical music, the surgeon she had been watching perform a similar procedure was delicately suturing the wound, almost as if gently conducting the orchestra. It was quite a contrast after the heavy metal performance next door.

Anyway, it was a very efficient way to discover how much punishment the human body can tolerate and, equally useful, that I'm not particularly squeamish. Interestingly, one of our astronaut instructors had passed this way a week earlier to review the training and had witnessed some surgeries without any untoward reaction. Then he had spent some time in A&E where he had been shown an X-ray of a patient who had just had a motorbike accident. The picture of all the broken bones was too much for him, and he nearly passed out. In this area, clearly, you don't know what's going to trigger you until you find out.

* * *

In October 2011, Rebecca and I found a nice, small town house in the suburbs of Bonn – a beautiful city with large green spaces, great public transport and miles of cycle paths hugging the Rhine. Our new home was just west of the river, a half-hour drive from the European Astronaut Centre. Bonn was a lively relief after the quiet of Uckendorf. Not that we would be hitting town very much for a while. Thomas was two and a half, and he was about to get a baby brother.

We had been in the new house for just two weeks. Rebecca's father had helped us move and then done a tag-team with her mother, who came over to help in anticipation of the new arrival. With Thomas, as panicky first-timers, we had clearly erred on the side of caution and arrived at the hospital far too early. But we were experienced parents now. There was no way we were going to make that mistake this time.

It was Sunday, just after lunch, when Rebecca casually mentioned that it was probably time to go to the hospital. She had been having contractions all morning, she said, but hadn't really wanted to bother anyone about it. However, she felt that they were coming 'a bit quicker now'.

Once again, the trusty 'go-bag' was ready. Rebecca's mother took the reins with Thomas while I drove Rebecca to the Johanniter-Krankenhaus, south Bonn, with her casually occupying the passenger seat this time, rather than on all fours in the back as before. I was thinking to myself, on the drive, that we really had this birth thing down to a fine art now.

The first inkling I had that things were much further ahead than anticipated was when we had to stop on our short walk from the car park to the hospital, not once, not twice, but three times. Each time, Rebecca winced and bent double until the pain passed and she could move again. By the time we were through the hospital's front doors, Rebecca was all but ready to lie on the floor and give birth there. With the help of some friendly German nurses, we managed to get her into the lift and up to the third floor where a bed was waiting for her in the delivery room.

Less than twenty minutes later we were cuddling a very pink-faced bundle which we named Oliver – who perfectly completed our family and who at that point had no idea how close he had come to being born in a car park.

I made the mistake once of saying that Oliver's birth was really very easy. Rebecca shot me a look.

'Those were the most painful twenty minutes of my entire life,' she said.

As a man, never comment on the ease of childbirth.

* * *

In the spring of 2012, Frank De Winne called me into his office. Frank was the Belgian astronaut and first European commander of the ISS who had recently appeared in the video backdrops used by U2 on their '360°' tour – a solid claim to fame, in my book. Slightly more importantly, he had also taken over from Michel Tognini as the Head of the European Astronaut Centre.

Frank explained to me that the opportunity had arisen to send a European astronaut to serve on NASA's NEEMO 16 mission, commencing in June. NEEMO stands for NASA Extreme Environment Mission Operations, which is a less clunky acronym than CAVES, but not by much. And just as ESA's CAVES had started using a Sardinian cave network, so NASA had for some years been using the ocean as a space analogue, creating a controllable extreme environment with the help of Aquarius, a unique undersea research station, 8 kilometres off the coast of Key Largo and 25 metres down on the ocean floor. This was a hugely exciting opportunity – the chance to live for twelve days under the ocean and qualify as an aquanaut – and no European astronaut had ever been selected for it. I would be the first.

One slightly awkward thing: I had only just returned to the family, including our young baby, after three weeks in Russia, where Thomas Pesquet and I had completed our training in the Russian Orlan space suit. The Orlan is a very different beast from the American suit that we would most likely wear if we ever got assigned to do a spacewalk from the ISS. But at that stage, with nothing certain about our future missions, it was about keeping our options open and being trained up and ready to go for all eventualities, including the offer of a spacewalk from the Russians, so off we had gone to Star City. (Verdict on that Russian suit by the way: comfy and easy to climb into, but not as tight fitting and therefore not as easy to work in as the American suit, or EMU as it's known – the Extravehicular Mobility Unit. Now, that's a proper title).

Anyway, after those three weeks away, I was looking forward to spending some time at home with the baby. I'm very sure Rebecca was looking forward to me doing that, too. But now it looked like I was going to be turning straight around and heading off to the States. Before I could do NEEMO I would need to go to Houston and do some spacewalk training in the American EMU suit. On the plus side, Rebecca and the boys would be able to come and join me for a few weeks before returning to Germany, while I stayed on, so at least we'd now get some time together. Rebecca was so adept at managing our hectic lifestyle that I probably never gave her the credit she deserved for travelling alone across the Atlantic with a toddler and a baby. The trip could also serve us as a bit of a recce. We knew that if my mission came through we would probably be moving out to Houston for a while. We were more than familiar with Texas, of course, after our years at Fort Hood, but it was good for all of us to be able to get a sense of what awaited us in and around the Johnson Space Center.

This wasn't my first time at JSC. The six of us ESA astronauts had flown out and been shown around during basic training, and had then travelled on to Japan to see the operation at the Tsukuba Space Center, in a kind of whistle-stop tour of the world's key space complexes. But no matter how many times you go to JSC as an astronaut, the history of the place will always hit you. You're aware that you're parking in the same slots where the Corvettes would have been lined up, back in the Apollo era. The Mission Control room from that time is preserved exactly as it was, as a museum piece, and it still smells of cigarette smoke when you first walk in.

My main business this time, though, was in the Neutral Buoyancy Laboratory where spacewalk training is conducted, a giant white-walled warehouse containing a diving tank which is five times the size of an Olympic swimming pool and 12 metres deep. Life-size mock-ups of Space Station modules and various other equipment sit on the bottom – impressive to see.

Ordinarily the beginner will go into the pool with an experienced astronaut. My partner on my first three-hour familiarisation run was a stocky American with a gruff demeanour – a little intimidating. That was my first encounter with Scott Kelly, my future commander on the ISS, had I but known it, and the best commander I could have asked for – a genuine leader and, when you got to know him, selfless and caring in ways that the initially blunt demeanour didn't prepare you for. Having him there to administer astute and constructive criticism during the spacewalk training was also a bonus.

This was just the basic course; later I would do an advanced course and beyond that there would be mission-specific spacewalk training, in which I would work extensively alongside Tim Kopra, who was eventually my EVA partner and the other excellent commander I had during my time on the ISS. But even this first stage was highly intensive. Everything was filmed and analysed up in the Mission Control room above the pool. When you weren't in the pool on a spacewalk, you were studying – really cramming to get your head around all the different tools, the different techniques, the tether protocols to prevent your equipment (or worse, yourself) from floating away in space, the airlock procedures, the communication systems …

You were taught to think of your space suit as a spacecraft and to familiarise yourself with its emergency procedures to the point where they became a reflex. What happens if the electrics fail? What if your cooling or ventilation systems fail, or your carbon dioxide removal system? There were procedures for all these outcomes which needed to be practised and absorbed.

And what if your partner on the spacewalk fails? That was another scenario we rehearsed for: the rescue of an incapacitated crew member. When you're out there in space, working separately, you're meant to be aware of what your crewmate is doing at all times and check in with each other every few minutes. 'How's it going, buddy? Whereabouts are you now? What are you working on?' And if you don't hear anything back, you need to go to them and find out what's up. And if they're unresponsive when you reach them, there's a whole procedure for getting them back into the airlock.

At some point in the training, your partner would be given the nod to fall silent on you. They would be watching how long it took you to notice. And once you did notice, you had twenty minutes to retrieve your apparently unconscious, and certainly unhelpfully dead-weight buddy, and haul them back to the safety of the airlock, making sure not to snag any tethers along the way.

Exhausting work, all in all. No one goes to the gym after spacewalk training. Even when you're not lugging other people around, it's a full-body workout down there in the water in that suit. And, as with the real thing, you realise that it's all about efficiency and saving your strength. If you can do something in a way that involves just one less tether clip, you'll do it. And when you're five and a half hours into a spacewalk and there's nothing left in your fingers, you'll be grateful that you made those tiny savings along the way.

I loved spacewalk training, just as I would eventually love spacewalking itself. It spoke loudly to the pilot in me. It had so many similarities to flying a sortie, in terms of the preparation you put in beforehand and the choreography of it, and it offered the same immense sense of satisfaction when you got it right.

After that course, I returned briefly to Cologne and then flew to Florida for the NEEMO mission, the objective of which was to develop tools and map out techniques and procedures towards a notional journey to an asteroid at some point in the future. (Asteroids were a particular focus of NASA's thinking during the Obama years.) There were a few days of initial training on the quayside and in the ocean, mostly learning how to use the Kirby Morgan diving helmet which we would be wearing when working outside the Aquarius habitat. And then it was splashdown.

With our small dry-bags of belongings, the four of us on the mission, along with our two instructors, put on scuba gear and dived down the 25 metres to the cylindrical Aquarius habitat. It was like something from that James Cameron film, The Abyss. We ducked in and onto the wet porch, walked up the steps, took off our scuba kit, dried off, and then had a cup of tea.

Aquarius was an extraordinary place – a confined cabin at the bottom of the sea. At the far end were six bunks, three on one side, three on the other – very much like submariners' quarters. I was on the top bunk with just a small gap between my head and the curved ceiling. Between the two sets of beds, on the back wall, was a lovely circular window – excellent for watching the passing marine life. There was a living area with a table that could just about fit six, and a separate cubbyhole which was also our Mission Control space, with a radio, a couple of laptops and all our electrical equipment. Our kitchen, such as it was, consisted of a microwave oven and a hot tap. Only meals that you could add water to or microwave would be consumed over these twelve days.

I was also pleased to notice, on arrival, the presence of a Portaloo. Practically the first instruction we were given, though, was: on no account use it. It wouldn't flush well, it blocked and clogged, and it would soon stink the place out. It was there solely for the eighteen-hour period at the end of the mission, when the hatch would be closed and we would be sealed in the habitat, going through decompression before we returned to the surface.

In the meantime, we were invited to slip out and pee off the wet porch and, for nature's larger calls, to avail ourselves of the Gazebo, a construction next door to Aquarius, a little like an upturned eggcup, more or less the size of a Soyuz capsule, and with an air pocket. This was also the designated safe space in which we could gather if there was an emergency in Aquarius, such as flooding or a fire, since swimming straight to the surface would risk potentially lethal decompression sickness.

In order to use the Gazebo as the outside toilet, the routine was to duck-dive into it, lower your trunks, poke your bum out of the back, and empty your bowels into the ocean. Simple in principle.

Alas, it turned out that by doing so, you would successfully attract the attention of every hungry fish in the vicinity, for whom your faecal matter appeared to constitute fine dining. Some of those fish had no compunction about coming very close in order to eat, either. Consequently, I can confirm that triggerfish, with their teeth designed for cracking shells, have quite a bite on them. You could reach back with your hand to try and flap away the frenzied shoals attacking your buttocks, but to no avail. Rare was the trip to the Gazebo from which you didn't return with either a blood-ied finger or a shredded butt cheek.

Brilliantly, somebody had the idea of introducing a pierced tube connected to an oxygen tank which the Gazebo-user could switch on to send up a kind of curtain of bubbles, behind which we could privately do our business. To our great delight, the 'bub-bler' solved the fish problem – but only for the forty-eight hours that it took the fish to work out what was going on. Thereafter the 'bubbler' itself became a kind of dinner gong, summoning the fish to supper in even greater and more persistent numbers. There was no escape. When people refer to the excitement and sup-posed glamour of the astronaut life, I remember the 'bubbler' and think, 'If only you knew.'

Alongside me in the mission crew of four were Dottie Met-calf-Lindenburger, a NASA astronaut and former science teacher who had flown the Shuttle to the ISS; Kimiya Yui, who was at the time an un-flown astronaut with the Japanese Space Agency, having been selected in 2009, at the same time as me; and Steve Squyres, a professor of Physical Sciences at Cornell University, who was on the NASA advisory board and was the Principal Investigator of the Mars Exploration Rover mission. That MER project had been staggeringly successful, landing two rovers on Mars – Spirit and Opportunity. Contact with Spirit had been lost in 2010, but Opportunity was still going strong (and would carry on until 2018). In the evenings, Steve would sit at a desk with a cup of tea, receiving the day's latest pictures from Mars and planning the rover's upcoming routes.

'Right, where do you want to go next, guys? What do you reckon?'

We were on the ocean floor, plotting where Opportunity would be roving on Mars in the coming days. It was surreal.

On one occasion we had a live link-up with the ISS as well, which was also very cool: people 400 kilometres above the planet chatting with people 25 metres below sea level.

The four of us performing the mission weren't the only people around. When we were out on an EVA, there were a whole host of support divers accompanying us, some to help, some to observe and some, seemingly, because it was fun to be down there seeing what was going on. The occasional visitor would pop up from the wet porch, too, and come in for tea and a biscuit, maybe bringing some fresh food down to us. You'd be pouring them a second cup and they would suddenly look at their watch and say, 'Blimey, I've been down here forty minutes. I'd better get back up before I need to decompress.'

Each morning, two of us would perform an EVA while the others were in the habitat, liaising with Mission Control on the Florida coast. Then in the afternoon we would swap around. The scenario was that we were learning how to spacewalk and conduct scientific research on an asteroid. The first week was fairly basic stuff, looking at different methods of translation and working with tools, and the second week got a little more exotic, trialling jet packs, for instance, which were great fun, and very *Stingray*, but also dangerous because you didn't want to get it wrong and get carried upwards too far. I think all of us were understandably wary about the threat of decompression sickness. We grew only more so after a near mishap, during one of our EVAs.

Near the end of the exercise, a support diver who had been working with us started to return to the surface with a bag of tools, not realising that he had snagged on an umbilical. As the diver rose, he was unwittingly pulling up Kimiya with him, and Kimiya couldn't do anything to stop him. Fortunately the diver realised before Kimiya was too close to the surface. That could have been extremely nasty.

One night, when we were relaxing, Steve and I decided to suit up and spend a couple of hours outside the habitat. We popped out and went and sat on the ocean floor where, very soon, we found ourselves surrounded by incredible marine life. During the daytime when we were out there, you might see a bit of stingray activity, but nothing else very much and, in any case, we were too busy concentrating on what we were doing to observe properly. At night, though, as we sat and watched, the scene illuminated eerily before us grew strange and magical – a nurse shark just swimming around us, enjoying itself, barracuda coming in to feed, goliath groupers swimming around the habitat. When you live underwater for twelve days, you come to appreciate how habitual these sea creatures are, doing the same thing at the same time. The nurse shark would sleep in the same patch of soft sand every night. You could set your watch by the moment she would come in and settle down.

I rank that NEEMO mission among the best experiences of my life. On the flight back home, I could only reflect that being an astronaut was taking me to some incredible places. Wouldn't it cap it all if being an astronaut could now also take me to space?

CHAPTER TWENTY-ONE

The long and the short of it, a grilling on the telly,
and vodka round the fire

In late 2012, Frank De Winne announced that ESA had managed to procure a seat for one of us on a short-duration Russian mission to the ISS, due to launch in the autumn of 2015. The selected astronaut would fly in the left-hand seat of the Soyuz, there would be a Russian commander, and in the right-hand seat would be the British soprano Sarah Brightman.

Good news? Actually, I'm not sure I'd ever had such conflicting emotions about anything.

Clearly, for Thomas, Andy and myself, any announcement of expanded flight capacity should have been a cause for celebration. It increased the chances that we would be flying soon, and it was a heartening sign that ESA were out there batting for us, grabbing flight opportunities for their astronauts when they could.

Also, for whichever one of us was selected, it would mean a confirmed trip to space. How could anyone turn their nose up at a trip to space?

At the same time ... a short-duration flight? That's ten days. Six-month missions were the norm. A six-month mission was what we had been training towards and what each of us had set our hopes on during these past three years. There would be no spacewalking in this ten-day mission, no working with the Space

Station's robotic arm – barely enough time to get used to weight-lessness. You'll fly to space, you'll be exceptionally busy, you'll do as much science as you possibly can, and squeeze in some outreach work. But, realistically, it's ten days in orbit – two and a half of which would be spent in the Soyuz during rendezvous, so probably six meaningful days on the ISS by the time you've completed your arrival procedures and done the prep for leaving. Six meaningful days in space, versus six months. There was no comparison.

And which of the three of us was going to get handed this glit-tering yet oddly compromised prize – this unexpected gift from Frank, which we couldn't help but feel was also the short straw? It was blindingly obvious to me, and I'm sure it was blindingly obvi-ous to Andy and Thomas also.

The astronaut who was going to be assigned to this trip was me.

We knew there was a 'ministerial' coming up in November. That's the crucial round of meetings in which ESA get together with ministers from the member states and hash out budgets and contributions and the expected industrial return from those investments. Future astronaut flights can be powerful negotiating tools during such meetings. Long-duration missions would cer-tainly be on the table during this upcoming session: in particular there was a newly-acquired six-month mission in the diary for November/December 2015, to which an ESA astronaut would need to be assigned fairly soon. (Most people assumed that mis-sion would go to Thomas.)

But ESA would now also have this ten-day mission to put on the table – the perfect opportunity to try and tempt the Brits to come in through the door and finally contribute to the Human Spaceflight programme. 'Here, look: we've got a guaranteed trip for a British astronaut we can offer you. Nice little package – the perfect starter kit.' Meanwhile the big boys in the playground could carry on planning for their long-duration missions.

I could see how it made sense from ESA's perspective, but some part of me was going to be a bit disheartened if this happened. Which sounds ungrateful, I know. Just a decade earlier, most ESA missions were two-week missions, either on the Shuttle or short Soyuz trips. For some of the senior astronauts, that was their only option. But the parameters had changed, and when everyone around you was spending six months in space, and you were the only one about to get the ten-day stint ... well, you couldn't help but feel a bit short-changed.

And then there was the added dilemma of flying with a British celebrity, Sarah Brightman. Now, space tourists were nothing new in 2012 – although amongst the space agencies the preferred term is 'spaceflight participant'. A private American company called Space Adventures was brokering these seats for spaceflight participants. Dennis Tito, the American engineer and entrepreneur, flew to the ISS in 2001, funding himself to do so. If you had $25 million at your disposal, why wouldn't you? Guy Laliberté, the Canadian founder of Cirque du Soleil, flew in 2009. Anousheh Ansari, the founder of Prodea Systems, flew in 2006, self-funded, and became the first Iranian in space. The Russian Space Agency loved the spaceflight participants. They brought in money and they generated public awareness.

Nevertheless, my heart sank a little further to think that I might be part of what was nicknamed a 'taxi-flight' for a spaceflight participant. And let me say straight away, that reaction had nothing to do with any feelings I might have had about Sarah Brightman. I would get to meet her later, when some of our training time overlapped in Star City, and she was wonderful: professional, dedicated, extremely capable and fun to be around.

What I was dreading was what it would look like if this mission came my way. After all that excitement in the UK about my selection back in 2009, and the gathering anticipation about the possibility that Britain might be sending someone to space for the first time in two decades, I could see only too easily how it would play out if I now ended up accompanying a British singer to the ISS. I had formed – and publicly talked about – a vision in which a British astronaut reached out to inspire a whole new generation of engineers, scientists and adventurers. I could only wince to imagine the British press now telling the story of 'Britain's Major Tim' hopping to space and back as a celebrity's taxi driver.

I had to take the pragmatic view: better ten days in space than no days in space. Better a flown astronaut than an un-flown astronaut. And better that the UK was participating in ESA's Human Spaceflight programme in some form than not at all. But I will admit, dealing with the disenchantment and concern over how it might be perceived by the press was difficult for a while. And if it was hard for me, it was possibly even harder for Rebecca. In so many ways, of the pair of us she had made the greater sacrifice, upping sticks to move to Germany, leaving her family behind, enduring those isolated first years, putting up with me being away for weeks and months on end. It was tough for her to think that we might have invested so much in this adventure, only for it now to become a source of anxiety.

The UK was going to be represented at the November ministerial by David Willetts, the Conservative Minister for Science and Universities, who also happened to be the MP for Havant, which is very near Westbourne where I grew up. I had met David in London once, to discuss the issue of UK funding for space and was very impressed by him. His nickname was 'Two Brains' and you could see why. He was clearly a highly intelligent person. But he was also someone who was a very intent listener, which is not a quality one always associates with politicians.

In that meeting he had asked me what I wanted from a mission and why I thought it was important. I set out three reasons why it would be good for the UK to be involved in the ISS programme. Firstly, a mission involving a British astronaut would help to grow the UK's burgeoning space industry, by offering the chance to get involved in human spaceflight, an area which was only going to expand as time went on. Secondly, a mission would benefit the UK's scientific community, which had a really strong cohort of expertise in life science and space biomedicine, yet wasn't currently getting access to space station science. And thirdly, a mission had massive outreach potential in terms of education and STEM learning – science, technology, engineering, maths – from the youngest schoolchildren upwards.

I left thinking the meeting had gone well. David Willetts had listened keenly and I sensed he could see where I was coming from. How he would get on in negotiations with Jean-Jacques Dordain, though, I had no idea. And whether he would think a short-duration mission might serve the UK as well as anything at this point, I also didn't know. I only knew that Jean-Jacques Dordain could be very persuasive.

The November ministerial came and went. The rumour was that David Willetts had made a great showing on the UK's behalf and had some very positive discussions, but there were no specifics that I could find out. An announcement on flight assignments was expected in the wake of the meetings, but it didn't come straight away. In fact, it didn't come for ages. We got through Christmas and the New Year without any statement being released. The silence dragged on, through Thomas's fourth birthday, through my forty-first in April. Those were edgy and agonising months for both Rebecca and me. We were desperate for the wait to be over – to have the inevitable confirmed so that we could get on and make the best of it. I kept talking to Matt Goodman, who was Head of Communications at the UK Space Agency, and a good friend of mine, to see if he knew what was up. And he did his best to keep me informed. But the reality was that nobody seemed to know anything.

I'll always remember the day. It was midway through May, bright and sunny. I had popped back to the UK and had dropped in to see my old Apache instructor mates at Middle Wallop – something which I managed to do only twice in my whole astronaut training phase. They had taken me out on an Apache sortie for the morning to do some flying training. I was very happy to be back in that familiar cockpit and I'd had an absolute blast. When I got back on the ground, there was a missed call on my mobile from Matt Goodman.

I was certain I knew what he was calling about and I rang him back straight away, standing there on the grass right beside the Apache that I'd just jumped out of. I'd rehearsed this conversation so often in my head by now. He would be calling to confirm, at long last, that I'd been selected for the short-duration flight, and I was ready to sound enthusiastic about that because right now I just wanted to get assigned and crack on with it.

'So,' I said. 'What's the news?'

'You got it, mate,' said Matt.

'2015, short-duration?'

'No,' said Matt. 'You've got the long-duration. The long-duration in 2015.'

There are times when you are so elated that the world swims before your eyes a bit. This was one of them. Never in my wildest imaginings had I thought that I would be handed that 2015 flight. In my mind, I was already climbing into that Soyuz alongside Sarah Brightman and coming back less than a fortnight later.

But no. I was going to space for six months, with everything that that entailed. Not for the first time in this story I was standing with a phone to my ear and a ridiculously broad smile on my face.

A huge weight had just lifted clean away. All the training, all the studying, all the doubt and uncertainty along the way, the decision to walk away from a job I loved, the upheaval inflicted on Rebecca and the family … it hadn't been in vain. I'd got the dream mission.

Two fantastic hours of flying an Apache round the skies, and then you get told you're going to space for six months. Mornings don't really get any better than that.

Incidentally, Matt Goodman had slipped me the news under the radar. When Frank De Winne from ESA called a couple of hours later to hand me the assignment officially, I would have to be surprised and delighted all over again. To be honest, I didn't find it that difficult.

I still have no idea what went on in the conversation between David Willetts and Jean-Jacques Dordain at that November ministerial and in the months that followed. I can only assume that Willetts produced a magic wand from somewhere. In the face of a recession which was biting everyone hard at that point, the UK increased its overall funding of ESA that year, including an all-important contribution into the Human Spaceflight programme. The UK still wasn't investing to the extent that Germany, France and Italy were doing, but the fact that they were increasing their spending sent a strong message, which the director general clearly heard. Indeed, Jean-Jacques Dordain welcomed the UK's investment, saying it was the most important news from the 2012 ministerial meeting. I will always be grateful to David Willetts. I'll always be grateful, too, to Jean-Jacques Dordain for his flexibility. He gave the UK a flight ahead of a French astronaut, which must have been a dispassionate decision for him, to say the least. And now I had my mission and Rebecca and I could take our lives off hold and ready ourselves for three exciting years in Houston and all the experiences beyond.

As for that short-duration flight, it went to Andy Mogensen. Sarah Brightman began her training in Star City in Russia in January 2015. I was there some of the time, and I know how she threw herself into it. She was so dedicated – up until one and two in the morning, studying her Russian, learning her Soyuz systems. Sometimes she would come down and socialise with us in Shep's Bar, which was our Star City watering hole, having a drink and watching a movie together.

And then, in May, she suddenly wasn't going to space any more. There were some stories in the papers sneeringly alleging that she had failed to make the grade and flunked out, but that wasn't the case. She was doing really well. There was a problem somewhere with the funding for her mission and the Russians simply pulled the plug on her training. One day she was there, the next she was gone. Her seat was taken in the end by Aidyn Aimbetov, a Kazakh cosmonaut. Sergei Volkov, the Russian cosmonaut, commanded the flight.

To this day I continue to think that mission was initially earmarked for me. It was acquired, I'm convinced, with a view to bringing the UK on board financially, only for the UK to come on board to an extent that ESA hadn't been expecting. Meanwhile, having spent a busy ten days in space in 2015, Andy has had to watch his classmates Alex Gerst and Luca Parmitano complete their second long-duration missions to the ISS and, at the time of writing, Thomas Pesquet is the next ESA astronaut scheduled to fly again. Acquiring the short-duration mission was no doubt good for ESA, but when they did so one of us was always going to end up being hard done by. I feel bad for Andy but, of course, he knows, as we all do, that this has nothing to do with his abilities as an astronaut, which are as good as anybody's in the class of 2009, and everything to do with funding and politics. That's just the nature of the game.

* * *

My assignment to Expedition 46 to the ISS was announced amid fanfare at a press conference in the IMAX Theatre at the Science Museum in London. Thomas Reiter, the German astronaut who had taken over from Simonetta Di Pippo as director of Human Spaceflight, was beside me on the platform, along with David Willetts, David Parker from the UK Space Agency and Ian Blatchford, the Science Museum's director.

In the opening remarks, Reiter told the gathered press and dignitaries that this was 'a remarkable moment for your country'. David Willetts spoke of the 'Apollo effect' – the inspirational value of the mission. 'I have high hopes it will interest a generation of students in science and technology,' he said. David Parker suggested that 'nothing inspires like human explorers at the final frontier'. The Prime Minister, David Cameron, issued a statement: 'I am sure Tim will do us proud, and I hope that he will inspire the next generation to pursue exciting careers in science and engineering.'

When it was my turn, I spoke about the privilege of being given the chance to fulfil what was clearly the high point of my career and I tried to give some sense of the scale of my gratitude to the people who had supported me. I also thought I had better get ahead of a couple of questions which I knew would be coming my way in the Q&A session. Earlier that month Chris Hadfield, the Canadian ISS commander, had marked his departure from the Space Station by performing David Bowie's 'Space Oddity', accompanying himself on the acoustic guitar he had been using up there in his free time. So to spare everyone the trouble of asking, I confirmed that, yes, I do play the guitar, but very badly; and that I wouldn't inflict my singing on anybody.

Perhaps I would have earned some kudos at this point if I had gone on to mention that Chris Hadfield and I once jammed together. This was in Shep's Bar in Star City early in 2012 – an occasion that will go down in the annals of rock history, I'm sure. Star City was always a lively place with astronauts coming and going at various stages of mission training. At this point, Sunita Williams and Akihiko Hoshide – Suni and Aki – were in the final months of preparation for their July launch to the ISS from Baikonur. Thomas Pesquet and I were there to do our Orlan space suit training and we had been reunited with Luca Parmitano, himself only a year away from his launch.

It was a Friday night, after a busy week of training, and Thomas, Luca and I were chatting with Suni and Aki, hearing their stories of what launch, spacewalking and living in space were really like. Chris Hadfield was there, too, and the bar was getting busier. As Friday night stretched into Saturday morning, the guitars came out and the singing began. Luca, Thomas and I all play guitar, albeit not to Chris's standard. In fact, my musical talent and my ability with the Russian language have a lot in common, all things considered. However, I cut my teeth playing Pink Floyd's 'Wish You Were Here', and I've got that number down pretty well.

Suni suddenly exclaimed that she wanted us to play her something that she could record and have as one of her three tracks to listen to in the Soyuz capsule before launch. (Each crew member gets an allocation of three songs to be piped in during those moments.) We could hardly refuse. I still have the recording of 'Wish You Were Here' that we laid down that night and, remarkably, it starts off quite well. But then the singing starts and things begin to unwind quite rapidly after that. True to her word, Suni gave the recording to her Soyuz instructor and had it played in the capsule before she and Aki were blasted into space. I imagine they were still laughing at our expense when they made it into orbit.

Anyway, I thought better of telling that story in the Science Museum press conference, and settled instead for quickly extinguishing any musical expectations people might have had for my mission, post-Hadfield.

When the Q&A session finished, and the event wrapped up, the publicity blitz began in earnest. I paused to talk to Chris Evans on Radio 2 from the Science Museum and then, chaperoned by Jules Grandsire from ESA's PR department, I set off on a day of appointments – newspaper, radio and television interviews in dizzying rotation. The spirit everywhere was celebratory. People really seemed genuinely enthused at the prospect of a British astronaut in space, even if I had to keep insisting that I wasn't the first, as some people got carried away and suggested. That was Helen Sharman, of course, and it always would be. But I was the first ESA astronaut from the UK, and I was happy to wear that title. At 7.00 p.m. there was an evening drinks reception at the Houses of Parliament to celebrate Britain's participation in the mission, with more introductions and handshakes. And then when that finished at 9.00, Jules and I drove to the BBC studios for the day's final hurdle: *Newsnight*, and an interview with Jeremy Paxman.

Paxman, of course, had a reputation as something of a Rottweiler. He was famous for asking Michael Howard, then Home Secretary, the same question twelve times, and for his general air of aggrieved contempt. By now I was shattered. It had been a long day and I seemed to have spent practically all of it talking, smiling and shaking hands with people. I was pretty much ready for bed, but I still had this final hurdle – and a slightly intimidating one where it appeared I might be required to call on some of the endurance methods I had learned on that Resistance to

Interrogation course during my Army days, and which might have grown a bit rusty from under-use in the meantime.

I was waiting in the Green Room when Paxman came in and said hello. I don't know whether he sensed the weariness in my face, but he was immediately reassuring.

'Don't worry,' he said. 'You're the good news story. This will all be very light.'

I thought that was decent of him, to go to the trouble of putting my mind at rest. I felt my shoulders relax a bit.

While the programme played an introductory film clip, I was ushered into a chair in the studio. And then we went live. Paxman's opener, in a fairly uncompromising manner, was: 'Now, what are we going to get from the sixteen million pounds that's being spent on putting you up there?'

I thought, 'OK, there's possibly a touch of side-spin on that question.' But that was fine; it was a question I was ready for. I knew people would want to raise the cost v. benefit issue. I had my answer prepared: that it was an opportunity to boost the British space industry, that the money would flow back to the UK in the form of industrial contracts, that it would assist British medical research, and that it would inspire the next generation of scientists and engineers.

'But what's the point?' said Paxman, as if I'd just told him I was going to spend six months throwing a tennis ball against a wall.

And that really set the tone for the next four minutes or so: me doing my best to lay out the value of the Space Station's pioneering work in areas such as osteoporosis and the development of vaccines, Paxman responding with withering scepticism.

'But what are you actually going to do? ... Yeah, but *what* science? ... Do you think it might be a bit boring up there? ... But you're just drifting around, aren't you?'

No facts or figures, since that opening question: no carefully researched probing. Just a series of fairly basic provocations. Perhaps, when he'd said this would be the light item, he had meant in terms of his own preparation. Lucky for me, I guess.

'They just seem to be up there nowadays playing the guitar,' Paxman went on. 'It's not what many people would recognise as a taxing job.'

Hmm. It was a bit rich of him to make that reference to Chris Hadfield, I thought. During his increment on the ISS, Chris's crew had set the record for the maximum amount of science done in a week. Frankly, if there was anybody busking on company time here, it was Paxman. Fair play to him, though. After a day of being asked the same things over and over, this was at least different. And it certainly woke me up.

Incidentally, NASA use that clip of me getting grilled by Paxman in their astronaut media training, as an example of how to cope under hostile fire. At least, I hope that's why, rather than as an example of how badly things can go wrong.

That summer, I flew to Houston to begin mission-specific training. Rebecca and the boys were going to come out to live in the States in October, so I also did some house hunting. The rental market in Houston seemed to me to be expensive and fickle, with landlords reluctant to give you a lease for longer than six months at a time – a year at a push. I was becoming discouraged and the real estate agent who was helping me, Maria Wicker, eventually said, 'Well, would you consider buying a *new* home?'

She meant very new: in fact, unbuilt – a house on a planned estate which at that point was still acres of untouched greenfield. I explained that we were only going to be there for three years and I didn't really want to spend the first of those years living on a building site.

'Oh, no,' Maria said. 'Whole estates around here get built in six months.'

I didn't really believe her, but she took me to a field where there was a solitary show-home surrounded by acres and acres of space. I liked the house – a typical wooden-frame American home with a brick facade. I also liked the price: $270,000 for a 3,000-square-foot detached four-bedroomed place. By comparison with European prices, that was a great deal.

'But I need it by mid-October,' I said.

'No problem,' said the guy in the sales office. 'By then your house will be built, your street will be finished, and we'll be out of here soon after.'

I was still pretty sceptical. But I was wrong to be. Come October the house and its neighbourhood were complete. There was still some work going on at the far edge of the estate, which would continue for another few months but when you've got a two-year-old and a four-year-old to entertain, easy access to diggers and tractors can actually be an asset. After years in married quarters and rental properties, having our own home was a huge treat. Maria ended up a great friend of ours too.

At this stage most of my Houston-based training was taking place in Building 9, the gigantic training hall in the Johnson Space Center, an astonishing facility. The biggest US flag I had ever seen covered a wall at one end, alongside poignant memorials to the lost crews from Apollo and the Shuttle. There was an old Space Shuttle trainer in there and a couple of lunar buggies and then some humanoid robots that NASA were working on – bits of the past, bits of the present and bits of the future, all of human space exploration in one building.

And then there was the mock-up of the ISS, occupying about two thirds of the available space and featuring all of the modules in life-size volume but with gaps between them to allow multiple access points. That was where I began for the first time to get a feel for the Space Station in its entirety, with its 820 cubic metres of pressurised space, essentially equivalent to the room on an empty Jumbo Jet.

I was now beginning to train more closely with Tim Kopra, my future crewmate. Some of that training was about honing skills you had already started to develop – doing things you had done before but at a higher level of fidelity. Robotic training pre-assignment, for example, had been done on a computer simulator, on your own with an instructor. Now you worked as pairs in the full Cupola mock-up at the Johnson Space Center, going through robotic captures of visiting vehicles while controlling the simulated robotic arm and seeing those vehicles approaching. We were also schooled over and over in the Space Station's emergency drills, particularly the three big ones: cabin de-pressurisation, fire, and toxic atmosphere, created by, say, ammonia leaking from the cooling system into the air supply.

I did some more formal medical training at this time, as well. If you're lucky, there will be an actual medical doctor who is part of your crew. Kjell Lindgren, the NASA astronaut and a good friend of mine, had been on the mission which flew just prior to ours. He had a doctorate in medicine behind him and a three-year residency in ER in Minneapolis – a comforting person to have around when you're a long way from your GP. But on our mission we weren't set to overlap with any professionals, so Tim Kopra and myself both trained as CMO – Crew Medical Officer – which included some dental training and lots on the use of drugs, defibrillators and resuscitation techniques. We did some of that training on parabolic flights, back aboard the vomit comet, because administering CPR in a weightless environment is a whole different skill. You need to be able to push against something in order to apply the necessary down-pressure. One technique is to do a handstand above the patient with your feet on the ceiling and your arms stretching down, adding a whole new layer of gymnastics to the heart-failure scenario.

And then there was the spacewalk training which was the most exciting aspect of it all from my point of view. First you would learn to suit up another crew member in the airlock and how to assist them to get out of the hatch for an EVA. And then another day it would be you getting suited up and easing yourself out. We got to wear the Class One spacesuits, not the training suits – the real deal, bringing it all that little bit closer. We were put in the vacuum chamber, learning to trust in our equipment as the air was slowly sucked from all around us. I had placed a bowl of water at my feet and it was amazing to watch it spontaneously boil and then freeze as both the pressure and temperature plummeted to what would be fatal levels should you be foolish enough to remove your helmet.

We were back in the pool again at the Neutral Buoyancy Laboratory, conducting six-hour spacewalks, only this time it wasn't generic training. By now, our instructors had a feel for what spacewalks Tim Kopra and I might have to perform during our mission and were putting us through our paces. We were slated to install an International Docking Adapter, due to arrive a few months before us on a SpaceX cargo vehicle. If all went to plan, it was going to make for an exhilarating and complex spacewalk. If you immersed yourself in those training sessions, and treated them as though they were real events, the anticipation would flood through you. I would catch myself in the middle of exercises thinking, 'Holy smokes, this is going to be absolutely incredible.'

During this time I was also going for four-week blocks to Star City to train on the Soyuz which would carry us to space. Star City is an extraordinary place. During the Cold War, as the nerve centre of Russia's developing space programme, it was a closely guarded secret. Nowadays, as you leave Moscow and its nightmar-ish traffic behind and draw closer to it, you feel you are gradually receding through time and going back to those days. You could be in some kind of Cold War movie, driving through the woods and then reaching the checkpoint at the gate before passing into the rather solemn compound, with its communist-era red-brick build-ings all intact and its numerous statues of Yuri Gagarin. There is nothing shiny or glitzy about Star City. It's all very functional and Soviet. And then, as you pass through, up pops a patch of purest New England – three semi-detached white houses in an area of lawn. That's the American compound.

ESA astronauts stayed next door in a building called the Pro-phylactory, known as the Prophy, close to Star City's very beautiful Russian Orthodox church. We all got basic en-suite rooms, with old rugs and heavy wooden furniture (Russians do seem to love a wooden display cabinet), where the heating was either on or off. We shared a kitchen but most evenings we would go round to the American cottages and cook up group meals. 'OK, I'm doing a chilli tonight. If someone can bring the nachos and someone else bring the cheese, and someone do salad or dessert …' It was all extremely sociable.

It was on an early rotation to Star City during this period that Tim Kopra and I had to do the Winter Survival course. At this point, the Russian cosmonaut and Soyuz commander assigned to our mission was Sergei Zalyotin, a former fighter pilot in his mid-fifties who had been to the Mir space station and had already completed one short mission on the ISS. However, just before we flew out to Russia, Frank De Winne had said, 'Don't get too attached to your training with Sergei. He's not going to be your commander.' It was all shrouded in mystery, but Sergei appeared to be a placeholder assignment, although this was never officially acknowledged or explicitly talked about between us during our training together. The point where it became apparent that Sergei himself knew that he wasn't actually going to be coming with us was when we were all sent out to do our Winter Survival.

This was one of our training's more punishing obligations and would probably be nobody's idea of a great holiday – two nights of nothing-inclusive self-catering in the forest surrounding Star City where the temperature was likely to drop as low as minus 25°C. On the positive side, at least it's dry cold, rather than wet and slushy. On the negative side, you'll be building and sleeping in your own shelter.

The exercise starts, however, in a Soyuz capsule, left in the woods, where the first challenge is to change out of the thin space suit with which you would have returned to Earth and pull on your thermal layers. With three people rammed into the tiny space, this is as challenging as you might imagine, with legs and arms all over the place.

Then you get out, bringing along your Soyuz survival kit: a hacksaw, some matches, a couple of signal flares and, yes, some more of those disgusting firelighter food tablets without which no survival exercise is complete. And off you go into the woods to chop down some trees, get a fire going and construct somewhere to sleep. Ahead lie two uncomfortable nights because even if you are a master shelter builder, your shelter isn't going to provide much warmth at those temperatures and even standing by the fire, the parts of you that aren't getting toasted would be frozen.

On the first night, the instructors came by at about 10.00 p.m. for their final check visit, and then went off somewhere warm, saying, 'See you in the morning.'

They had not been all that long out of sight when Sergei suddenly said, 'I'm just going for a walk.'

Tim and I were a little concerned. He wasn't about to do a Captain Oates on us, was he?

'Don't worry,' Sergei said. 'Back in ten minutes.'

Tim and I watched him go, setting off into the frozen forest. Ten minutes later, true to his word, Sergei was back, carrying a large rucksack. He put the rucksack down on the ground, opened it up and began pulling out its contents. Out came a pillow, a duvet, a plastic tub containing hot stew, and two bottles of vodka.

'Jeez, Sergei. What the …?

I don't know whether it was his wife, his brother or a friend – but someone had arranged to meet him somewhere on the wood-line and make the drop.

'It's supposed to be a survival exercise, Sergei.'

'Yes,' said Sergei, philosophically. 'But any fool can be uncomfortable.'

He cracked open one of the bottles of vodka.

'We'll just have a little drink,' he said. 'For good health.'

Tim and I looked at each other. We were cold, it was late at night. We weren't fools.

A while later, when all of our stomachs were warmed by delicious hot stew and all of our healths were preserved by the best part of a bottle of vodka, Sergei wrapped himself in his duvet, put his head on his pillow and went sweetly to sleep. Tim and I, happily inebriated, took it in turns to keep the fire fed.

The next morning, I felt absolutely horrendous. It's one thing to have a hangover, quite another to have one at minus 25 when you've barely slept and you're dehydrated with no water around to quench your thirst. But what are you going to do? The next night, the same thing happened. With the instructors out of the way at 10.00 p.m., Sergei trudged off to meet his supplier and collect the stew, which we put away along with the second of the bottles of vodka. This was rapidly becoming a real exercise in survival – just not quite the exercise that Tim and I had prepared for.

The usual debrief following that exercise was extremely awkward. The instructors had clearly found some incriminating evidence – maybe a feather from Sergei's pillow, maybe some drops of stew in the snow, I'm not sure. Either way, we sat before the Russian panel, and the question came, in Russian: 'Is there anything you would like to tell us?'

Our innocent expressions would have won awards.

'Did you have anything with you during the exercise that you shouldn't have had?'

Again, we feigned ignorance which, given the state of my spoken Russian, wasn't all that hard for me to do. There was no way that Tim and I were going to drop Sergei in it. He knew he wasn't going to fly, he had done this training before in any case, and he hadn't been interested in being cold and miserable again for nothing. He left the Astronaut Corps not long after this, and Yuri Malenchenko was assigned as our Soyuz commander.

Technically, the panel could have failed us on that exercise and made Tim and me do it again. We got away with it, though, largely because the circumstances surrounding Sergei's assignment had put the pair of us in a slightly awkward position, but also perhaps on the grounds that we had shown unbending loyalty to our commander even so. After all, if your commander tells you to drink vodka and eat stew ...

CHAPTER TWENTY-TWO

Carry-on luggage, Christmas in Star City,
and the long goodbye begins

My bags were packed for space eight months in advance. That wasn't just me getting over-excited. It was the rules.

Packing for space turns out to be more complicated than just shoving a few things in a holdall. A well-oiled NASA machine whirrs into operation and there's a sequence of deadlines to be met for the delivery of your belongings which starts from at least eighteen months before launch. Your clothes get sorted earliest – your day clothes, your underwear, your PE kit, including the cycling shoes you're going to need to clip into the pedals on the cycling machine (or the ergometer, as it's called) and the running harness to tether you to the ISS's treadmill. These were going to travel up well ahead of me on a cargo craft.

The white fabric duffel bag in which you are allowed to pack your personal items is about the size of two shoeboxes. I had been thinking carefully about the outreach element of my mission since I got assigned. I knew that I wanted to run the 2016 London Marathon from the ISS's treadmill, in parallel with the actual event in April. I wanted to do that on behalf of the Prince's Trust charity, for which I was an ambassador. That meant I had to get a Prince's Trust running top cleared for spaceflight, with an ESA logo in place and its material checked for flammability.

I wanted to take an England rugby shirt with me. That item, too, had to be vetted, as did the Raleigh International T-shirt that I wanted to fly with to acknowledge my Operation Raleigh trip to Alaska, and as did the T-shirt with the tux-and-bow-tie design on the front of it, which I had a hunch might spare me a 'nothing to wear' wardrobe dilemma at some point. And Rebecca had cut the corners off two blankets that Thomas and Oliver had slept under as babies, which I also wanted to make room for.

I borrowed an idea from Kjell Lindgren and took 1,000 tiny fabric keyring fobs which I would be able to hand out to people at outreach events so they would own something that had spent six months in space and had been around the planet 3,000 times. On the keyring fobs were the words 'Principia – Flown in Space'. Every time ESA sends an astronaut to space it holds a competition inviting people to name the mission. Four thousand people submitted suggestions for mine, and I chose Principia, after Sir Isaac Newton's seminal physics text. As for the mission's patch, which would be sewn to the right arm of my flight suit, that was thrown over to the viewers of Blue Peter. Three thousand of them sent in designs and the winner was a thirteen-year-old, Troy Wood, who put a bright red Newton's apple and a rocket over a green map of the UK.

You have to sign a disclaimer stating that nothing you take to space will be used for financial gain. This stipulation followed some incidents during the Apollo era of stamps and other collectibles being flown in space and then sold off. Moreover, everything that travels with you, potentially monetisable or otherwise, has to be carefully checked. I wanted to take the watch commemorating the first UK Apache squadron, that Rebecca had given me as an anniversary gift. I needed to certify that the battery was a certain type, and because the watch had a glass face, which could shatter, it had to be sealed in a plastic bag with a promise never to remove it on the ISS. All the family photographs that I took had to be bagged up for travel as well. There's a broad rule against removing those too, on account of them being flammable items, but everyone quietly forgets that and decorates their crew space with a few mementos of home.

Obviously the weight of your luggage was subject to restrictions. When my items were bagged up, Bernadette Walls, who was looking after the crew packages, shook her head solemnly.

'Tim, in all the years I've worked with NASA, you're the only one who's exceeded your weight limit.'

Clearly I was trying to cram in as many outreach items as possible. Fortunately for me, Tim Kopra was under. In a scene parallel to the one you'll see on the floors of airport check-in halls the world over, I took some of his lighter things and he took some of my heavier things and we balanced it out.

We were given an additional 1.5 kilograms allowance by the Russian space agency, which flew with us in the Soyuz and could therefore be finalised much later. Among some other personal effects, I made room in that extra luggage for a book – an already quite well-travelled one. It was a signed copy of Yuri Gagarin's autobiography, *Road to the Stars*, loaned to me by Helen Sharman. It was the volume that Helen had taken to the Mir space station and which would now accompany another British astronaut on another pioneering trip to space.

Meanwhile, I was completing my final training tasks. I flew to Japan for one last refresher on the science I'd be conducting in the Japanese segment of the ISS, in particular installing a new furnace for combustion experiments. And I flew to Moscow for my final exams – a formal and old-fashioned process, like so much about Russian spaceflight, where you are grilled verbally by a panel of about fifteen stern-faced instructors and officials in a hall that is open to seemingly anyone living in Star City who wishes to pop along. It was a bit like being in the dock for a criminal trial. Moreover, I was being tried in Russian, albeit you are entitled to have an interpreter on hand. This is widely believed to be a smart tactic because the interpreters, who sit through these exams day in and day out, tend to know the answers better than you do and, if you're struggling, they can sometimes help you out a little by making sure that something is, shall we say, gained rather than lost in translation. Either way, I got through. I was qualified by the Russian board as suitable to fly. But then, at this late stage, it would have been pretty embarrassing for everybody concerned if I hadn't been.

In the big finish to that exam process, Yuri, Tim and I approached the commission members in our soft Sokol space suits and Yuri stepped forward to select one of a number of sealed envelopes laid on a table in front of him. That envelope contained a randomly chosen list of emergency situations. Yuri now handed it, unopened, to the instructors. Then we climbed into the Soyuz simulator and the instructors went to their training room, and we began the mission simulation – every aspect of the Soyuz journey ahead of us, from launch to rendezvous and docking, to un-docking and re-entry – into which were injected the emergencies from the list, which we had to cope with. As I recall, we drew a depressurisation on launch and a fire on the way back down, along with a sprinkling of other less major incidents. All of this took a few hours, with a break in the middle for lunch, and then afterwards we went straight into a big debriefing, again open to the public, in which our performance in the simulation was critiqued.

There was quite a bit of toing and froing during this assessment. It was expected that your Soyuz commander would fight their corner and even haggle for points, bringing the panel down from a proposed twenty-point deduction, say, to a five-point deduction. Yuri did his share of scrapping on our behalf, but there had been at least one senior commander in the past who had simply shrugged and let the instructors do their worst – a mild protest against the haggling charade – knowing that they were hardly going to stop the crew flying to space at this point. I was glad that I had witnessed a few of these debriefs already, having wandered in as a rookie astronaut, just to find out how they worked. It meant it didn't feel quite so bewilderingly arcane when it was my turn.

Another great landmark when flying with the Russians: getting fitted for your personally moulded seat lining for the Soyuz. This involved a trek from Star City across Moscow to the factory of a company called Zvezda. There I was required to change into a tight, white-hooded onesie which made me look like a character that hadn't quite made it into the final cut of the movie *Elf*. I then had to lie in a kind of ceramic bath and have warm gypsum poured around me, keeping very still until it hardened around my rear torso. Then I was levered out and a craftsman in a white coat went to work to plane off the unwanted plaster. Two further fittings, with adjustments, would follow until the craftsman was satisfied. It was certainly a good fit, but a couple of inches of extra headroom had to be built in. If all went to plan, I would be needing this seat to come home in, and I would be coming back taller. With my spine offloaded in microgravity, the discs, tendons and ligaments in my back would relax during those months in space and cause me to elongate from 5 foot 8 to 5 foot 10. Sadly, gravity would quickly restore me to my original condition.

These final months were a period of mental preparation – not just for me but also for the family. Back in Houston, I took the boys to Building 9 and walked them through the mock-up space station. 'This is where Daddy is going to eat, and this is where Daddy is going to sleep …' And, of course, they were more interested in how I was going to use the loo. I was trying, as best I could, to normalise for them what was about to happen, but it brought the imminent reality of it home to me, too.

In June, six months before our launch, Tim, Yuri and I flew to the Baikonur Cosmodrome in Kazakhstan to act as backup crew at the launch of the Expedition 44/45 crew. Our duties there were to mirror and support our three colleagues, Oleg Kononenko, Kjell Lindgren and Kimiya Yui, in the build-up to launch, but the chances of us being required to replace the prime crew at the last minute were remote. If anything had happened to the prime crew, the launch would have been scrubbed and the backup crew might perhaps have flown, say, a month later. There was no question of us having to leap into the rocket with a few minutes' notice. So more than anything this was a great opportunity to witness the build-up to a launch and experience it all calmly from one step back. ESA thoughtfully invited our partners to come, too, so Rebecca had a chance to familiarise herself with Baikonur and the launch process. The situation would be just that little bit less alien and intimidating for her when we came back for our launch in December, allowing her to focus on supporting the boys. There were tours of the museum, summer drinks in the garden of the small but rather nice Seven Suites hotel at the Cosmodrome where they put the prime and backup crews … It was all very civilised and unpressured.

It was also good for me to have that experience with Tim and Yuri. After all of our training together in Houston and Star City, the three of us felt like a unit now. Tim and I had a great under-standing. We were from the same mould in many respects. He was a US Army officer and an Apache pilot who had been to test pilot school. We had been living two miles apart from each other in Houston, and Rebecca and I had been spending time with Tim and his wife, Dawn. He was nine years older than me and an already flown astronaut, and I was fascinated to watch him at work. When he disagreed with something, he would sit there quietly, unruffled, formulating his argument and waiting for the moment to make himself heard. He was never volatile, never outspoken, always calm and logical, but also opinionated enough to be able to take things in the direction they needed to go. Those were qualities that made him an excellent ISS commander, the role he assumed halfway through our mission when Scott Kelly returned to Earth.

The launch we were witnessing in Baikonur had been delayed by six weeks. A Progress cargo vehicle had made it into orbit only for something to go disastrously wrong during separation from the rocket's third stage engine. The Progress vehicle was sent spinning into space and tumbled out of control for two days before re-entering the Earth's atmosphere and burning up. It didn't pay to think too hard about what would have happened if that had been a crewed spacecraft: two days of spinning, aggressively enough to make you horribly nauseous but probably not aggressively enough to cause you to pass out, and then an uncontrolled, fiery re-entry. It was awful to contemplate. And although thankfully it was only a cargo vehicle, the technical problem was with the rocket – the same one employed to fly astronauts. Flights were necessarily grounded while the incident was thoroughly investigated.

This was on top of two other incidents in the year prior to my launch involving cargo vehicles. A Cygnus exploded seconds after lift-off and a SpaceX supply vehicle blew up over the Atlantic. The reminders were mounting, if we needed them, that this was a risky business with no guarantees.

I had a personal reason to rue the explosion that wiped out that SpaceX capsule. My space suit was on board. There were only two sizes of suit on the ISS – a large and an extra-large. I needed a medium. It wouldn't just be mine, of course. Several crew members on future missions would also be needing a size medium, so it made sense for NASA to fly one up. But it had been scheduled to arrive in time for my mission, and now it was lying somewhere on the bottom of the ocean.

Now, if it meant the difference between getting a spacewalk and not getting one, I would more than happily have worn any sized suit that was going – crammed myself into a small or stretched myself into an XL. Alas, EVA is complicated and dangerous enough without the extra challenge of a suit that doesn't fit. Serious injury can result if your shoulder joint doesn't happen to match the suit's hinged joint, so astronauts are only allowed to spacewalk in a suit size that the flight surgeon has approved. And in my case, that suit had just blown up. Worse still, the International Docking Adapter which Tim Kopra and I were hoping to install on an EVA had blown up with it.

'There goes my space suit,' I thought. 'And there goes my spacewalk.'

This was another occasion when I owed so much to the people on the ground. Within that EVA community, if they think you're a good operator who can get the job done, they will go the extra mile for you. That was the reason my medium-sized suit had been sent up in the first place, and that was the reason that a replacement was now prepared and sent, too. The NASA and ESA team understood the fact that nobody from the UK had ever done a spacewalk and they knew how big a deal this was. There was still no guarantee I would get an EVA, but my replacement suit was packed into the next scheduled Cygnus vehicle. It docked five days before we arrived.

When it was time for the Expedition 44/45 crew to launch, Tim, Yuri, Rebecca and I sat out on the roof of the search-and-rescue tower, just 1.5 kilometres from the launch pad. It was 3.00 a.m. on a warm, perfectly clear summer night. We watched as the

ISS passed directly overhead and seconds later, as the rocket's engines lit up, I was astounded by the intensity of the noise that came roaring out towards us at that distance.

But that wasn't even the half of it. There was the briefest of pauses and then the engines opened up to full power – a cacophonous thunder of bass notes that punched outwards from the launch site and directly into your chest. As the Soyuz lifted and began its climb, the rumble gave way to an awe-inspiring, visceral crackle.

I think all of us on that rooftop were grinning. In a very short time, Tim, Yuri and I wouldn't be experiencing that noise from 1.5 kilometres away. It would be coming from 50 metres beneath us.

* * *

When does the countdown start? I could say, in my case, it was from two and a half years out, maybe – on the grass beside that Apache, the day I was assigned to my mission. Or perhaps it started even further out than that – six and a half years before launch, in the kitchen, with the call from Paris telling me I had been selected as an astronaut. Or maybe it was even a year before that, when my eyes first fell on ESA's advert. Maybe the clock had been running since then.

But wherever you dated it back to, we had now definitely entered the closing stages. At the end of November, when I had finished my final exams, Rebecca and the boys flew out to join me in Moscow. They would stay from now until the launch in three weeks' time, but I would have to spend the last two of those weeks in quarantine, with limited face-to-face contact with Rebecca and with (according to the rules) no contact at all with anyone under twelve, children being considered too much of a viral threat. So we knew this would be the last chance we would get for quite a while to have some at least semi-normal family time. We were given two rooms next to each other in the Prophy at Star City which became our hotel for the week, and we turned that time into our Christmas. We went for walks, built snowmen, messed about in the woods, skated on the frozen lake.

Well, I say 'skated': I shuffled around very carefully in my boots because to twist an ankle or break an arm at this point would have been unimaginably disappointing. The closer the launch came, the more nervous I was – not about the launch itself, but about the possibility of something occurring that might prevent the launch from happening. For God's sake, don't let me be sick, don't let there be a technical issue, don't let me injure myself in a dumb skating accident …

Still more than two weeks out, and over 2,000 kilometres from the launch site, the extensive Russian pre-flight traditions and ceremonies now began. Tim, Yuri and I went to the Kremlin wall to lay flowers at the graves of Yuri Gagarin and Sergei Korolev, the father of Russian space travel. Back at Star City we attended the pre-flight wreath-laying ceremony at Gagarin's statue and then formally signed the visitors' book in his office, which is preserved exactly as he left it before his fatal flight in a MiG-15 training jet in March 1968.

Then, on the morning of our departure for Kazakhstan, we assembled for the traditional farewell breakfast. Here, to my humbled amazement, we were addressed by Alexei Leonov, twice Hero of the Soviet Union and the first man to walk in space (18 March 1965), a pioneering adventure which came close to killing him. He was now eighty-one but still commandingly strong of voice and firm of handshake. A table piled with meats, cheeses, fruit, bread and pastries went largely ignored as glasses of vodka were raised around the room in a long succession of toasts. Each of the crew members had to propose a toast, too, which meant I had spent the past few days facing probably my greatest fear: public speaking in Russian. But, after much careful rehearsal in the build-up, I managed. And then it was time to leave, although only after another solemn Russian ritual: the tradition of sitting down before you go anywhere. We all sat, the room fell silent for about thirty seconds while we all calmed our minds, and then we stood and left.

Out by the bus, the crew said a very public goodbye to our families, with officials and VIPs gathered around us, and with the press present and cameras pointing. It was cold and dark and my younger son, who had just turned four, was notably distressed by all the camera flashes, so Rebecca took the boys away from the hubbub and stood close to the bus to wait for us to board. Thankfully, I had said my proper goodbye to the boys earlier, before we walked out of the Prophy. The next time I saw them, in Baikonur, there would almost certainly be glass separating us, so I took what I figured was my last chance to hug them properly in our room before we left for the breakfast. And so began a series of protracted and mostly public farewells: goodbye, but I'm not actually going to space yet, and I'll see you in just over a week, except not really in circumstances where we'll be able to be together properly ... Hard for a four-year-old and a six-year-old to get their heads around this. Hard for their father, too.

The bus took us to the airport where, along with the VIPs, the officials and the backup crew, we boarded a gloriously battered old Russian Tupolev plane for the four-hour flight to Kazakhstan. In place of seats, the plane had a sofa running lengthways down the aircraft, so we sat side-on with lap straps minimally holding us in place. Once we were airborne, out came the vodka and the toasts began again, with the officials and VIPs clearly keen to drink us on our way in a spirit of high revelry of which Tim, Yuri and I were frequently the focus, and yet really quite detached from. Our minds were already beginning to be elsewhere.

On the tarmac in chilly Kazakhstan, a formal greeting awaited us from local officials and we were directed to one side briefly to be introduced to a large gathering of Kazakh schoolchildren. And then a coach took us across the huge, flat, brown expanses of the Kazakh steppe to the history-soaked Baikonur Cosmodrome, the largest and oldest operational space facility in the world, scene of the launch of Sputnik 1 in 1957 and of Vostok 1, the first ever crewed spacecraft, in 1961. It was stirring to reflect that our Soyuz would lift off from the same pad from which Yuri Gagarin had been launched into space.

The coach passed through the gates and deposited us at the now familiar Seven Suites hotel. The supporting entourage would be staying in the larger Cosmonaut Hotel next door. From the moment we passed inside we were effectively in quarantine. Since the pandemic of 2020, everyone would recognise the chief features of the astronaut's quarantined life: social distancing, strictly limited interaction, copious applications of hand sanitiser. Essentially, we were in lockdown. Over the next few days there would be a few final lessons on operating the Soyuz. We would do some manual docking training on a portable Soyuz simulator. We would cover one last time the search-and-rescue plan in the event of an aborted launch. And we would do a lot of last-minute admin: signing pictures, filling forms, firming up final arrangements for some of the outreach aspects of the mission. But there was also time to draw breath, gather your thoughts and ready your mind for the journey ahead. It would be the last time for some while that I would have that luxury.

The pre-flight traditions and rituals began to ramp up again. In the second week of quarantine, Tim, Yuri and I planted a tree in Cosmonaut Grove, on an area of grassland behind the Seven

Suites, going down towards the Syr Darya river. It's a historic avenue paved with bricks and, as you pass down it, the trees graduate from the ones planted by Gagarin and Leonov, now tall and stout, down to saplings from the modern era. Here I found the trees planted by my fellow Shenanigans, Luca, Alex, Samantha and Andy, and where I now dug a hole for my own.

Five days before launch, there was the official flag-raising ceremony outside the Cosmonaut Hotel and then a formal signing of books and mementos in the Cosmodrome museum. Two days before launch, tradition, again dating back to Gagarin, would stipulate that we all got our hair cut. This was my last chance for a while to be trimmed by a professional: in space I would be shearing myself with vacuum-suctioned clippers.

Meanwhile, Rebecca and the boys and our small group of family and friends had arrived from Moscow, just in time to see the official roll-out of the Soyuz rocket, carried horizontally by train to the launch pad – a journey which the crew are required *not* to see because it's considered bad luck for them to do so. However, it's considered good luck for the crew to fly to space with coins flattened on the rails by the rocket's train. I had duly handed over some Russian small change to Yuri Petrovich, ESA's head of Star City who had looked after us so well during our years of training in Russia. Yuri had got them crushed and returned to me.

Each crew member is allowed to invite a total of fifteen guests – not that we could mix with them. They were staying in the Sputnik Hotel, just outside the compound where we were quarantined. Along with Rebecca, my parents were allowed, after a medical check, to be in a room with me a couple of times. We sat, well spaced out, round a large table and talked, commonly under the watchful eye of Dr Savan, the medic in charge of quarantine, who was known to everybody, somewhat unfairly, as 'Dr No'. After all, if you were the one person responsible for ensuring that no viruses arrive on board the International Space Station, you would probably be heard saying 'no' a few times too. My access to everyone else who I had invited was at a designated time in a conference room, with me on one side of a large glass window, on my own with a microphone, and everybody else in chairs on the other side.

It wasn't the most natural of social settings. For one thing, I felt like a goldfish. For another, in front of me sat people from so many different parts of my life: my friends from Test Pilot School, Andy Ozanne and Dave Marsden; my best man Ian Curry and his wife Claire; Rebecca's parents and aunt and my sister and niece … I was so grateful that they were all there, and so pleased to see them. They had come in full cheerleading mode, with banners and flags and Union Jack bobble hats. Yet the only person linking these people was me, and I was stranded behind glass. It was strange talking to just one person at a time as the microphone was passed around – like I was holding a press conference for my friends.

We only had thirty minutes to chat, but I soon discovered that they were having far more fun than we, the crew, were. Having quickly bonded with Tim and Dawn's family group, they had already found themselves invited to a Kazakh wedding reception and established some friendly rivalry involving drinking games at the bar.

Let's not forget the presence of Thomas and Oliver, two small boys who had been dealing remarkably well with an exhausting schedule, strange surroundings and a building sense of anticipation for something they couldn't fully comprehend. It was agonising at this stage to be so close to the boys and so far away from them at the same time. But my luck was about to change in that regard. Rebecca, who had been allowed scheduled visits during these three days, had been absolutely scrupulous about playing by the rules. This had definitely put her on the right side of Dr Savan. He had also had the chance to observe our children and see that they were healthy. So, on that last night, when I asked him if Rebecca, the boys and I could slip out quietly for what we would now call a socially distanced walk in Cosmonaut Grove, he let us do it.

It was pitch dark and bitterly cold as the four of us set off between the trees. The boys skipped down the path to the Grove saying, 'This is amazing, Daddy. It's just like normal.' I loved how resilient and endlessly positive they were, no matter what we threw at them. Rebecca and I would get the chance to hold each other and say a last goodbye in the morning. For now we talked to the boys in the ways we always had, trying to keep it light, trying to shield them from the stress of knowing that what I was about to do was risky and that things could go wrong. We talked about when I would be back. Six months is pretty meaningless to a four-year-old, and 186 sleeps would feel like a lifetime for a child, so I explained how it would be summertime when they saw me and we would be spending lots of days together then.

Rebecca and I had prepared ourselves well for this. It was no rash decision of the heart. We were a good team and we were in this together, as we had been for every major decision in our lives. Rebecca had the time well planned out and the boys were used to me being away for extended periods. We were determined that this was going to be a really positive experience for us as a family, despite the obvious risks and challenges.

People wonder how you can do it – how you can step away from your own young children and put yourself in the way of danger. It's hard – in fact, it's the hardest thing I have ever had to do. The morning after this walk, I would look out of the window of the bus that was about to take me to the launch site and see Oliver on the shoulders of Yuri Petrovich and Thomas on the shoulders of my boss Frank – the pair of them held up right beside me, at face level, just beyond the glass. And I would smile out at my sons as hard as I could while thinking, 'Don't let this be the last time that I see you.'

But I would still stay on the bus and climb into the rocket.

I have seen astronauts described as selfish because they can do this. And maybe all exploration *is* selfish, at some level. But what if people didn't push boundaries and explore? And if that curiosity and drive and that urge to probe the limits are inside you, should you go out in the world and be who you are, or should you try and hide it?

Maybe what you are doing at such moments is about something greater than yourself, in any case. I think back to that ESA press conference in Paris in 2009, after the six of us were selected, when we each had to say something about what we thought it meant to be an astronaut. I said I thought the privilege was being part of an international team that could make a difference, and getting to be at the cutting edge of something so much bigger than any of us. I believed it then, and I still believe it. We need people to push the boundaries, and if I'm not prepared to do it myself, why should I expect anyone else to?

In Cosmonaut Grove on the eve of the launch, we walked the length of the avenue of trees and went down a small set of steps into a sheltered area that was completely out of sight of the complex. Did I then defy Dr Savan's quarantine regulations to hug my sons as hard as I have ever hugged them? I couldn't possibly comment on that.

I can only say that, as we made our way back through the Cosmonaut Grove to the hotel, there was a sense of quiet contentment between us all. We were ready. The next day I would be flying to space.

CHAPTER TWENTY-THREE

Too much borscht, not enough Lady Gaga,
and engines at full thrust

Tim Kopra enters the habitation module first, so I have a moment or two to take in the view. From this height, I can see the crowds off in the distance, my family and friends in there somewhere, and then beyond that, a broad swathe of the Kazakh steppe, stretching away for miles. Another entry for the list of sights my boyhood self would never have expected to see: Kazakhstan from the top of a fully fuelled rocket.

It's late afternoon. I have completed the strictly regulated launch-day routines. I have breakfasted well – eggs, bacon, Russian tea – knowing that my next proper meal won't be for many hours and won't actually be all that proper, in the normal sense. I have completed my medical checks, having already been badly unnerved to discover that my pee this morning has a slightly red tinge to it.

Please, no. Nothing wrong at this stage.

Then I remembered last night's borscht. Thankfully just a case of too much beetroot.

I have been given the choice of a US or Russian enema, and have chosen the Russian without really knowing what the difference is, but assuming the effect is largely the same. I have showered with anti-bacterial soap and then called for the flight surgeon who has dried me with sterile towels. I have donned my sterilised white long-johns and my blue flight suit and I have made the short walk to the Cosmonauts Hotel next door.

There, along with Tim and Yuri, I have attended the final management meeting with, among others, Frank De Winne from ESA and representatives from NASA and Roscosmos. The mission has again been toasted, this time with champagne although in the case of Tim, Yuri and me, only with orange juice. Our wives have also been there and we have each been given a minute in our own private room to kiss and hug and say goodbye. Following that, I have left my signature in black marker pen on the door of the crew room – a tradition, not an act of graffiti. And I have been copiously blessed by a Russian Orthodox priest, olive-tinctured holy water drenching my hair and flight suit.

Then, outside the hotel, I have boarded the bus for the thirty-minute ride to Building 254, cheered on by my flag-waving friends and family, clad in their Union Jack bobble hats and almost certainly destined for the record books as the rowdiest astronaut support party in Cosmodrome history. In Building 254, the three of us have changed out of our flight suits and into our Sokol space suits and had them pressure-checked, behind glass, while our close family members, among the melee of assembled press and VIPs, looked on. Eventually, suited up, with grey wellies on and carrying the cooling unit that looks like a packed lunch, I have climbed onto the bus which will bear me to the launch pad. A very large crowd of well-wishers has surged forward to send us off. Yuri Petrovich and Frank, being well used to this, have grabbed my sons and elbowed their way through the pressing mob so that I have seen them through the window one last time. After a week of goodbyes, this has been our final and most public one.

When the bus has gone a certain distance, I have got off and undone the laces and rubber pressure seal on my airtight, leak-checked and carefully sanitised space suit in order to pee against the bus's back wheel. When Yuri Gagarin made this same comfort stop in 1961 would it have occurred to him even remotely that more than half a century later all launching astronauts would still be emulating him? And would he have held on if he had known?

Back on the bus, I have looked out of the window to see the sun low on the horizon behind the waiting rocket, making it seem to loom even more gigantically against the sky. I have stepped off the bus and back into a mayhem of onlookers, VIPs and ambassadors, in which I have been grabbed under the arms and assisted towards the rocket, as if I weren't able to do it myself or mustn't be allowed to expend any precious energy. Finally deposited at the foot of the steps, I have stood and waved. And because I was not sure how long we were meant to do this for, I have anxiously looked over my shoulder several times to check that Tim and Yuri haven't left me behind.

And then I have made the 50-metre elevator ride to the capsule, looking in awe at the frosting on the side of the rocket, in all its immensity, and with clouds of vapour pouring off it.

And now here I am, looking out across Kazakhstan until it's time for the ground crew member to help me into my seat. I climb through the tightly packed habitation module, stuffed full with cargo, and then wriggle downwards, feet first, using the footholds, into the tiny descent module below, arriving in the commander's seat first and then gingerly moving over to the right, going slowly and taking care not to knock against anything or snag and tear my suit. Once again I find myself grateful to be an experienced caver. I plug in the two electrical cables, one for my communications headset, the other attached to the medical harness next to my chest which will monitor my heart rate and breathing throughout the flight. I connect the two hoses – one for piping in air for cooling and ventilation, the other supplying oxygen in the event of an emergency. I attach the knee braces and the five-point harness, which the ground crew member tightens using his full force before handing me my checklists.

Yuri follows me in. After all the ceremonies and the hoop-la, the emotions and the protracted farewells, there is something enormously clarifying about settling down to work in the capsule, a space which, after all the hours of training, we are very used to sharing. It's the three of us now, with the spotlight off, concentrating on doing what we know we can do.

There is built-in slack in the schedule, in case of any snags, and when our checks and drills are complete we still have fifty minutes left to wait. While we do so, our pre-made choice of mood-enhancing music is piped into the capsule by Ground Control. Tim Kopra really hates Lady Gaga so I have asked our Russian instructor to make sure there is plenty of Gaga in our final hour.

As the first notes of 'Bad Romance' ring out, Tim gives me a narrow look.

My three personal choices: 'Don't Stop Me Now' by Queen; 'Beautiful Day' by U2; 'A Sky Full of Stars' by Coldplay. Yuri hasn't picked anything. I think he's been to space so many times that he's run out of tunes.

Next thing, we're listening to Europe, 'The Final Countdown'. 'Who the hell chose this?'

A little joke courtesy of our instructor, it seems.

It all passes the time. Indeed, I'm quite startled when Europe's tacky synth-rock anthem fades out and I see the clock showing just three minutes to launch.

The engines start, a lovely, deep rumble way below us. No final countdown, in fact. No '10–9–8 …' at the moment in your life when you would most expect one. Instead we hear our instructor calmly announcing the stages of the final launch sequence.

'Engines at full thrust.'

Five seconds to go. The rumble from below becomes a thunderous roar. The rocket sways to the right and then to the left, as if buffeted by a sudden wind, and then, almost imperceptibly, it lifts and continues to lift, slowly, steadily, as all that pent-up power begins to expend itself.

The noise is the main thing at this point – the incredible thunder and that muscular crackle that only rockets make. And then, as we begin to gain altitude, the acceleration really starts, as 9 million horsepower drives us skywards. My stomach muscles are clenched, and I'm concentrating on breathing deeply, checking our systems and watching the clock, knowing that the first-stage separation is going to come at two minutes.

We're accelerating hard, just over 4 g. There's a bang as the launch escape system is jettisoned and then, a few seconds later, an abrupt jolt as the four first-stage boosters separate. It's a strong motion but less powerful than I've been anticipating: it was more aggressive when we practised it in the centrifuge. This is followed by a rapid deceleration and a slightly disconcerting sensation of falling. But then the second-stage rocket engine picks up again and the acceleration builds, more gently this time. Now we're flying smoothly at 1.5 g. Yuri switches the internal camera to the one pointing in my direction and I give a grin and a thumbs-up. Rebecca and the boys won't see it yet, but maybe someone out there will. I'm not faking it. This is the most exhilarating ride of my life.

And it's only just beginning. Around 85 kilometres above the Earth, the protective nose-fairing splits in half and is discarded, unblocking the windows, which means I can now look up and get a tantalising glimpse of the blackness of space approaching, though I'm strapped in and too low in the seat to see out directly. There's another big jolt as the second stage decouples. Now all that's left is the third stage of the rocket with our spacecraft perched on top. And it's time to get a move on.

If the first stage was all about power, noise and vibration, the third stage is all about speed – jaw-dropping speed. As the third-stage engine fires, we accelerate back up towards 3 g but we're flying almost horizontally now, rather than directly rising, and the sensation of pure velocity is simply overwhelming, blasting us forwards in an astonishing rush that seems to last an impossibly long time. The thrust is consistent but we're burning through our fuel and the craft is becoming lighter so we're on an exponential curve, getting faster and faster. For someone passionate about speed, this is about as good as it gets. The acceleration is mind-blowing and I've long since gone past the point of being able to comprehend just how fast we're going. Surely we've got the altitude now, so what's all this thrust about? How much longer can this go on for? About four minutes is the prosaic answer, and this ride won't stop until we reach a staggering twenty-five times the speed of sound.

At the end of those four long minutes we will reach the third-stage separation. This is going to be a sensitive moment. It was the point of malfunction for that cargo vehicle recently, the rocket not coming clean away but pressing on behind the space-craft and sending it tumbling. Is that in our minds? It's certainly in mine and I suspect it's in Tim's, too. If we don't get an auto-matic separation, it's Tim's job in the left-hand seat to separate us manually from the third stage. Tim's gloved hand now moves slightly in the direction of the relevant buttons, readying himself, just in case. Almost at the same time, Yuri calmly extends his own hand and almost lays it on Tim's in a manner which seems to say, 'Not yet.' Things are still looking OK.

And straight after that there's a massive jolt, the largest by far, as the third-stage engine cuts out, bringing me lurching forward against my straps as the acceleration drops from 3 g to zero in one stunningly emphatic instant.

And then ... stillness. The capsule is eerily quiet and the mascot chosen for this precise moment by Yuri – his daughter's plastic pen in the shape of a rocket – floats on its chain before our eyes.

It's only eight minutes and forty-eight seconds since we left Earth and we are just over 200 kilometres away, in orbit.

I loosen my straps and float up slightly to look out of the window. Yuri is concerned about me moving around in case it's too much too soon and I become nauseous. But actually I'm glad to do it because my legs are sore and I need to relieve my knees. Also I am desperate to see what's out there.

What's out there is the Earth that we have just left behind, and specifically the Sea of Japan which I can make out clearly. And even while I'm looking down at this already incredible sight, the moon rises – just pops up on the horizon, out over the Pacific, one of the most beautiful things I've ever seen. The difference about moonrises in space is that they happen without warning: there's none of the preliminary fanfare that precedes a sunrise, with the atmosphere gradually brightening. The Moon just arrives in the sky. So to catch one the first time you look out of the window feels extraordinarily lucky – charmed, almost.

I reluctantly sink down again and join the others in beginning to work through the rendezvous procedures as we settle into the six-hour journey that still lies ahead. We have been hard at work in orbit for about forty minutes when I notice the light in the capsule changing. I rise up to the window again and watch the Earth gradually become illuminated – my first sunrise from space.

I go back to work. Before long some gold solar panels appear outside the window. We've made it to the International Space Station. Now all we need to do is dock.

What could possibly go wrong?

CHAPTER TWENTY-FOUR

Climbing aboard, learning to walk, and the view from up there

The Soyuz idles 90 metres back from the Space Station's docking port. So near and yet so far.

We retreated to this point the first time when the capsule suffered a thruster sensor failure and aborted our automatic docking. This time, Yuri has brought us out here manually, knocking his own mis-aligned re-approach on the head and backing away.

Now he steadies the craft and gets ready for another attempt. I have never in my life been so keen for the law of 'third time lucky' to apply.

Once again, Yuri hunches forward to peer into the periscope. And once again, Tim and I are largely powerless to help him, unable to see what he's seeing. All I can do is continue to hold the communication stick out of Yuri's way, and wait in a state of high apprehension.

The capsule begins to move in and almost straight away I can sense the difference in Yuri's demeanour. The tension leaves him, he is steady-handed. He has very clearly got this under control.

Yuri brings the capsule in for a textbook manual docking.

We were relieved and jubilant, of course. But there wasn't time to savour the moment. The delay had put us several minutes behind schedule and ahead of us lay a long series of drills before the hatch could open. We had to complete the pressurisation checks between the two modules, start to put the Soyuz into hibernation mode (we hoped we wouldn't be needing it for six months) and to ascend, one by one, into the habitation module to get our Sokol suits off and change into our flight suits. All of this would take nearly two hours.

During this process, Yuri was talking on the radio to the Russian crew members on the ISS, Sergey Volkov and Misha Kornienko, and coordinating some of the checks with them. After a while, though, I heard a very familiar New Jersey accent come through the intercom.

'Welcome to space,' said Scott Kelly. 'What would you like for dinner?'

I had no idea that the ISS doubled as a twenty-four-hour fast-food drive-through, but I was delighted to hear that it did. Scott had obviously been rummaging through some of our 'bonus food' containers – the special treats that had been sent up ahead of us. He asked what we would like him to warm up for us. I requested the bacon sandwich.

While the ISS's commander got on with the important work of heating my snack, we completed our final checks. With the pressure equalised, the hatches could be opened and we floated up the narrow tunnel and, at long last, into the Space Station.

Waiting to greet us with hugs were Misha, Sergey and Scott, who was capturing the moment on his camera. No time to chat, though. We were behind schedule for the video-link down to Kazakhstan and our arrival conference, so we were now swept straight through to the service module where we clamped on a headset and stood in front of the camera talking to family and the press back in Kazakhstan. After witnessing the launch, our friends and families had returned to their hotels and had then reconvened at an old theatre in Baikonur to watch the arrival, meaning they had all endured the intensity of the docking drama. It was great to be able to reassure them in person that all was well with us, but it also felt highly surreal. We had been on the station fewer than five minutes and here we were, cheerfully having a group video chat with friends on Earth.

With the live link completed, the pace finally slowed and I was at last able to begin to absorb my surroundings – and also my bacon sandwich, warmed for me by Scott. I gratefully opened the can and bit into it, but then noticed Scott and Misha looking on rather longingly. They were both three quarters of the way through year-long missions on the ISS and neither had seen a sandwich, nor smelt the enticing aroma of warm bacon, for a very long time. I gave them each a bite.

Scott then took us on the commander's tour, pointing out the emergency equipment and showing us around the various segments. I was all hands and feet at this point, adopting the classic star-jump shape of the rookie in weightlessness, trying desperately not to collide with things and kick cameras and equipment off the walls. I was grateful to get into the Russian segment, which is smaller in diameter than the US segment and where I could grab the handrails more easily and pull myself along. By contrast, stranded in the middle of the US segment, I felt like a puppy on a shiny floor. It would take a few hours of experience before I could trust myself to move without barging into something, or, indeed, somebody.

Even while my body was trying to work out what the hell was going on, it struck me how familiar everything felt. All those mock-ups and those hours in the simulators had prepared me for what to expect in precise detail of the look of the Space Station and its layout. As for its atmosphere, the place had a slightly antiseptic, faintly hospital-like smell to it. It was lit with fixed-frequency 'white light', which could be dimmed, but was quite harsh when turned up full, and there was a constant hum from the ventilation fans. (The station got fitted with LED lighting after we left, allowing for softer lighting in the evenings to aid sleep, which will have made a big difference.)

But what no amount of time in a simulator could have prepared me for was my first look out from the Cupola, the Space Station's observatory, with its multi-paned, clear-glass cathedral-style window. Its permanently jaw-dropping view of the world extends for 1,600 kilometres in any direction, the vista from the UK to Greece lying within a slight turn of the head. Here was the planet as very few of us are privileged to see it. Going to the window at random moments in the day (and while I was cleaning my teeth would become one of my favourite times to do this), you might catch sight of the wind-sculpted sands of the Sahara, a gently smoking Siberian volcano or the lights of a thousand night-fishing boats glimmering in the Gulf of Thailand. I hadn't been a photographer before I went to the ISS but I quickly realised I would have to become one to capture some of this. The Cupola became my favourite place to go in any spare moment that I had.

In those first hours, Yuri had some unpacking to do, bringing out of the Soyuz the more urgent cargo and some laboratory items that we had brought up and which would need refrigerating. He did this as casually as someone who had just got home with the shopping and was putting the cold things straight in the fridge. I watched him twisting around, turning upside down, passing in and out of places with no problem at all, adapting so fast that you would assume he had never been away from the place.

While Yuri did that, Tim and I checked out our quarters – in my case, a cosy enough space in the floor of Node 2, about the size of a shower cubicle. I also made my first use of the ISS loo which, as I would later discover, is the part of the station which attracts the most curiosity back on Earth. So, again, for the record: the ISS boasts a pair of loos, both of them the approximate size of a phone box. For the next six months, with foot restraints to keep me steady, I would be peeing into a fan-assisted hose (very important to remember to switch that fan on) and pooing into a self-sealing rubberised bag stretched across the opening of a solid-waste container ('*Please be sure to leave a fresh bag in place for the next crew member*'). The urine would be recycled back into drinking water, and the waste container, when full, would be removed for disposal to a cargo craft where you would very much hope that no passing astronaut would accidentally kick off its lid. But we'll come to that story later.

When Yuri had finished his duties with the Soyuz, we all came together for a meal in the Russian segment. After the bacon sandwich, I wasn't especially hungry, but it was great just to be there. My new colleague Sergey, as he always would, played the Master of Ceremonies and the chief toastmaster. Misha, although quieter, also had a wicked sense of humour. Following the palaver of the launch build-up, the drama of the flight and the spotlight of the cameras, it felt reassuringly normal for just the six of us to be floating around a table, sharing the experience. It would become our Friday night routine.

Incidentally, I also noticed during that meal that the ISS stainless steel cutlery was engraved '*SHUTTLE*'. Good to know that NASA's multi-use items were being properly recycled.

By then it was about 2.00 a.m., and Yuri, Tim and I had been awake for a solid twenty-four hours. It was time to experience the one major aspect of ISS life that training can't really prepare you for: sleeping in weightlessness. Some astronauts like to be tightly tucked up; others go completely detached and simply float around in their quarters all night. My preferred method was to clip my sleeping bag to the wall and zip myself into it with my arms inside and held across my chest to prevent them rising up of their own accord in the night and knocking me awake. It took a couple of weeks to get used to, but even then, going to bed would never feel quite as satisfying as it does on Earth, with the luxury of a pillow and some gravity to bring your head down onto it.

Waking up that first morning and remembering where I was felt pretty thrilling. My head was a bit fuzzy, as though I had a mild hangover and a blocked nose, but there were no other after-effects and I was soon having breakfast and then cracking on with the first day's tasks, which mostly involved unpacking things from the Cygnus cargo vehicle, trying to find my clothes, my medical kit and generally getting sorted bit by bit.

The Space Station runs to Greenwich Mean Time and everyone keeps themselves to that circadian rhythm despite the fact that in any twenty-four-hour period, night would come and go sixteen times as we orbited Earth every ninety minutes. You might break for coffee at 11.00 a.m. and it would be dark outside and you would be somewhere over New Zealand. Or you might be getting ready for bed and looking out onto broad daylight in Hawaii. But if you made sure to eat your meals at regular times and stick at it, the rhythm would soon establish itself.

From now on I could fire up my iPad each morning and see my day mapped out entirely by the planning team in Mission Control, complete with a live red timeline working its way across the screen. All of us had a conflicted relationship with that red line. When you rode with it or even got ahead of it, it was your best friend. When you fell behind, it was your sworn enemy. From Monday to Friday, the twelve-hour working day would be broadly split between fulfilling the science tasks, maintaining the Space Station, and exercising. The exercise was vital to counter some of the depredations of weightlessness, to keep up muscle mass, bone density and cardiovascular fitness. A two-hour workout was scheduled into most of our working days, for me normally from 5.00 p.m. and using the station's treadmill, its static bike and its Advanced Resistive Exercise Device, a kind of piston-driven multi-gym.

The largest part of our work was conducting science in the laboratories, assisting with the pioneering research enabled by a microgravity environment. Sometimes you were commencing experiments, sometimes you were continuing to process experiments that had been running for years and would continue long after you were gone. And sometimes you yourself were the experiment. I signed up for twenty-five life science experiments during my mission, monitoring the reactions of my own body to an extended period in weightlessness. I spent a lot of time siphoning off my blood, separating it in the ISS's miniature centrifuge and then storing it to be shipped back to Earth and analysed. All in all, we would complete well over 250 experiments in my months on the ISS.

We worked a five-day week in the main, and then Saturday mornings were devoted to cleaning tasks. Yes, astronauts hoover. Indeed, astronauts hoover with style. One of my most pleasurable tasks was flying the perfectly regular, off-the-shelf vacuum cleaner through the station on its massively long extension lead. Saturday afternoons I set aside for my education outreach projects – recording messages, managing students' science experiments, talking to schoolchildren on ham radio, hosting events from space.

On Sundays we would have a scheduled video link-up with our families. Those were great for me, and a real morale boost, although they were not without their technical complications. There was only so much bandwidth for video calls, so we staggered them. Sometimes you would get the slot you wanted, other times not. Sometimes the lag made conversation really difficult and occasionally the audio was dreadful and you would have to

417

wait while Mission Control re-patched it through another route. But it was contact, and a sight of the boys, and that was the main thing. In addition to that, I called Rebecca nearly every day, normally in the evening after we had done our work and our prep, which was late afternoon for her, in Houston. If she was free, that was great and we would chat, and if she was busy, that was fine too, and we would say, 'Talk later.' It felt perfectly natural.

So that was the general shape of the ISS week, but I had only been on board a handful of days when a problem emerged to disrupt it. At the daily planning conference, which took place every morning at 7.00, Ground Control reported that a CETA cart on the outside of the Space Station had become stuck halfway between two docking stations. A CETA cart is a sort of shopping trolley on rails, which runs along the truss of the Space Station. Bits of equipment can be bolted to it and sent out by remote control to the station's further reaches. Mission Control had discovered that they couldn't move the cart and had reached the conclusion that its brake was on. This was by no means a trivial matter. The CETA cart is quite heavy and it needs to be securely docked before the Space Station can shift its orbit by using its thrusters to perform a re-boost. If the brake were to fail and the CETA cart were able to move around on its rails during that manoeuvre, it could gather momentum and slam into a docking station, causing damage. And if the Space Station can't do a re-boost to change its position, it's not able to avoid any incoming space debris, which could smash into the ISS and damage it. For understandable reasons, Mission Control likes to maintain that re-boost option.

Scott knew straight away what had happened. On his recent spacewalk with Kjell Lindgren, shortly before our crew arrived, he had been working to secure the CETA cart brake handle, and he must have accidentally locked its brakes. We debriefed that later

– it had been such a trivial task during Scott and Kjell's mammoth spacewalk, a five-minute get-ahead, and yet nowhere had there been a note in the procedure to check that the brake was off on completion. As is so often the case, it's the small things that will bite you. Now somebody needed to go out and release the brakes. Scott and Tim would have to stage an emergency spacewalk.

We had arrived late on Tuesday, it was now Friday, and the EVA was set to happen on Monday – an incredibly tight turn-around. Spacewalks are normally planned and rehearsed for weeks in advance, with all the 'translation paths' meticulously mapped out, and with development runs practised and modified in the pool on the ground before being sent up to the crew. Now, with less than a week on board, and still orientating ourselves, Tim was going to have to suit up and accompany Scott, and I was going to have to serve as IV, helping them into their suits and getting them into the airlock, being ready to support if anything were to go wrong, and then helping them safely back in at the end.

It was a lot of responsibility to be given, and it meant a week-end of full-on prep. Scott would have had a right to feel anxious about a rookie being handed these key duties so early in their mission. But I was determined to prove that I was up to it, and that it hadn't come too soon.

All went smoothly with the prepping of the suits over the weekend. On the morning of the walk, I helped Scott and Tim get dressed and assisted them into the airlock; soon after that they were out in space. It was extraordinary to look through the window and see them out there. It was equally extraordinary to be inside and to *hear* them out there. The clinking of their tethers and equipment on the handrails as they were clipping themselves on reverberated through the Space Station. Back at work in the lab, I could follow their progress by listening to where the tinkering noises were coming from outside.

Scott successfully released the cart after forty-five minutes. He and Tim then stayed out for another two and a half hours finishing up some other tasks, mostly routing cables. Then they came back in, grinning from ear to ear because everything had gone smoothly, and we did the clean-up, sorting out all the tools and tethers, cleaning and refurbishing the space suits. That venture was a huge confidence boost for me, very early on in the mission. You always want responsibility and sometimes you can get frustrated because you feel you could handle more of it than people are willing to give you. But this was a situation where I had definitely been handed responsibility, with people senior to me depending on me to do a good job, and it was hugely satisfying afterwards. It also left me longing hungrily for the opportunity to do a spacewalk myself. I wouldn't have to wait long.

After the hectic few days brought upon us by that unforeseen spacewalk, it was suddenly Christmas. The chance to relax a bit felt very well timed. The ISS Christmas tree – all of one foot tall – was dug out from somewhere, Santa hats were in evidence, and our pre-packed Christmas stockings from Mission Control were hung up with care. Our orbit on Christmas Eve took us directly over northern Europe, making the Space Station perfectly visible from the ground, so if you were scanning the skies for Father Christmas that night, you might also have picked out the ISS. Missing Christmas with the family was going to be tough for all of us but, in my case, the fact that being in space was still such an all-consuming novelty at this stage definitely helped. It also helped to know that Rebecca and the boys were spending the holiday at a big family gathering in Comrie, being well looked after and having a good time.

Quite late on Christmas Eve, I decided I would ring my sister and wish her a happy Christmas. I had been very organised about bringing phone numbers to the ISS, knowing that we could place calls (although we couldn't be phoned). I'd known I wouldn't be taking my iPhone to space so, in quarantine in Baikonur, after consulting the internet on the best method, I had exported my phone's contacts onto an Excel spreadsheet, and made a careful list of all the people I wanted to call from space. Lots of them were friends and family, and others were people I had worked with or who had been instructors or mentors along the way, whom I wanted to thank. The plan was to work through the calls in any spare time I had over the 186 days.

It all seemed to be going well. People appeared to like getting a call from space. Sometimes you would get their answering machines – indeed, it happened with my parents a couple of times. But that worked just as well, in some ways. A missed call and a recorded message from the ISS was in itself quite a novelty.

The Christmas Eve call to my sister didn't go to answerphone. It got picked up – but unfortunately, not by my sister. A woman answered whose voice I didn't recognise.

I said, 'Hello, is this Planet Earth?'

The woman said, 'No, it's not,' and hung up.

That was a little embarrassing. I then rang my parents – on the right number this time. After we'd had a conversation, I put out a tweet apologising to the woman I had phoned in error at that late hour, and explaining that it was a genuine mistake and not a prank call. The tweet went viral, the media picked up on it and eventually even CNN were reporting on the astronaut who had called a wrong number from space.

I blame the software. My sister emailed me her proper number after this episode and I compared it to the one on my list. And I was able to work out that, somehow, in the process of converting all my numbers from the phone to the spreadsheet, any phone number on my list ending in a nine, such as my sister's, had been rounded up to a zero. Which is not helpful. So now, any time I came across a number which ended in a zero, I had to ask myself, 'Is that actually right, or could it be a rounded-up nine?' So much for my beautifully prepared pre-flight admin. There were a couple more wrong numbers along the way.

We celebrated Christmas on the ISS by watching *Star Wars: The Force Awakens*. Scott had managed to get the movie sent up by NASA as a data file, which was just as well because the ISS's painfully slow crew Wi-Fi certainly didn't lend itself to streaming. Scott had also at some point managed to charm NASA into sending up a white-screen and a projector. It was 'for briefings', apparently. It also made a very good home cinema. During the day we all had packages from home to open. Rebecca sent up, among other things, a calendar for the twelve days of Christmas, with each day containing letters from a member of the family, or a neighbour in Westbourne, or an old schoolfriend. It was the most thoughtful Christmas present I had ever received, and I was dumbfounded to reflect that all of this would have needed to be planned out and packed up eight months in advance, to be ready to fly on the cargo vehicle.

Care packages would come up with every cargo or Soyuz space-craft, and receiving them was always a great moment. I remember the same feeling from when the post arrived during military operations. The care packages were normally near the top layer of any cargo vehicle that arrived, but we would squirrel them away and get on with unpacking everything else. Then in the evening everybody would disappear to their crew quarters, open the packages and read their letters.

Chocolate always went down well. So did pistachio nuts – pre-shelled, I probably don't need to say. I had to develop a method for eating those, and found it worked best if I tore the corner off the packet and put it in my mouth. Then I would float to the top of the crew quarters and push myself back down off the ceiling. The nuts would still be going upwards in the packet as I was coming downwards, so I would get a mouthful before landing on the floor, taking the bag away from my mouth and holding it closed while the contents settled again. Repeat until full of pistachio nuts.

The ground crew would normally put a small supply of fresh fruit in with the arriving cargo, and just briefly the ISS would fill with the surprisingly powerful aroma of oranges. Once Rebecca put in a handkerchief doused in her perfume; another time a picture of my grandparents on their wedding day. What would they have thought? I imagined being able to burrow back through time and tell them, 'You know, one day your grandson will be looking at this picture in space.'

* * *

Right after Christmas I discovered the special misery of falling ill on the Space Station. I will spare you the grimmer details, but as a consequence of being required to deliver multiple urine samples in a short space of time for a life science experiment (and take it from me, urine samples are never easy in weightlessness at the best of times), I landed myself with an infection. That let me in for three days with a shivery fever and a temperature of 102° until I finally got the right antibiotic prescribed for me from the station's very impressive onboard medicine cabinet, and knocked it on the head. A horrible experience, though. A fever in microgravity has very little to recommend it and I realised that I never wanted to be ill on the Space Station again.

Meanwhile we got the news that I had been fervently hoping for: Tim and I had been scheduled for a spacewalk. On 15 January, my freshly shipped medium-sized space suit would be the first with a Union Jack patch on it to walk in space – but only, of course, if it worked.

That was a pretty tense hour or two. Scott was in the role of IV for Tim's and my walk, so he was unpacking and building the suit prior to pre-positioning it, along with Tim's, in the airlock. The suit was basically complete, with its battery installed and its lithium hydroxide canister, used to scrub our exhaled carbon dioxide, charged and ready. All that remained now was to flick the switch on the fan, which controls the air circulation and the cooling system and is essentially the closest thing the suit has to an on-button. Scott called me over.

'Tim, you need to be the one to flip this switch,' he said.

We looked at each other. This was the moment of truth. No noise from the fan, no spacewalk. We did a '3–2–1' countdown and I flicked the switch. The fan began to spin and hum. It sounded like music to me. The spacewalk was on.

Originally Tim and I had been prepped to do a spacewalk in which the prime task would be fitting to the outside of the Station an International Docking Adapter, a module which would help pave the way for the arrival of new commercial spacecraft at the ISS in times ahead. But that plan had exploded, along with the docking adapter and my original space suit, in that malfunctioning rocket less than a year ago.

The great consolation was that our job instead became to repair a solar panel which had broken on the furthest edge of the Space Station – a challenging spacewalk, right out to the station's limits. We would have to carry with us something called a Sequential Shunt Unit, an item about the size of a small fridge (thank goodness for weightlessness), take out the old one and plug in the new. That was the primary task, and the secondary task was to lay some cables along the outside of the pressurised modules in preparation for the eventual arrival of that ill-fated docking adapter. That part of the walk we had already trained for in the pool. The other part I now trained for with multiple virtual run-throughs. These days spacewalkers on the ISS have access to cutting edge virtual reality systems. Back then we made do with a pair of goggles and a basic laptop which you turned upside down and wore strapped to your head. It sounds primitive, I know, but the engineering model was very high fidelity. You looked a bit of an idiot while you were doing it, I don't doubt – but then who doesn't look a bit foolish in a VR headset? I must have rehearsed our spacewalk about ten times in that makeshift VR environment.

There's a lot of apprehension in the build-up to a spacewalk but very little of it tends to be about the dangers of the hostile environment that you are about to venture out into – a merciless vacuum where temperatures can vary from minus 160°C to plus 120°C, depending on the whereabouts of the Sun. All of your focus beforehand is on the practical execution of the walk itself. Those translation routes – the literal matter of where your hands and body are going to be at every stage of the walk – are meticulously mapped out in advance. Every spare moment I had, in the days leading up to the walk, I would be visualising those translations, much in the way that I had always pre-visualised aircraft sorties, carefully eliminating the unknowns and replacing them with plans for every step.

One of the things that weighed most on my mind in the run-up was how crowded the airlock was going to be when Tim and I set out. The easiest way to get a spacewalk off to a bad start is to get snagged up in the airlock. And bear in mind that everything in there has to be tethered so that it doesn't float out of the hatch and off into space. We had that bulky new SSU in there. We had a tool bag each. We had a huge bag of cables and a large external light, the fitting of which was going to be one of Tim's tasks later in the spacewalk. Essentially there was a spider's web of tethers in that airlock waiting to get tangled, and since Tim would be first out of the hatch, it was my job to organise everything so that it could be passed out efficiently. The night before the walk, I spent an hour sitting on the floor in there, thinking about it, and then ultimately changing the entire configuration – working out which piece of equipment needed to be in which location, and tethered to which handrail, for the smoothest and least effortful exit.

On the morning of the walk, Tim and I ate hearty breakfasts (our last food for many hours) and got ready to suit up. Getting into a space suit while in space is a major physical challenge. On the ground, people used to remark on how easily I got in and out of mine, which I put down to those early potholing expeditions. In weightlessness, with nothing to push your feet against, it was a proper battle – and one which I had almost managed to lose a couple of days earlier. During my suit-fit check, I adopted the necessary position for getting myself into the hard upper torso section (arms straight above head, palms outwards) and was clambering in as best I could when I felt my shoulder twist. In the hours afterwards, my left forearm and my left thumb turned worryingly numb. I had obviously pinched a nerve, which I then had to report to the flight surgeon on the ground. I could still grip with my left hand, so we decided it wasn't going to impact on my ability to perform the spacewalk. But imagine how gutting it would have been to get that close to a spacewalk and then put yourself out of action in the innocent act of trying on your suit.

On the day, however, I managed to climb into it without doing myself any further mischief. Some people, I'm sure, would find it horrendously claustrophobic to be inside that helmet and enclosed by that stiff, restricting armour, and to know that, without assistance, there is no escaping from it. But for me it was the moment to get into the zone. It was like those first minutes spent quietly in the darkness at the start of a caving trip. I told myself to embrace it. 'This is your new environment now. This is where you are, and you're actually perfectly comfortable with it.'

We spent a few hours in our suits before the walk, breathing pure oxygen and flushing the nitrogen from our systems. And then we passed into the airlock and closed the hatch, sealing ourselves off from the rest of the Space Station. Once again I was struck by how tiny that space was – more of a cupboard than a room, and stuffed with baggage. I was in a position with my head towards the hatch leading back into the Space Station, my feet down by the hatch that goes out into space, and Tim's feet directly in front of my visor. We were topping and tailing in there, as if in some overdressed game of sardines. I opened the valve to begin the depressurising of the airlock and once we were down to vacuum, Tim opened the hatch at the other end.

At that point sunlight flooded the airlock, an incredible sight. People sometimes wonder if the moment when the hatch opened was the scariest part of the walk – but on the contrary, that was the point at which I finally felt myself relax. The build-up to a spacewalk is all pressure, stress and loss of sleep: checks, drills, painstaking procedures. When that hatch opened and the sun came in, it was as though a voice had said, 'OK, gentlemen. You know what to do. Over to you, now.'

Tim went out and got himself secured, and then I started passing out the bags of equipment in that carefully rehearsed order. Once they were all out and pre-positioned, which took about ten minutes, it was my turn. I pushed myself down the airlock feet first and slid into space.

That was the moment when Scott, monitoring our progress from inside, came on the radio and said, 'Tim, it's really cool seeing that Union Jack going outside. It's explored all over the world. Now it's explored space.'

Which was a tremendous thing for him to go to the trouble of saying – and an extremely proud moment for me and, I hope, the many people who had helped me get there.

I acknowledged Scott and then, having taken a moment to get acclimatised, followed Tim as we began our long climb out to the solar panel at the end of the truss. Tethered to me was the bag containing the SSU. I moved slowly and steadily. Most of the route was across areas of the ISS designed for spacewalkers, with handrails, where you could kind of bounce yourself along and find a rhythm. But there were also a couple of harder points – such as passing under the CETA trolley, which was more like a rock climbing crux – involving bigger reaches and more awkward finger points.

I was making good progress when Reid 'Tonto' Wiseman, our IV person in Houston and an experienced spacewalker himself, came on the radio and asked me to retrace my steps a short way and check out a suspected tether snag which someone on the ground had picked up on the camera. In a spacewalk, every hand-hold and every motion of your body is a small victory, so going backwards was a bit of a blow. I was also pretty certain there were no snags. But this was not the time to argue, and from my own experience of being in Mission Control, I knew they wouldn't be asking if it weren't important. I retraced my steps, confirmed that the tether was actually fine and then carried on.

I was getting into the flow now, starting to have a better feel for how the space suit moved, which was quite different from training in the pool where the viscosity of the water provided some resistance. In space you moved so much more easily and gained momentum so much more quickly – momentum which was then far harder to stop. It was both liberating and cautionary at the same time.

Even with the short delay to check the tether, Tim and I arrived at the end of the solar panel with time to spare. We couldn't begin unbolting the old SSU until the Sun had completely disappeared and there was no residual sunlight hitting the panels and threatening us with an electric shock. So that meant we now had a few minutes to wait and simply enjoy the extraordinary – and, indeed, unheard of – luxury of just hanging out in space. Attached by a short tether, I pushed off and allowed myself to float where I was.

They were the most incredible few minutes. What surprised me was how entirely serene I felt. I was weightless, with no forces exerting themselves on my body. I was not too hot, not too cold, and the only sound was the gentle hum of the ventilation fan. To my left was the Space Station. In front of me was Tim. Down below me, gradually going into shadow, was the Earth. And over my right shoulder was the universe.

I was aware of a sharp contradiction. Part of me, taking all this in, was thinking, 'This is just wrong. You shouldn't be here. You are an imposter in this hostile environment. This is so far beyond the limits of what you were intended to witness or comprehend. Looking down on the planet while floating in space – this is not something a conscious human being was made to do.' And yet, at the same time, it felt so serene and completely natural. I still don't think I've fully processed that experience. Perhaps I never will.

It was during those moments of hanging out that I suddenly heard some cracking and noticed some very faint scratches appear on my gold sun-visor. I wondered for a moment whether I had been hit by a micro-meteorite. In fact, slightly less dramatically, the rapidly dropping temperature had shrunk the visor and caused this mild cracking effect. But it only added to the eeriness of it all.

When the eclipse came, we went to work, our helmet lights illuminating the work site. The bolt-on SSUs had proved tricky to shift in the past. Tonto, our IV in Houston, would attest to that, himself having struggled to release a stubborn SSU during an EVA just over a year earlier. It was because of this incident that I had been tasked with fashioning some makeshift tools prior to our spacewalk, including a toothbrush taped to the end of a socket extension bar in case we needed to clean and lubricate the threads. Thankfully, everything went smoothly. We got the old unit out and got the new one in flush. And thankfully I later found myself a replacement toothbrush. Houston was happy. Job done.

I put the failed unit in a bag, tethered it to myself and set off on the journey back along the truss. I would stow the old SSU in the airlock, collect my bag of cables and then climb out again to commence the second phase of the tasks. At one point on the way back, I had to pass along the CETA spur, a stretch of polished titanium alloy tube, with a few handrails spaced along it. It was like taking a shortcut, spanning a gap between the more comforting large structures of the truss and airlock. There was nothing around you – you simply held on to the spur and pulled your way along it, with your feet dangling.

I chose that moment to look down – which perhaps, on reflection, was unwise.

Below me I could see my boots. And below my boots I could see Western Australia.

I was hit by a sudden wave of vertigo. Not surprisingly, I guess: 400 kilometres is an awfully long way down, and especially when you're hanging from a pole. My fingers in their thick gloves instinctively tightened their grip. Something that Chris Cassidy, a NASA astronaut and former Navy Seal, had told me flashed into my head. 'The best cure for vertigo is to wiggle your toes. And when you wiggle your toes, your fingers will relax.'

I wiggled my toes and, indeed, just as Chris had assured me they would, my fingers relaxed and the vertigo passed. I put the moment behind me and moved on.

The work site where I would start attaching the cable was on the underside of the ISS, and the subsequent wire routing was not without its challenges. At one point I found myself squeezing head first into the narrow gap between an external platform and the side of the lab and wondering whether I was actually about to get stuck. I had practised this move in the pool and it was tight then, but it seemed even narrower in reality. It took me five minutes of wriggling and twisting before I managed to find the sweet spot and get my suit through the gap. That took me into an area known as the Rat's Nest, a notorious mess of cables where you have to be super-careful not to get your tether snagged. All along this tortuous route, I had been immersed in the unique challenge which is laying cable in space, a highly intensive workout for the fingers involving the twisting of copper ties with gloved hands while controlling a bundle of wire whose every instinct is to float off into space rather than lie flat against the Space Station's outer shell. Nevertheless, it's very satisfying when you get it right, and I was thoroughly into the flow of it when Tim, who was working separately from me, round at Node 3, came on the radio.

'There's some water in my helmet.'

At the first mention of the word 'water', I knew that we would be concluding this spacewalk very quickly. Two years earlier, my classmate Luca was on his second spacewalk from the ISS when he noticed liquid entering his helmet from the ventilation duct at the back of his head. The water stuck to his face and covered his eyes and nose. It also cut out his communications system. Unable to hear, see or speak, and wondering whether his next breath would bring him a lungful of water rather than oxygen, Luca used the small tug from his retractable safety tether to guide him back to the airlock. His crewmate, Chris Cassidy, helped him inside where they discovered about 1.5 litres of liquid had entered his helmet. He was safe, but it was one of the most serious emergencies in the ISS's history. It hadn't slipped our notice that Tim was wearing the same suit that Luca had worn on that ill-fated spacewalk.

Unsurprisingly, Mission Control were soon on our radio. 'Guys, you can start opening your cuff checklist to page 7. We are in a Terminate case.'

By the time I got to Tim and looked into his visor, there was a bubble of water inside his helmet the size of a golf ball. He wasn't in any immediate danger, but that golf ball wasn't going to get any smaller and we needed to get him inside the Space Station and out of his suit.

Mission Control instructed us to switch positions: Tim was now EV2 and I became EV1. The significance of that was that EV1 is the last person into the airlock and the one who has to close the hatch. It wouldn't ordinarily be the job assigned to the rookie astronaut in the pair. It's hard work to get into the airlock when there is already somebody in there, and it's hard work to close the hatch. And if the hatch doesn't get closed then both of you are in trouble. Of course, I had trained for this. But only once. And about two years ago. So yet again, it was time to step up.

We didn't waste any time bringing all our bags inside. Instead, we secured them as best we could – they would have to be picked up on a later spacewalk. With Tim inside, I turned around and entered the airlock, feet first. I realised I couldn't get in fully until he had moved his legs but we quickly sorted that out. I then had to take myself halfway out again to close the thermal cover – and then back out once more to close it properly, as the Velcro hadn't fully sealed. Then I got the hatch locked and we were safely in.

The focus now was all on Tim and getting him out of that helmet. Scott pressurised the airlock, striking a careful balance between the need to get Tim's helmet off soon, but not at the risk of damaging our ears by increasing the pressure too quickly. Misha and Sergey were there too, waiting with towels ready, as they had been for the last forty minutes since hearing the call to terminate the EVA. The atmosphere in Mission Control, where Rebecca and Tim's wife Dawn had been following the spacewalk, was tense. A couple of experienced astronauts always accompany partners watching an EVA in Mission Control, in case something should happen. As soon as things had started going wrong, those astronauts would have begun an unspoken process to reassure and support the families but also to be ready to activate contingency plans should the situation get more serious. As soon as the hatch to the Space Station opened, and with everything under control, that tension dispersed. Scott took pictures of the water in the helmet for Mission Control to examine in the investigation that was bound to follow.

When both of us were out of our suits, we all had a cup of tea and talked about how it had gone. It was turning into quite a mission for Tim: an emergency spacewalk just after arrival, and then an emergency *during* his spacewalk a few weeks later. Still, the pair of us had been out in space for four hours and forty-three minutes. It would have been lovely to have stayed out there for longer, but at least the walk had achieved its prime objective and had been declared a success; the call to come in had definitely been the right one.

We then went back to our crew quarters to get changed properly. It was at that point that I took a look at some emails, and I was knocked backwards. My support crew had sent up some of the messages of encouragement that had been posted during the day, among them a tweet from Paul McCartney, '*wishing you a happy stroll outdoors in the universe*'. I had no idea of the buzz back on the ground that this spacewalk would create. I didn't even expect it to get picked up on very much. Only now, in the wake of it, did I discover that the entire walk had been televised in the UK. It was all pretty surreal.

But of course, the Space Station is a great leveller. I don't know what Paul McCartney was doing that night, but for me, an evening of clearing up lay ahead – drying the sweat out of the space suits, cleaning them down with antiseptic wipes, taking off the carbon dioxide scrubbers and setting the batteries to recharge.

I was proud as I could be, though. Later that night I removed the Velcro'd Union flag patch from my suit and stuck it to the wall of my crew quarters. It stayed there until I left.

CHAPTER TWENTY-FIVE

The chocolates that went everywhere, gorillas in space, and fond farewells

In February 2016, I was getting myself ready for *Cosmic Classroom* – a live link-up from the ISS, with Kevin Fong, doctor and television presenter, hosting from Liverpool in front of a large audience of children ready with questions, and with other schools around the country following on the internet. My plan was to show everyone what water looks like in space, the way it coalesces into a translucent bubble. And then I wanted to put an Alka-Seltzer tablet into a water bubble so they could see how, rather than fizzing out as it would in gravity, the gas would expand entirely within the bubble, creating a kind of liquid snow globe.

I asked Scott whether there might be any Alka-Seltzer on board the Space Station.

'Try the toy box,' he said.

The toy box? The Space Station had a toy box?

It turned out that, down in Node 2 at the very front of the Space Station, there was an unmarked bag that nobody on the ground seemed to know about. It had been handed down from crew to crew and contained the accretion of twenty years of daft stuff that astronauts had brought up for outreach. I unzipped this bag and looked inside. A treasure trove! There were a couple of

435

Nerf guns in there, there were water balloons, face masks of all kinds, an inflatable planet Earth, a mini basketball and hoop ... and there was Alka-Seltzer.

The experiment with the gas bubble passed off smoothly – which was just as well, given that there were half a million schoolchildren in the UK watching. It certainly went better than my project with the M&M's. On the ISS was a ball the size of a small football, made of two thin, clear plastic halves, which someone had at some stage patiently filled with M&M's from the Space Station's snack cupboard. This ball of coloured candy was a brilliant device for outreach events and for demonstrating movement in weightlessness. You could shake it up and get all the M&M's moving randomly as a way of describing Brownian motion. And if you moved down, all the M&M's moved up inside the ball, and when you moved up, they all moved down, and so on.

I set up the cameras in the European laboratory, used this ball to make a little outreach video, and then took it back to where it lived, in Node 1. By now I had mastered the fine art of moving around in weightlessness. The rookie days of feeling like a puppy were far behind me. If this had been skiing, I'd be making a black run look easy. So it was perhaps with an air of overconfidence that I came flying through the hatch from the lab to Node 1, holding the ball lightly in my hand and causing it just to clip the top of the hatch.

Disaster. The two halves of the plastic separated and 150 M&M's, now free of confinement, carried on down into the Russian segment. Which didn't especially please the Russians, I have to say. I did my best to clear up, but we were finding M&M's all over the Space Station for days after this and I would bet good money that there are still some stray ones on the loose up there.

Still, better a ball of M&M's than a ball of what Scott found one morning. Noticing a strong smell of urine in the loo, he decided to explore. Tentatively removing a panel at the back of the loo, he uncovered ... Well, words hardly begin to convey the full horror of it. It was like some kind of terrifying creature from *Alien* – a monstrous, horribly dark ball of urine mixed with pre-treat. Liquid from the loo must have been leaking out behind the panel for a number of days while this sticky, viscous ball of stale pee and chemical grew and grew. One could only recoil in horror – and then, in my case, take a photograph of it.

Luckily this monster showed no inclination to go anywhere. But it still needed to be exterminated. Scott's solution was to use a long succession of towels to soak it up – applying one towel until it was saturated, then bagging it and applying another. It was easily the worst clean-up operation I witnessed in my time in space. Bear in mind that, as commander, Scott would have been well within his powers to delegate this problem away. It's a measure of his leadership that he didn't do so.

There was nothing any of us could do about the second-worst lavatory-related incident of the mission. That was when Tim Kopra was rooting around in the depths of an overstuffed storage module. Somehow, as he worked his way through the confined space, his trailing foot managed to kick the top off a KTO, one of the Russian solid-waste containers holding many weeks' worth of individual bags of human waste on its way to incineration. A small explosion announced his error as the build-up of foul-smelling gas escaped. By the time Tim had located the offending item and jammed the lid back on, the stench had permeated to all corners. The station's air system was pretty good at masking smells in general but it was entirely overwhelmed on that occasion. We could only sit it out and wait for the atmosphere to clear.

Late in February I finally found the excuse I had been looking for to wear my tuxedo-print T-shirt. The producers of the Brit Awards approached ESA to ask me to do a remote presentation to

Adele of her Global Success Award during the Brits ceremony at the O2 Arena in London. I was told later that my appearance on the screen was greeted with a gale of screaming from the audience and, on the stage, a visibly moved Adele said, 'My kid is going to think I'm so cool.' With retrospect, my time on the ISS often takes on the appearance of a succession of entirely surreal moments, designed to make my younger self tip his head back and honk with laughter, and this one was definitely up there.

In the competition for best novelty outfit on the ISS, my T-shirt was, I am sad to say, trumped by Scott Kelly. He had an entire gorilla suit up there. I really thought he must just mean a gorilla mask when he told me about it. But no. It was the entire suit – a whole gorilla. I think his twin brother and fellow astronaut Mark had managed to get it packed and sent up. Scott bided his time before deploying it. But eventually the moment came, and the chosen victim was Tim. Scott hid in his crew quarters and I went to tell Tim that he needed to make a call to the ground. He set off for his quarters where a gorilla jumped out and mugged him. You can't really be expecting that to happen when you're in orbit, and his reaction was pleasingly extreme.

Still, Scott wanted to do something more with the suit. His girlfriend, Amiko, worked at NASA in Houston and knew when the cameras on the ISS would be going live in the US lab. At a time when Scott could be sure that it would be caught on the screens at Mission Control, he donned the suit and floated casually the length of the US lab. Then, for good measure, he floated all the way back.

He returned to his crew quarter and rang Amiko.

'How did it go down?'

Amiko had to break it to him gently. Nobody had said anything. Not a single remark. It seems you can let a flying gorilla loose on the ISS and nobody on Earth will notice.

Eventually Scott and I made a little video with the suit. I pulled an innocent-looking bag into the lab, and Scott/the gorilla burst out of it and chased me around the station. Somebody in the US edited the resulting footage to the Benny Hill music which worked perfectly, and I now frequently use the clip as an ice-breaker when speaking with groups of children.

Meanwhile, I was continuing to work through my phone calls. In addition to speaking with friends and family, each crew member has a special allocation of two or three calls to someone they might want to speak to from space. I had already called Chris Evans on his Radio 2 breakfast show early in February, on Rebecca's birthday in fact, to surprise her with a message. I had wanted to talk with Captain Eric 'Winkle' Brown, an extraordinarily skilled Royal Navy test pilot who had flown more aircraft than any person in history, but was greatly saddened to be given the news that he had passed away on 21 February, before our scheduled call.

However, I was delighted to have the chance to speak with someone who had been an inspiration to me and whom I deeply admired: Professor Stephen Hawking. I knew that a conversation with Stephen was not going to be without its challenges and was quite nervous before the call. I shouldn't have been. Stephen had prepared some answers to questions I had sent earlier, and his wonderful daughter, Lucy, was there too. I would ask a question, Lucy and I would chat and when ready, Stephen would answer. Whilst we were talking, we had the most amazing pass over the Bahamas which I was able to share with Stephen and Lucy. It was one of the most memorable moments of the mission.

Shortly after this, I spoke with another hero of mine and close friend of Stephen's, Dr Brian May. Having earned a PhD in astrophysics from Imperial College London, Brian takes an enormous interest in space and it was wonderful to have the opportunity to give him a tour of the Space Station. It was also a pretty unique experience for Tim, Jeff and me, all fans of Queen, to listen to Brian jamming on his guitar for us, having just finished a gig in Brussels.

On 29 February Scott rang the bell to signal the formal hand-over of command to Tim Kopra. The following day, Scott, Sergey and Misha left the station and returned to Earth. It was hard to say goodbye to them. You form a powerful bond with anybody you spend time with on the ISS. You have shared the thrill of it, and the risk of it, and you know that there are not many people in the world who have that experience in common. In Scott's case, the bond went back to our training in Houston as well.

He and Misha had each completed a year in space, an astonishing feat of endurance by both of them. But it was interesting to sense their different attitudes to it. Misha, a quiet, considerate person, seemed to be counting the days down until he left. He would say, 'I've only got sixty-eight days to go now.' Scott, by contrast, was always adding the days on: every day was another to add to the tally, as far as he was concerned – another day that he had been in space. And, much though he was looking forward to seeing his family, Scott seemed to find it hard to leave. I think he knew that he would be retiring as an astronaut after this mission, and wouldn't be coming back. So this wasn't just farewell to us, it was farewell to a whole stage of his life. And it was hard for him because he was genuinely attached to the ISS. On that last day, he was taking pictures in the Cupola for ages. We practically had to drag him out of there, make him put his camera down and get into his Sokol suit.

We watched their Soyuz capsule push back and descend slowly beneath the Space Station. We could continue to see it as it set out on its first orbit, a bright light gradually disappearing until it was unrecognisable as a spaceship. And then we saw the streak as the craft re-entered the atmosphere, and they were gone.

It was a strange feeling, watching them disappear. I thought about how I would have felt if someone had said to me, at the very last minute, 'If you get changed quickly, you can climb in with them and go home.' Obviously, the thought of returning to loved ones and Earthly comforts would have been tempting: comforts such as a shower, which I was beginning to long for, and the ability to go for a run outdoors and feel rain on my face, which I was probably longing for even more.

At the same time, I had clearly become attached to life on the Space Station, too. I knew that if someone had said, 'The door's open – leave if you wish,' I wouldn't have entertained the idea of not staying, completing my mission and squeezing this experience for everything it was worth. In a way I couldn't have imagined, but which seems common to everyone who spends time there, the ISS was definitely exerting a strong hold on me.

Going down to three people on board the station seemed to change the tempo. I don't think our individual work schedules altered much at all, in fact, but the general pace just somehow seemed to slow. Suddenly there were half as many of us floating around in this 747-scale place, where sometimes you could go a whole day and not really see much of anyone. And then a fortnight later we were watching a tiny point of light getting bigger and bigger, eventually revealing itself as a spacecraft closing in on the docking port, and the next thing we were welcoming on board Jeffrey Williams, Oleg Skripochka and Aleksey Ovchinin.

I knew Jeff well from training with him at Johnson Space Center. He was an enormously experienced NASA astronaut, on his fourth mission to the ISS. Oleg was on his second mission and Aleksey was a rookie like me, though he would eventually return as a Soyuz commander in 2018. We bonded well as a team in no time at all and Friday nights in the Russian segment continued as ever.

Not long after Jeff, Oleg and Aleksey arrived, I was back at the Cupola window and watching another spacecraft draw closer to the ISS. But this time I was observing its approach even more intently. It was a Dragon cargo vehicle and I was at the controls of the Space Station's robotic arm and getting ready to grab hold of it.

I'm sure any astronaut who is tasked with performing a robotic capture on the ISS will tell you that they experience some apprehension in the build-up. Essentially, for those few minutes, the responsibility for connecting a $100 billion Space Station with a few hundred million dollars-worth of spacecraft and cargo lie, literally, in your hands, and your hands alone. Leaving aside our docking episode on arrival, I would probably have to rank the critical ninety seconds of this operation as the most intense of my mission.

Also watching closely was Jeff, standing with me in the Cupola and looking over my left shoulder.

'Two metres,' he declared, loudly and clearly. This was not the time for any misunderstanding between us.

Two metres was the distance between the end of the Space Station's robotic arm and the grapple pin on the cargo vehicle, which was at that point holding position 10 metres beneath the ISS. The robotic arm had been steadily closing the gap from 5 metres and we were now entering the most delicate phase in the process of capturing a visiting spacecraft.

It would have been comforting to know that Jeff was measuring these critical distances using a highly sophisticated piece of equipment – a laser rangefinder perhaps. In fact, he was judging it by comparing the size of the vehicle to pictures of Dragon at various stages of approach on a piece of A4 paper he had printed out ear-lier. Wet finger method then. As incredible as it may seem, that's the way it's done on the ISS.

Any second now, Dragon would be put in 'free drift', meaning that the spacecraft's station holding and attitude control system would be deactivated. It would be gently tumbling in space. At the same time, the Space Station's thrusters would be inhibited. The last thing you wanted during this delicate task was any unex-pected movement from either vehicle.

The robotic arm moves a little bit like a helicopter. The left hand controls translation: forward-backward, left-right, up-down. The right hand controls pitch, roll and yaw. All I needed to do was to 'fly' the robotic arm, exactly matching Dragon's gently tumbling motion whilst gradually closing the distance, bringing the arm over the grapple pin and finally pulling a trigger to catch the biggest (and most expensive) fish I would ever catch in my life.

What I wouldn't have given for some of the Apache's technol-ogy at that moment: laser rangefinder, linear motion compensator, image auto-tracker and a decent set of hand-controllers. This task would be a doddle if I were in that helicopter. But we had none of this. As it was, I was going to have to earn my supper tonight. Months of training had gone into these next few moments and, quite simply, the consequence of failure didn't bear thinking about.

'Go for free drift,' I called to Jeff.

I steadied my hands on the controllers, gripping them more tightly than I had in training, and watched to see how Dragon was moving. In previous simulations, our instructors had thrown every scenario imaginable at us, including a spacecraft tumbling so dra-matically that we would have to abort the capture. Thankfully, that wasn't the case today. Dragon was only gently drifting. I kept closing the distance whilst matching its motion – some roll, a bit of yaw, moving left and up.

Jeff had been calling down the distance: one metre, half a metre …

'In the envelope,' said Jeff.

'Trigger,' I called.

I had captured a Dragon. I made a radio call and then went to call Rebecca, who had been watching the whole thing in Mission Control. I don't think I had ever focused so much attention on one single task in my life. As a pilot, you learn to maintain situational awareness, not to become too fixated on one thing at the expense of missing something vital elsewhere. This was different. My entire attention had been on that spacecraft. Frankly, the Earth could have imploded beneath me for all I would have noticed. The coffee afterwards tasted very good.

On the morning of Sunday 24 April, I put on my running kit, climbed onto the ISS treadmill and set off to run the London Marathon on behalf of the Prince's Trust. It wasn't the first time I had run it. That was in 1999, when I was twenty-seven. (Three hours and fifteen minutes, if you want to know.) But it was certainly the first time I had run it in weightlessness and while 400 kilometres above the start line in Blackheath. Mission Control helpfully livestreamed the BBC's race coverage to a screen in front of me, so I was able to feel more of a part of it as I ran, and also able to enjoy at least a virtual version of that traditional marathon experience – getting overtaken by someone dressed as a Smurf.

I needed that visual inspiration, as it is a lonely run up there, no crowds cheering you on, although I knew there were teams from ESA and the UK Space Agency supporting me on the ground. Because I was running with only 70 per cent bodyweight, the maximum force the bungees could muster, my legs were having a relatively easy time of it, but on account of the harness, my waist and shoulders were getting seriously punished. Meanwhile, the droplets of sweat from my face were migrating to the top of my head and coalescing into a wobbly bubble in my hair which I would periodically have to soak up in a towel. I'm pretty sure that doesn't happen on the streets of London, even to the people dressed as Smurfs. Anyway, I pounded on, actually speeding up towards the end of the 26.2 miles out of sheer desperation to be free of the harness. (Three hours and thirty-five minutes, in case you're wondering.)

These efforts of mine were rather put into the shadow by Eddie Izzard who, at roughly the same time, was completing twenty-seven marathons in twenty-seven days across South Africa on behalf of Sport Relief. I rang to congratulate him from the ISS the night before he was due to finish with a double-marathon. I have no idea how he managed that much running, but I knew I wasn't going to try and emulate him from the ISS. After just one marathon in space, it was three days before the bruises calmed down enough for me to put the harness back on again.

In late May, I was in the middle of my morning routine. Having washed and shaved, I was brushing my teeth and looking out the Cupola window. As well as taking in the view, it was my habit to devote a few moments to looking over our hibernating Soyuz spacecraft. I thought of it as the equivalent of doing a pre-flight walk-round of an aircraft on the ground. I got to know every scratch, tear and bump on that Soyuz's thermal covering over that period. However, on this occasion my examination was interrupted. Out on the horizon, against the completely black and starless sky, I could see a bright light.

That was odd. In the daytime, usually you only saw lights in space when it was another spacecraft coming towards you, and nothing was due to arrive.

Then I saw another light, quite close to the first. Odder still. And then a third. The lights were evenly spaced and flying slowly in formation out in the far distance.

I called Jeff over.

'What do you make of that?'

Jeff was as puzzled as I was. There was now a neat line of lights drifting across the darkness.

How to explain it? Surely not …

Were we witnessing extraterrestrial activity?

In fact, no. We were witnessing frozen urine.

It took us a while to work that out. Eventually we could see that these distant lights were actually individual droplets of liquid – and not far away, as they appeared to be by a trick of perspective, but very close to the window. The droplets were venting at regular intervals from the docked Progress vehicle, whereupon they were instantly freezing in space and floating gently away.

I found Yuri inside the Progress, hard at work with a pair of pliers and a spanner. I told him about the escaping fluid.

'Ah,' said Yuri. 'Well, that would explain the drop in pressure from the urine tank.'

Mission Control had already identified that something was wrong on board the Progress and they had sent Yuri in to explore. Unlike the US segment, the Russian segment doesn't recycle its urine. It transfers it as liquid waste into tanks on the cargo vehicles and disposes of it with the rest of the trash. One of these tanks, carefully monitored on the ground, had shown a pressure drop. No wonder. It was leaking into space, sending a gentle stream of frozen pee past the Cupola windows and doing a very convincing impression of a visiting alien life form. For the record, that was as close as I got to spotting any such thing during my six months.

As the end of the mission neared, things started to appear on your timeline which were geared to going home. There was bag-packing to be done. Yuri was increasingly tasked with reawakening the Soyuz. Tim and I had to go and sit in our seats and check the liners still fitted us. It was very strange to go in there after all this time. Yuri, as commander, had been going in twice a day. One of his tasks every morning was to put a piece of paper in the craft showing where the emergency landing sites were on that day's orbits and the timings for the engine burns in case we needed to leave in a hurry. Fortunately we had never needed those pieces of paper and Tim and I had had no cause to enter the capsule for half a year. I now eased my newly extended frame into the seat. Before launch I had been able to get three fingers above my head. Now I could barely get one: a perfect, snug fit.

Life back on Earth began to subtly reassert itself. Emails were coming from the ground team about what would be happening when I got back. Can you make the Farnborough Airshow? Can you make a reception at Number 10? All very exciting, of course, and an indication to me of how well the mission had been received at home. Yet in many ways, I didn't want to turn that tap back on. I'd been in this blissful isolation for six months where I hadn't had to think about my schedule, and where my biggest personal decision on any day was what I was going to have for dinner. The rest was taken care of.

In other words, there was slowly more to think about. And in space, I think it's fair to say, you don't get an awful lot of time for thinking. You're always on your game, always concentrating, always switched on – out of necessity. It was an extreme environment, and in order to function, you had to normalise it. That was the way I had always done things, on an operational tour, in a cave, underwater – you just made it normal. You got into a daily routine. And that made you function efficiently. The downside of that was, you lost perspective. There was no time in the routine to stop and wonder. A lot of the time you forgot that you were actually in space.

And that, I realise, is why I loved the Cupola so much. We spent our days being chased by the red line, thinking about procedures. What's up next? How much time have I got? What tools do I need? How much prep time do I have? And then you would take a moment to go into the Cupola and stand at the windows. And when you were there, with the world spread out below you, there was no question about it: you were most definitely in space. And there it was: your invitation to wonder.

In the last two weeks of my mission, because of the leaking Progress, we had to close the shutters on the Cupola windows. There was potential window damage from the debris. We could only open the shutters briefly to do a monitoring task outside the station. It made the ISS a very different place to be. Now you worked all day without seeing the sunlight come and go, and, of course, nearly all of the beauty of being in space lies in the view. It made it easier to detach myself.

At the same time, two days before departure, I went frantically round the ISS taking hundreds of pictures. I had been pointing my lens out of the window at the world and had taken almost no pictures at all of this amazing place, this implausibly ambitious international construction that had sustained and supported me all this time. And I knew I wouldn't want to forget it, if I didn't get the chance to return.

On the last morning, a Friday, I unclipped my sleeping bag, having already stripped down and cleaned out my crew quarter for the next person. We were due to get into the spacecraft at 2.00 a.m. on Saturday but there was a full day of work ahead. Friday morning found me in full PPE, in the Japanese module, bringing in an experiment from outside on the external platform where aerogels were being used to collect micro meteoroids flying around in the solar system. These were being sent back to be analysed on Earth, where apparently organic compounds were being found in some of them. Incredible, groundbreaking science, like so much that goes on at the ISS. And also quite fiddly. Work overran and I missed my scheduled fitness slot. Eventually Jeff had to come and take over for me on the final clean-up so that I could finish packing, get in the spacecraft and go.

Working up to the last, then. But that was OK. That was what it was about. That was the routine. Yuri, Tim and I stood at the entrance to the Soyuz while Jeff, Oleg and Aleksey took the traditional farewell photographs. And then the hatch closed and we left.

*   *   *

We undock and begin to back out slowly between the solar panels, and immediately I feel vulnerable. The Soyuz seems so small – the Soyuz *is* so small. But also, I've come to think of the Space Station as my protection, my secure place. I have never felt the sheer threat of space so much as I feel it at this point. It is as though the three of us have jumped off the ocean liner and are rowing away in a tiny lifeboat.

The thrusters push us out into space. I'm knackered. It was nearly 6.00 a.m. by the time we undocked and we've been up all night. There was a crew sleep in the schedule for the previous afternoon, but I don't think any of us was able to persuade our bodies to sleep at that point. For the first orbit, the capsule is very quiet and I'm struggling to keep my eyes open.

I'm wide awake for the second orbit, though. We do a burn, and it's the first time that we have fired up the main engine – a moment of truth, in the circumstances, like starting your car after a long cold winter. There's a comforting low noise, like a dog's growl, and we feel a firm push on our backs. The engine still works.

The burn has to be precise. Burn it too long and we'll be too steep when we re-enter. Too short and we'll skim through the upper layers of the atmosphere and be out in space again. We need to burn for four minutes and thirty-seven seconds exactly.

My eyes are glued to the clock. My job if the engine fails at any stage in the burn is to work out how long we will need to fire the back-up thrusters to keep us on course. It always makes me slightly anxious to think that a highly critical, high-stress task such as this is being entrusted to my less than notable powers of mental arithmetic. Thankfully my maths will not be tested today. The burn goes flawlessly. We have less than halfway round the planet to our point of re-entry, on the right trajectory and starting to fall.

A quarter of the way around the planet from that initial burn, and descending all the time, we separate the spacecraft. Jeff Williams has warned me about the noise at this point. It's through the wall right next to your ear from where we're sitting, and it's fourteen pyrotechnic bolts going off, staggered, like a machine gun firing. With each explosion you feel a small jolt inside the module. The habitation module, depressurised, goes one way, the service module goes the other and the descent module, containing us, continues on its way.

We've been gradually descending all the time. By now Earth is less than 200 kilometres away and the descent module is in a gentle tumble, end over end. It's a good time to look outside. I see the Persian Gulf pass across the window followed by black sky as we cartwheel through space.

That tumble continues until we hit the upper layers of the atmosphere where, thanks to the beauty of natural aerodynamics, the craft orientates itself, heat shield forwards. At this point I notice the first gentle onset of g — not a slam, just a persistent, building force. I push myself back into the seat, tighten the harnesses and stow the checklists in the pocket beside me, alongside the map. Yes, a folded paper map of the world, showing numbered orbit paths. Only in a Russian spacecraft.

One or two sparks come past the window, then three or four, then four or five. The g force is steadily building – 0.5 g, 1 g, 1.5 g … There are more sparks. We have curtains in the Soyuz, little circles of fabric that cover the windows to dim the Sun's glare, but I'm lifting mine up because I want to watch the sparks fly by, the bits of burning spacecraft. Inside, it starts to get hot. Given that the temperature outside is rapidly increasing to a high point of around 1,700°C, it's perhaps not surprising. Before we left, Yuri had asked me if my ventilation fan had been working during launch. It had been. But for some reason it wasn't when we re-boarded. Yuri and I have been sharing one fan between us, meaning there is much less airflow coming through our suits. This is not going to do either of us any favours in the increasing heat.

Now the windows are starting to scorch and turning brown. Pretty soon, I can't see out through the charring – but I don't want to look in any case, because I'm really starting to feel the g's now and I need to be back in my seat. I re-tighten my harness. As we drop like a brick, the g's mount: through 3, to 4 and on to 5 … All this time Yuri is reading out the parameters: the time, the g force reading, the crew condition. Nobody apart from Tim and I can hear him because this is the communication blackout phase, but it's his duty to report to the spacecraft recorder which is capturing the information for later analysis.

The noise from outside, originally a whispering of air, has grown to a constant rush, as if we're aboard a jet. I continue to watch the clock. Now we have passed through the upper atmosphere, the g profile starts to drop off slightly and I'm bracing in advance for the braking chutes because I know that's going to be tough. We are still travelling just a bit faster than the speed of sound when two pilot chutes pull out the drogue parachute, and the effect is wildly aggressive. The chutes are off-axis with the capsule, so immediately the capsule, still dropping, flings around wildly, like a windsock in a hurricane. This belligerent motion lasts for twenty seconds and if you didn't know that this is what it's supposed to do, this is the moment, I think, when you would swear you were going to die.

Then the wild motion stops, which is an enormous relief. And now I'm waiting for the main canopy to open. Jeff has told me that it happens with a big jolt, and to be ready for it, but I'm feeling nothing. Surely it should have occurred by now. Has it failed to deploy? It's another Yuri moment: I look across at him and somehow he seems to know what I'm thinking and just calmly raises a hand. We're fine. The canopy has opened but it just happens to have done so softly.

There now follows about fifteen minutes of relative tranquillity under the canopy – a nice slow drift groundwards after the meteoric descent. However, my eyes are stinging with sweat and I can't rub them through my visor, so I have to try and blink it away, and I'm also getting ready for the landing, which I know will not be soft. Yuri has an altimeter on his wrist which is now what we're going on. Meanwhile our antenna will already be sending signals to the Search and Rescue helicopters.

From 500 feet, we're braced for the landing at any time. We know the rules: arms across your chest, neck firmly back, tongue well away from your teeth. Do not lift your head or look out.

The soft landing thrusters fire about half a metre off the ground: bang-bang. It's just enough time to ready yourself before we slam into the deck.

Imagine you are sitting in a wooden chair, and somebody behind you has a baseball bat and permission to smack the back of your chair as hard as they like. This is what that impact feels like.

And then we bounce, roll over and skid, and the parachute drags until Yuri cuts one of the risers from the control stick between his legs. One riser, but not both. If we have missed our landing site and are in the middle of nowhere, we may be grateful to have that parachute material for shelter.

And then finally we are still.

We have come to rest on our side, with me at the top of the craft, desperately using my arm to hold all the documents (including the map) on the shelf above me and stop them falling down onto Tim's head. Only now that we have come to rest do I begin to feel disorientated for the first time. Suddenly I am static and I feel gravity again and, to be perfectly frank, gravity sucks. I am baking hot, drenched in sweat and every time I move my head even slightly the world spins violently.

We have precisely hit our landing target on the Kazakh steppe, so it's not long before there are noises outside and the hatch opens. Suddenly there is summer sunlight in our battered tin can and an influx of air. Most of all, though, there is the smell of scorched grass and, still more powerfully, the smell of scorched spacecraft, which has a nasty, bitter tang to it. Yuri is out first, then it's my turn. A crew member reaches in for me and drags me out. You can add the smell of burly ground crew member's sweat, after a long day lingering in 30° C heat, to my first olfactory impressions of Earth. So much for those romantic accounts of the great and restorative smell of our planet.

I am placed in a carry-chair. I want to walk because I very badly want to be in control of my own motion, and not get jolted around. But this is the way they do it, so a chair it is. I am carried off across the Steppe, my bearers tripping and stumbling on the tussocks of grass. Cameras are pointing at me. The BBC seem to have made it to the site. I smile broadly, but the truth is I feel utterly awful.

Sergi, my flight surgeon, is there. Sunglasses are placed on my nose. A blood pressure cuff materialises on my wrist. It joins the watches that I have put there – future eighteenth birthday presents for certain boys of my acquaintance – along with my own. I realise that I have come back to Earth looking like a dodgy watch salesman. I ask Sergi to look after the watches for me. My one other piece of precious carry-on luggage – the Union flag patch from my spacewalk suit – is safe in my knee pocket.

The BBC want an interview. I somehow manage to say a few words. Somebody is trying to reach Rebecca, who is watching this from EAC in Cologne, on a satellite phone but it isn't working. At least she knows I'm back and safe.

I am carried about 100 metres to a medical tent. Inside, there are three fabric-screened partitions, one for each of us. The medical staff help us out of our suits. Any movement of the head is horribly provocative at this point, so lowering my head to push it out through the chest opening is a real challenge. After shucking off the suit, I need about four minutes to recover. I sit and look fixedly at the wall of the tent.

An IV drip goes in. I am then allowed to walk a short way to the waiting BTR-60 Russian armoured personnel carrier which is going to take us to the helicopters. I feel perfectly strong enough to do this, but am aware that my balance is not quite right. I also suddenly appreciate, from the tenderness underfoot, how soft the soles of my feet have become from lack of use.

The BTR-60 provides another bumpy ride across the Steppes to the fleet of Mi-17 helicopters that will fly us, with our support teams, to the airport. I smile to think that the last time I was in the back of one of these helicopters I was manning a general-purpose machine gun with a big grin on my face. At the airport, Tim, Yuri and I cross the tarmac together and are taken to a back room for more medical checks. A Kazakh reception committee awaits us, but Tim isn't feeling too well. I'm not feeling so great, either. I have been unable to pee, probably on account of the Kentavr – a tight anti-g garment worn under our space suits. Sergi has given me an ultrasound and discovered that my bladder is full, and distended like a rugby ball. I'm going to need to have a catheter inserted in order to empty it, which rather rules me out of the Kazakh reception. Tim can't quite muster the wherewithal, either. Poor old Yuri, who seems to have shrugged off a descent from space as if it were a raincoat, has to attend on his own.

Tim and I weather it out in the back room. Someone hands me an iPad to complete a spatial awareness test. I'm not ready for the weight of it at all and it drops a few inches before I remember: there's gravity here. I will shortly fly by jet to Norway with Tim, while Yuri flies to Moscow. There is only time for the briefest of partings, but I manage to tell Yuri that it's been an absolute honour to have flown in space with him and to have had him as my Soyuz commander. Yuri just quietly smiles and nods.

Rebecca has somehow managed to get a package for me onto our NASA jet. It contains Yorkshire teabags and a KitKat. I am able to have tea and toast on the plane and I start to feel better. I then fall asleep, and stay asleep until we land in Norway.

It's midnight there, yet still light – which would have thrown me, if my body clock hadn't been comprehensively thrown already. Another plane has flown up from Cologne to meet us here. Frank De Winne is there, and so is David Parker, now ESA's director of Human Spaceflight, and Jan Wörner, ESA's new director general.

And so is Rebecca. It feels so good to see her that, for a while, the rest of the world stops spinning in the corners of my vision.

The ESA aircraft which will fly us back to Germany has a big Union flag draped up its steps. It's after two in the morning by the time we arrive at Cologne airport, but my parents are still up to see me. I know I've put them through a lot of worry these past six months – in fact, these past forty-four years – but it seems they've forgiven me. In fact, they seem elated. And despite the late hour, there's even a welcome committee at the European Astronaut Centre, where so many friends and colleagues who have supported my mission have turned out to welcome me home.

We've been given family rooms in the EAC crew quarters. I take my first longed-for shower which is bliss, but I find I can't stay in there long because the water running past my ears seems to increase my feelings of vertigo. It's gone 3.00 a.m. when I fall into bed. Luckily the room doesn't continue to spin when my eyes close and I sleep deeply.

The next thing I know I am being woken by two very excited small boys leaping into our bed.

How quickly normal life resumes. And how brilliant it feels when it does so. One day you're an astronaut, falling out of the sky after six months in space. The next you're just any other dad, feeling like you have a stonking hangover, getting jumped on by his kids on a Sunday morning.

AFTERWORD

Beyond

'Savour every moment in space.' That was the advice from Helen Sharman as we chatted in a Costa Coffee at Heathrow when she handed over Yuri Gagarin's book. I'd heard similar words from fellow astronauts in the weeks before launch.

'Don't forget to look down on Earth whenever you can.' (Jean-François Clervoy)

'Take some time for yourself every day.' (Thomas Reiter)

And perhaps the most prescient advice:

'Be ready for life never to be the same again.' (Frank De Winne)

Living in space is truly remarkable for so many reasons. Even if you strip away the surreal nature of your surroundings, it's hard to beat the basic practicalities: you can phone anyone but no one can call you; the only emails you receive have been vetted and are highly relevant; every day is fully planned for you, with dedicated time for exercise; there's no cooking, shopping, laundry … and no commute. An enormous team involving several space agencies around the world is dedicated to making your life on the ISS as easy and efficient as possible, allowing you to focus on work.

Frank was right – life wouldn't be the same again. In fact, life couldn't have been more different when I returned from space. There was a schedule, of course, and ESA worked hard to try and protect their crew member as much as possible in the weeks immediately after landing. But the reality is that everybody wants a piece of you – often quite literally.

If you're not donating blood, urine, saliva or faeces, or having muscle biopsies, then you'll most likely be found inside an MRI scanner, having an ultrasound, an ECG, an EEG or multiple X-rays – all in the name of science. I made the mistake once of asking the clinician what radiation dose I was being exposed to with all these X-rays.

'The equivalent of about three minutes of your time on the ISS,' she answered.

Point taken.

Our flight surgeon would prefer that, once back at our home base, we don't travel anywhere for at least three weeks, not just in order for us to be around for the collection of medical and scientific data, but also to give our bodies a chance to recover. All that exercise in space is great for our major muscle groups and our cardiovascular system, but all the small, stabilising muscles that hold the body together are weak. Our core strength is compromised and although our overall bone density loss may not be too bad (around 2 per cent after six months), some areas will suffer worse. I had lost 11 per cent bone density at the neck of both femurs, so rehabilitation was vital. There were sit-ups, twists, lunges, intense workouts with medicine balls and on the treadmill. I still remember when I was told I could run outside on my own for the first time. Running in the fresh air! It was bliss – almost as blissful as the first trip Rebecca and I made for beer and pizza at a *bierhaus* on the Rhine.

Unfortunately, rehabilitation has to compete with the many other demands on a returning astronaut, not least of all public appearances and multiple interviews with the press. And not all the science can be completed in Europe. I found myself on a long-haul flight to Houston, less than the stipulated three weeks after landing, in order to use a specifically calibrated X-ray machine at the Johnson Space Center. I didn't ask about the dose. On the way back, I stopped off in London to attend the Farnborough Airshow and then soon after that I was on my way to Moscow for debriefs. And so began a whirlwind six-month 'post-flight' period, bouncing between London, Cologne, Houston and Moscow. We also managed to squeeze in a ten-day speaking tour around the UK, where Tim Kopra and his wife Dawn joined Rebecca and me for the final three events in Edinburgh, Belfast and the Royal Albert Hall in London.

If space had been surreal for bringing a sense of tranquillity, wonder and a unique perspective on our place in the universe, the post-flight tour was surreal for exactly the opposite reasons. It was often madly chaotic, stressful, exhausting and frustratingly impossible to satisfy every demand. But despite this, the warmth and enthusiasm we received everywhere never failed to surprise and lift me. The mission had been a huge success – not just in the sense that I had ticked off the big-ticket items of the spacewalk – the science, the maintenance and working with visiting vehicles – but far more widely for the UK. The outreach programme had been immense.

Even in the build-up to the flight, the public response had far exceeded that of any other European mission, with thousands of people entering competitions for the mission name, mission patch, guessing my playlists and even cooking my food with Heston Blumenthal. When you are isolated on the ISS, you really don't get a sense that what you are doing is such a big deal down on Earth. I would be snatching a few minutes here and there to talk with a school over ham radio or run competitions, from computer coding to growing rocket seeds – just hoping that it would be helping to inspire some young people to take a look at space and science in a different way.

I was genuinely amazed when I returned to discover that the impact had been so dramatic. A staggering 33 million people had engaged with the mission, with over 2 million school children taking part in some of the thirty-four projects we had run. I thought back to that first meeting with the UK Space Agency's Jeremy Curtis and Libby Jackson, who were coordinating the outreach programme, and to the feeling we had that if only half of these projects came off we would be able to call it a success.

It was clear to see that the legacy of mission Principia would be one of inspiration for a new generation of scientists and engineers and I couldn't have been more delighted or grateful to the hundreds of volunteers who had supported these programmes on the ground.

I was also honoured that the mission had received such wide recognition. In the Queen's Birthday honours in 2016, announced eight days before I landed, I had been made a Companion of the Order of St Michael and St George (or CMG). This award is given on the recommendation of the Foreign Office, usually for distinguished services overseas. Well, I certainly spent a fair share of my time in orbit travelling over seas. On 1 December 2016, I met Her Majesty the Queen to receive the award, which I dedicated to the hundreds of people who had made mission Principia possible. I was also delighted that Eddie Redmayne was receiving his OBE on the same day and magnanimous enough to draw away any media attention.

I would have another opportunity to meet Her Majesty a few months later. On this very special and private occasion, Rebecca and I had been invited to dinner at Windsor Castle, along with other guests for the Easter Court – and to stay the night. Bed and breakfast, then. This was also the perfect opportunity to present Her Majesty with the Union flag which I had worn on my space suit during my EVA, the first to be worn in the vacuum of space, and subsequently smuggled down to Earth in my knee pocket. Ever since I had visited Buckingham Palace in 2011, at a reception celebrating exploration and adventure, and viewed the remarkable Royal Collection of artefacts from returning British explorers throughout history, I had dreamt of being able to add something. I'm quite certain that Sir Walter Raleigh wasn't running around the equivalent of Hobby Lobby in 1585, frantically trying to get his flag framed for the monarch, but needs must.

By Christmas 2016, it's fair to say that I was knackered. The pace of the post-flight tour, on the back of a six-month mission to space, itself on the back of an intensive two-and-a-half-year training regime and punishing international travel schedule, was catching up on me. I caught a cold (man flu, if the truth be known), which was a good excuse to go nowhere and do very little for a week. It was also a time to shift priorities. As a family we had embraced everything about the space adventure – different countries, different cultures, moving house, changing schools and making new friends. But one thing was undeniable – although our two boys had always been incredibly positive, I had been away an awful lot. Thomas was turning eight, Oliver five and, if truth be known, I had probably been around for only 50 per cent of their lives. That had to change. It was time to cut back on the travel and be more of a permanent feature at home, as opposed to a bowling ball that came crashing through, interrupting the routine for a week or two before disappearing off again.

That February we set aside some time for ourselves and did a Disney cruise. It was so unlike us. We were campers and hikers. We loved the great outdoors. This was the last kind of holiday we would have ever imagined signing up for. But our friends across the road talked us into it, and it was absolutely brilliant. The boat left from Galveston, the port just down the road from Houston. The boys had the run of half the ship and Rebecca and I could sit in the sun while drifting slowly across the Gulf of Mexico. It was the break we all needed.

My boss, Frank, had agreed that we could stay in Houston until the summer of 2017, so that our boys could at least finish the academic year. Then, I'd be needed back in Cologne to take over from Alex Gerst as head of astronaut operations – managing the team at ESA that provide all the mission support, organising launch and landing campaigns and ensuring that everything ran as efficiently as possible from the European point of view. As an astronaut, there is an undeniable draw to the Johnson Space Center, the nerve centre for space station operations, so it was hard to be leaving such a dynamic environment. It was also extremely hard to say goodbye to the many close friends we had made, in particular for Rebecca and the boys who had embraced Houston as their home for the past four years. But we knew Germany well, we knew that we could settle back into life in Bonn and that we would make new friends – it was all part of the ex-pat lifestyle that Rebecca and the boys had adapted so well to. We moved back to Germany, and I went back to work.

*   *   *

It was 30 May 2020, and once again I found myself smiling at a rocket launch. This time, however, I was at home, sharing the view with over 10 million people watching live, and the rocket wasn't just going up. Part of it was going to be landing again.

It wasn't the first time that I had watched a SpaceX rocket launch, but seeing the first-stage booster come back down under its own control, using its remaining thrust to return to a tiny landing pad attached to a moving vessel out in the Atlantic Ocean, was a sight to behold – the stuff of science fiction. I couldn't completely relax for a few more minutes though, not until the second stage had successfully delivered my NASA friends Bob Behnken and Doug Hurley into orbit.

This was the first time a crew had launched in a commercially built and operated spacecraft and the first launch from US soil since the Shuttle retired in 2011. It was a big deal for America, but it was a big deal for all of us with an interest in spaceflight, too. Despite the inevitable media hype, this really felt like the dawn of a new era of space exploration. And not just for SpaceX. As I write this, Boeing is preparing to launch its new crewed vehicle, Starliner, in a matter of months, and we are not far away from the first launch of NASA's Space Launch System – the most powerful rocket ever built, which will carry the Orion spacecraft to the Moon. Contracts are being awarded to build modules for a new space station, called Gateway, destined for lunar orbit, and NASA are furiously working on returning astronauts to the surface of the Moon by 2024. And Europe is very much a part of this plan. For nearly fifty years we have been confined to low Earth orbit, watching crews fly to various different space stations. But this is all set to change.

So many thoughts were running through my mind that Saturday in May as Bob and Doug made it safely to orbit, wearing some very futuristic-looking space suits inside a very futuristic-looking capsule. It couldn't have been more different from the Soyuz that carried me to space. I was also wondering if I had another part to play in this grand adventure – some more chapters to write in the days up ahead, telling of a return to the Space Station or, who knows, even a journey further afield. The European Space Agency is certainly keen to fly all its class of 2009 twice, which I hope will include a long-awaited six-month mission for Andy. And I hope that Matthias, the new addition to the Shenanigans who I first met in Paris all those years ago, before my interview with Jean-Jacques Dordain, will get his first mission soon.

Whatever happens, it's clear that for space exploration the limits are about to shift all over again. Nowadays, when I meet young people and see the awe and wonder in their faces when I talk about space, it fills with me deep satisfaction to reflect that they are the generation who will see us take the next big step for humankind. And how amazing to reflect, too, that maybe, just maybe, I have already met the first person who will set foot on Mars.

ACKNOWLEDGEMENTS

I have to start by thanking my wonderful parents for their endless support and encouragement, for giving me that childhood nirvana of a 'normal upbringing' and for maintaining such a great sense of humour throughout my various adventures, which were seldom within their comfort zone. And sorry for the sleepless nights.

I'm enormously grateful to Giles Smith for his warmth, humour, guidance and support during the writing of this book – thank you for sharing this trip down memory lane.

Huge thanks to my editor, Ben Brusey, for encouraging me to write this book and for his patience in waiting for it to happen. Thank you also to Jess Ballance in editorial, Susan Sandon for her ongoing support, to Klara Zak and Charlotte Bush on publicity, Sarah Ridley, Natalia Cacciatore and Rebecca Ikin on marketing and the dedicated team at Cornerstone for all your help and enthusiasm.

Thank you to my literary agent Julian Alexander, and to Sophie Laurimore and the team at The Soho Agency for keeping all the plates spinning, even during a global pandemic – you guys are amazing.

There are so many inspiring role models out there sharing their knowledge and enthusiasm, selflessly giving their time to support young people. My personal thanks to two teachers in particular, Anthony Forrest and Gina Griffiths, for giving me and so many others the opportunity to grow, assume responsibility and develop resilience.

A career in the military forges special bonds and lifelong friendships. I've had the good fortune to have worked hard and played hard with many great characters along the way. Special thanks to Ian Curry, Philip McCabe, Toby Everitt, Lindsay Morris, Jimmy James, David Amlôt, Jim Richards and Andy Ozanne for sharing your stories in this book, and for the fun, camaraderie and crazy times along the way. And to Matt Sills – we miss you, buddy.

To my colleagues at ESA, NASA, JAXA, CSA and Roscosmos, whose relentless work behind the scenes makes the extraordinary feat of sending humans into space possible, thank you for your passion, your dedication, your support and for continually striving to push the boundaries of what is possible. And to my fellow astronauts, thank you for your inspiration, friendship, advice and endless shenanigans.

There are hundreds of people behind every mission to space, but I owe a debt of gratitude to several individuals in particular for making my spaceflight a reality, namely Jean-Jacques Dordain, David Willetts, David Parker, Simonetta Di Pippo, Bernardo Patti, Frank De Winne and Thomas Reiter. And I'd like to make a special mention to the hard-working PR and outreach teams at ESA and the UK Space Agency, who helped make mission Principia the incredible success that it was.

Thank you to the many great friends that Rebecca and I have made along the way, who make our frequently hectic, nomadic family life so much fun. To our families, thank you for your steadfast love and support at every turn. And to my wonderful in-laws, Maddy and John King, thank you for always stepping in when I was not around, to help out on various long journeys and numerous house moves.

To Thomas and Oliver, you have experienced so much and still take everything in your stride. I'm immensely proud of you both – thank you for keeping my feet on the ground. And finally, to my soulmate Rebecca, this has been an adventure that would not have been possible without your love, encouragement, positivity and ability to embrace every new experience head-on. Thank you for being there every step of the way. We make a great team – where to next?

INDEX

PHOTOGRAPHY CREDITS